Philosophy and Medicine

Volume 153

The Philosophy and Medicine series is dedicated to publishing monographs and collections of essays that contribute importantly to scholarship in bioethics and the philosophy of medicine. The series addresses the full scope of issues in bioethics and philosophy of medicine, from euthanasia to justice and solidarity in health care, and from the concept of disease to the phenomenology of illness. The Philosophy and Medicine series places the scholarship of bioethics within studies of basic problems in the epistemology, ethics, and metaphysics of medicine. The series seeks to publish the best of philosophical work from around the world and from all philosophical traditions directed to health care and the biomedical sciences. Since its appearance in 1975, the series has created an intellectual and scholarly focal point that frames the field of the philosophy of medicine and bioethics. From its inception, the series has recognized the breadth of philosophical concerns made salient by the biomedical sciences and the health care professions. With over one hundred and twenty five volumes in print, no other series offers as substantial and significant a resource for philosophical scholarship regarding issues raised by medicine and the biomedical sciences.

Jason T. Eberl
Editor

Emerging Issues in Catholic Bioethics

Springer

Editor
Jason T. Eberl
Albert Gnaegi Center for Bioethics
Saint Louis University
St. Louis, MO, USA

ISSN 0376-7418 ISSN 2215-0080 (electronic)
Philosophy and Medicine
ISBN 978-3-032-08861-1 ISBN 978-3-032-08862-8 (eBook)
https://doi.org/10.1007/978-3-032-08862-8

This Springer imprint is published by the registered company Springer Nature Switzerland AG
The registered company address is: Gewerbestrasse 11, 6330 Cham, Switzerland

Acknowledgments

Volumes like this are a collaborative enterprise in which the authors of the individual chapters contribute far more to the overall quality than the editor. I am thus extremely grateful to the sixteen scholars who responded to my invitation to share their perspectives on the seven topics discussed herein. You would not be able to learn these perspectives, however, if not for the editorial and production team at Springer, in particular Floor Oosting, Travis Holloway, Immaculate Jayanthi, and Christopher Wilby, as well as Lisa Rasmussen and Søren Holm, editors of Springer's "Philosophy and Medicine" series who have graciously invited me to serve as Associate Editor of the series. I am most grateful for their support and work, as well as that of Christopher Tollefsen, in bringing this volume to fruition.

This project was conceived and developed during my service as the Hubert Mäder Chair and Director of the Albert Gnaegi Center for Bioethics at Saint Louis University. I continue to be humbly honored by this position and deeply appreciative of the scholarly opportunities it affords me. My formation as a bioethicist benefitted tremendously from my previous faculty positions at Indiana University-Purdue University Indianapolis and the Marian University College of Osteopathic Medicine. Throughout my academic career, I could not sustain my own efforts without the love and support of my wife, Jennifer Vines—who first prompted me toward bioethics as an area of specialization—and our daughter, August Eberl, who has carried on the family legacy as public health professional.

Contents

About the Contributors

Michael Baggot is Associate Professor of Bioethics at the Pontifical Athenaeum Regina Apostolorum and Invited Professor of Theology at the Pontifical University of Saint Thomas Aquinas.

Rachelle Barina is the Chief Mission Officer and Senior Ethicist at Hospital Sisters Health System and an affiliate faculty member of the Albert Gnaegi Center for Bioethics at Saint Louis University.

John F. Brehany is Executive Vice President and Director of Institutional Relations at the National Catholic Bioethics Center.

Peter J. Cataldo is a retired health care ethicist and former Catholic health care ethics executive, currently serving as Manager of Cataldo Consulting Services, LLC.

Jason T. Eberl is the Hubert Mäder Chair in Bioethics, Professor of Philosophy, and Director of the Albert Gnaegi Center for Bioethics at Saint Louis University.

Mary E. Homan is a region vice president of theology and ethics at CommonSpirit Health and holds an adjunct associate professor appointment in the Department of Community Health, Institute for Health & Humanity, Medical College of Wisconsin.

Daniel P. Horan is Professor of Philosophy, Religious Studies, and Theology at Saint Mary's College in Notre Dame, Ind.

Mary Jo Iozzio is Professor of Moral Theology at the Clough School of Theology and Ministry, Boston College.

Bernard G. Prusak holds the Raymond & Eleanor Smiley Chair in Business Ethics at John Carroll University.

Mark Repenshek is the Vice President of Ethics and Church Relations, Ascension.

Matthew P. Schneider is a professor and formator at Holy Apostles College and Seminary.

John Paul Slosar is the Senior Vice President of Healthcare Ethics, Ascension.

Erica L. Smith is the Executive Director of Hospital Sisters Mission Outreach.

Claudia R. Sotomayor is Chief of the Ethics Consultation Service in the Pellegrino Center for Clinical Bioethics and Associate Professor of Medicine at Georgetown University Medical Center.

William F. Sullivan serves as the Joseph P. Kennedy Sr. Chair in Bioethics, Kennedy Institute of Ethics, Georgetown University, and as Professor of Family Medicine, Department of Family Medicine, Georgetown University School of Medicine.

Henk ten Have is Research Professor at the Department of Bioethics of Anahuac University Mexico and Professor emeritus, Department of Global Health Ethics, Duquesne University, Pittsburgh, USA.

Christopher Tollefsen is Professor of Philosophy at the University of South Carolina and Interim Executive Director of the South Carolina Center for American Civic Leadership and Public Discourse.

Chapter 1
Introduction

Jason T. Eberl

The Roman Catholic Church has a tremendous global influence on health care practice, public policy, and bioethics. For example, in the United States alone, comprising "more than 600 hospitals and 1600 long-term care and other health facilities in all 50 states, the Catholic health ministry is the largest group of nonprofit health care providers in the nation. Every day, more than one in seven patients in the U.S. is cared for in a Catholic hospital" (Catholic Health Association 2024). Much of the Church's teachings to date have focused on issues such as abortion (CDF 1974; John Paul II 1995), euthanasia and end-of-life care (CDF 1980, 2020), assisted reproductive technology (CDF 1987), contraception (Paul VI 1968), justice in health care access (USCCB 1993), and recent advances in genomic biotechnology (CDF 2008).

Even with respect to these bioethical topics about which the Church's magisterium has spoken, interpretive debates and disagreement concerning how to manage pastoral implications and develop prudential policies have inspired considerable discourse among Catholic bioethicists. A previous volume I edited for this series, *Contemporary Controversies in Catholic Bioethics* (Eberl 2017), offered point/counterpoint analyses of seventeen distinct issues surrounding the beginning and end of human life, advances in genomic biotechnology, and health care policy. The present volume is intended to complement the previous one by offering again dual perspectives on seven topics not treated in the previous volume, but which have emerged at the forefront of scholarly debate among Catholic bioethicists, health care providers, and policymakers. In some cases, the paired chapters offer a true *sic et non*: one chapter affirming, for instance, that life-sustaining treatment may be unilaterally withdrawn under crisis standards of care—as during the COVID-19

J. T. Eberl (✉)
Albert Gnaegi Center for Bioethics, Saint Louis University, St. Louis, MO, USA
e-mail: jason.eberl@slu.edu

J. T. Eberl (ed.), *Emerging Issues in Catholic Bioethics*, Philosophy and Medicine 153, https://doi.org/10.1007/978-3-032-08862-8_1

pandemic—and the other denying the moral permissibility of unliteral withdrawal. In other cases, there is not strict disagreement about what is morally permissible or impermissible, but rather diverse analyses of the topic at hand: for example, different approaches regarding the understanding of, and appropriate care for, persons who identify as transgender. Finally, there are chapters which complement each other by one offering a more theoretical moral analysis and the latter providing more practical guidance for clinical decisions or policymaking, as in the case of incorporating artificial intelligence within Catholic health care.

The aim of this volume, as with the previous *Contemporary Controversies* volume, is to provide up-to-date ethical analyses that summarize available magisterial instructions and salient scholarly discourse, while also arguing for specific theoretical conclusions or delineating practical guidelines for enacting the moral principles that ought to guide the work of Catholic health care professionals in the clinical, research, and public health arenas.

The first part addresses how we ought to understand and incorporate into our moral theorizing the lived experiences of persons with diverse identities, specifically, persons who identify as *disabled* or as *transgender*. Disability identity is woefully underexplored in Catholic scholarly discourse, with sustained attention only recently having been occasioned through, for example, special issues of the *Journal of Moral Theology* (Romero and Iozzio 2017) and *National Catholic Bioethics Quarterly* (Eberl 2020). Transgender identity is, of course, a socially and politically fraught topic that merits careful consideration, with vastly divergent perspectives having been articulated by various Catholic sources alongside a recent statement by the US Conference of Catholic Bishops on "moral limits to technological manipulation of the human body" (USCCB 2023) and a magisterial denouncement of "gender theory" (DDF 2024).

Christopher Tollefsen proposes some new directions within a natural law bioethics that would show a greater concern for the role that beauty, work, and play might manifest in the lives of persons with disabilities. Tollefsen notes that biological impairments can create a cascading series of effects that negatively impact access to and enjoyment of the whole range of human goods. He then elucidates the ways in which attending to the goods of work, play, and aesthetic experience, specifically in terms of how such goods are pursued and realized within the context of friendship and community, might reverse the cascade. Tollefsen concludes that, to the extent that persons with disabilities work, play, and create amongst people who treat them with dignity and respect, to that extent will the Church, by welcoming and offering hope to all, fulfill her mission in the world.

Mary Jo Iozzio observes that dependence is often shunned by the temporarily hale who bely the reality of many people's lives that will be challenged by sickness, accidents, and ageing, which, along with other challenges, are normative for persons with disabilities and their caregivers. In recognizing the personhood and dignity of all members of the human community, Iozzio contends that the point of disability bioethics is to become aware of and attentive to human flourishing in the health care arena as well as in the commons of cultural, political, and environmental space. While many have gained cultural competence in matters related to race, gender, and ethnicity, Iozzio concludes that we must bring the cultural competence concerns of people with disability to the mainstream.

In discussing transgender identity, **Peter Cataldo** attempts to understand the real difficulties of people with gender dysphoria who wish to be treated in Catholic health care, and how Catholic health care can respond to their suffering in a way that is consistent with Catholic teaching on the creation of the human person as female or male. Specifically, Cataldo examines whether medical and surgical interventions on secondary sex characteristics are morally consistent with Catholic teaching. His analysis of the ontological and anthropological components of the Catholic moral tradition shows that the biological distinction between a human person's primary and secondary sex characteristics is significant with respect to the moral permissibility of interventions on secondary sex characteristics in the treatment of gender dysphoria.

Daniel Horan frames the denial of care for transgender persons as a form of epistemic injustice. In developing a course toward remedying this injustice, Horan offers a constructive intervention by arguing for the primacy of what medieval Christian philosophers called "intuitive cognition" over "abstractive cognition" in contending that one ought to believe transgender persons in their self-avowal about gender, affirming their first-person authority and proximate knowledge about themselves. He then argues for health care providers and ethicists to cultivate the virtue of epistemic humility, which involves a commitment to ongoing intellectual and moral conversion, as well as relativizing one's own abstractive knowledge in relationship to the self-knowledge of a transgender person.

The second part shifts the focus from caring for individuals to the wider moral concerns that uniquely arise in the *public health* and *global health* arenas. Both have suffered from a lack of sustained attention within Catholic and secular bioethical discourse until the outbreak of the COVID-19 pandemic. Due to the recent pandemic, however, there has been an increase in Catholic scholarly engagement with public health concerns (e.g., Eberl and Rozier 2022).

Mary Homan notes that how we define and understand public health shapes the theory and practice of public health ethics, which often reduces to a utilitarian justification for interventions that infringe personal rights and values, often to the detriment of the most vulnerable. Public health, though, implements evidence-based interventions with a concern for social justice to achieve health equity, thereby fulfilling a duty to address health concerns that individuals cannot solve but require collective action. Homan contends that the Catholic social and moral traditions provide a new vision of public health ethics focused on human flourishing and the common good, supporting public health's mandate to promote health and prevent disease, disability, and injury in the population. She concludes that an enriched ethics capacity, grounded in virtue theory, helps to clarify, prioritize, and justify possible courses of public health action while fostering respectful deliberation and alleviating ethical tensions.

Claudia Sotomayor and **William Sullivan** explore how Catholic bioethicists can mobilize the perspectives and resources of Catholic social teaching to critically engage public health decisions by attending to the basic and normative activities of human knowing, deciding, and loving. They propose that an adequate approach to public health ethics must attend not only to principles and concepts, but to a method by which agents reach concrete decisions in unique, complex, and evolving public health situations. The method they elaborate is G-O-A-LS®: intending and pursuing some Good; selecting appropriate and effective Options that align with this good; developing and exercising our intellectual, moral, and theological virtues as Agents; and attending to the particularities of complex-adaptive Life Systems. Sotomayor and Sullivan contend that ethical principles that can guide public health decision-making are generalizations of wise decisions reached through this integrated method in the key of agapeic love.

While public health ethics expands the sphere of moral concern from the bedside to the broader community, it remains largely limited to delineated communal contexts (city, state, nation, etc.). Global health ethics, on the other hand, takes in the entire human community, especially those who are made vulnerable by virtue of their socioeconomic or geographic contexts, recognizing that addressing global health concerns requires an intentional focus on the "cry of the poor" (Martins 2019; Block et al. 2023) in line with the concept of "integral ecology" emphasized by Pope Francis (2015).

Henk ten Have describes how global health ethics has emerged from acknowledging health as a global concern, articulating new issues such as the health impacts of environmental degradation, climate change, pandemics, hunger, inequality, and poverty. Ten Have contends that global health ethics requires a delicate balance between applying and implementing a universal ethical framework and the need to recognize cultural and moral diversity as well as different local traditions. He concludes that global health ethics requires a rethinking of the ethical framework for addressing global issues, as argued by Pope Francis, going beyond individual concerns and taking into account social and environmental concerns and emphasizing the harmony between the Creator, humanity, and creation, as well as the common human responsibility for individual, social, and environmental well-being.

Erica Smith and **Rachelle Barina** focus on faith-based global mission work, illuminating how interventions aimed at meeting the needs of low-resource communities across the globe often come with a range of unintended consequences which can be ineffective, counterproductive, wasteful, and even harmful. They call for maturing the actions of faith-based missionaries, including the spirituality that inspires such actions as an expression of faith. Smith and Barina consider the stories of two Catholic missionary figures who embody laudable and relevant virtues and values, such as humility, sacrifice, solidarity, and relationship. Based on these examples, they counsel missionaries to approach every engagement with the desire to deepen a sense of solidarity and act with reverence for the wisdom, experience, and daily life of people, which will prompt considerations, questions, and concerns about the quality and long-term impact of missionary actions.

As noted above, public health and global health concerns became acute during the COVID-19 pandemic, during which specific issues regarding clinical care and public policy were predominant among Catholic bioethicists and the public at large: namely, *triaging* of limited health care resources, such as mechanical ventilators, and the development and use, sometimes involving institutional or governmental mandates, of novel *vaccines*. In the third part of the volume, the authors reflect on the lessons learned from the recent pandemic in order to help prepare for future infectious disease outbreaks.

Bernard Prusak describes two distinct Catholic schools of thought on the permissibility of unilaterally withdrawing life-sustaining treatment from patients with poorer prognoses and reallocating it to patients with greater expected survivability. He proposes that the arguments proffered by these respective schools have not adequately engaged a deeper disagreement regarding the fundamental question of who does the physician work for—that is, whose agent should they be? Prusak proposes that, while it seems physicians ought not to engage in unilateral withdrawal of life-sustaining treatment, that position must reckon with the question of when, under crisis standards of care, patients present at a hospital, whose patients are they? Is the patient primarily the moral responsibility of the physician or of the health care system in which the physician is employed, which arguably has a wider social responsibility than just to the individual patients under the physicians' care? Both questions illuminate the tension underlying maintaining "ethics as usual" under a putative "state of exception" presented by a pandemic.

Jason Eberl defends the school of thought that justifies unilateral withdrawal of life-sustaining treatment under crisis standards of care, clarifying the applicability of double-effect reasoning, as well as employing just war reasoning analogously with respect to the strategic allocation of resources. While Eberl takes seriously the questions Prusak raises regarding the physician's responsibility *vis-à-vis* the patient(s) under their care, there is a legitimate basis for moving resource allocation decisions to a higher level of responsibility—i.e., a triage committee—thereby not forcing physicians to violate their fiduciary duties to their patients, but rather restricting—under criteria of justice and fairness—the resources available to the physician to care for their particular patient(s).

Turning to the ethics of developing and using certain types of vaccines, and in some cases mandating them, **John Brehany** argues that debates surrounding vaccine ethics within the Church have not been fruitful because disputants appeal selectively to ethical considerations that best support their own position, thereby neglecting other, more relevant resources in the Catholic moral tradition, including the virtues of prudence and charity and the principles of ordinary and extraordinary means. He thus argues that disputes over vaccine ethics are best addressed in light of these virtues and principles, as well as the ethical requirement of informed consent and situating proper regard for the principle of the common good in relation to the good of individual persons. Brehany concludes that, within this framework, people can more appropriately and effectively address other important moral goods, such as advocating for higher standards of safety, more individualized choices, and ending the role of abortion-derived cell lines in vaccine products and biologics research and development.

Matthew Schneider explores both the ethics of existing mandates and principles for possible future mandates. Schneider notes that, to force someone to act for the common good, there must be a great advantage to the common good compared to the little danger or disadvantage to them. He argues that vaccines and vaccine mandates can clearly serve the common good, especially when a mandate can achieve herd immunity or the end of endemic disease spread. Schneider concludes that most if not all existing mandates in the US and Canada serve the common good to a degree that is justified; however, not every mandate serves the common good, thereby allowing limited exemptions to vaccine mandates.

The last part of this volume treats the most novel topic to have emerged within Catholic bioethical and broader societal discourse: the integration of *artificial intelligence* (AI) into not only our daily lives but also within our health care delivery systems. In light of a recent document from the Dicastery for the Doctrine of the Faith and the Dicastery for Culture and Education, *Antiqua et nova* (DDF/DCE 2025), **Michael Baggot** offers theoretical and practical guidance from the Catholic tradition on cultivating the virtues most helpful for employing AI well in physical, mental, and spiritual health care. Baggot notes that, while AI systems can speed diagnosis and prompt new forms of treatment in physical health care, anxieties remain regarding the trustworthiness of programs notorious for hallucinations as well as how algorithmic biases could exacerbate current social injustices, mitigate patient autonomy, and damage empathetic doctor-patient relations. He further discusses how increasingly realistic companion systems in mental health care promise to fill in the gaps for those who lack access to human assistants, but cautions that reliance on such tools could create emotionally disordered and harmful attachments to digital substitutes that detract from authentic interpersonal relations. Finally, Baggot elucidates the ways in which new AI systems focused on spiritual health offer engaging access to enlightenment for seekers and those unsure about how to grow spiritually; yet, the same technologies raise fears about an excessively disembodied approach to the spiritual life and an unhealthy neglect of communal practices.

In the final chapter, **Mark Repenshek** and **John Paul Slosar** outline the insufficiencies in various guidance documents that have been put forth for ensuring generative AI (gAI) is adopted and utilized in a way that is consistent with the Catholic moral tradition, specifically with respect to ensuring that this technology does not disadvantage individual patients or certain classes of people. They argue for a casuistic method of analyzing potential uses of gAI technologies in the specific context of health care delivery, adopting a practice of piloting any and all gAI systems before widespread adoption. Drawing on existing directional and aspirational guidance, including *Antiqua et nova* (DDF/DCE 2025), Repenshek and Slosar identify the central values of concern and the risks to health care systems' Catholic identity presented by various applications of gAI. They conclude by describing the methodology, organizational tools, and processes they have developed to ensure that a Catholic health care organization is employing gAI in a manner that addresses these issues, evaluating gAI not just based on how well it does what it was designed to do, but also on its moral character and virtue.

References

Block, Jenny Weiss, M. Therese Lysaught, and Alexandre A. Martins, eds. 2023. *A Prophet to the Peoples: Paul Farmer's Witness and Theological Ethics*. Eugene, OR: Pickwick.

Catholic Health Association. 2024. Catholic health care in the United States. https://www.chausa.org/about/about/facts-statistics.

Congregation for the Doctrine of the Faith (CDF). 1974. Declaration on procured abortion. https://www.vatican.va/roman_curia/congregations/cfaith/documents/rc_con_cfaith_doc_19741118_declaration-abortion_en.html.

Congregation for the Doctrine of the Faith (CDF). 1980. Declaration on euthanasia. https://www.vatican.va/roman_curia/congregations/cfaith/documents/rc_con_cfaith_doc_19800505_euthanasia_en.html.

Congregation for the Doctrine of the Faith (CDF). 1987. *Donum vitae*. https://www.vatican.va/roman_curia/congregations/cfaith/documents/rc_con_cfaith_doc_19870222_respect-for-human-life_en.html.

Congregation for the Doctrine of the Faith (CDF). 2008. *Dignitas personae*. https://www.vatican.va/roman_curia/congregations/cfaith/documents/rc_con_cfaith_doc_20081208_dignitas-personae_en.html.

Congregation for the Doctrine of the Faith (CDF). 2020. *Samaritanus bonus*. https://press.vatican.va/content/salastampa/en/bollettino/pubblico/2020/09/22/200922a.html.

Dicastery for the Doctrine of the Faith (DDF). 2024. *Dignitas infinita*. https://press.vatican.va/content/salastampa/en/bollettino/pubblico/2024/04/08/240408c.html.

Dicastery for the Doctrine of the Faith (DDF), and Dicastery for Culture and Education (DDF/DCE). 2025. *Antiqua et nova*. https://www.vatican.va/roman_curia/congregations/cfaith/documents/rc_ddf_doc_20250128_antiqua-et-nova_en.html.

Eberl, Jason T. 2017. *Contemporary Controversies in Catholic Bioethics*. Dordrecht: Springer.

Eberl, Jason T., ed. 2020. Perspectives on disability. *National Catholic Bioethics Quarterly* 20(2): 217–219.

Eberl, Jason T., and Michael Rozier, eds. 2022. Perspectives on public health ethics. *National Catholic Bioethics Quarterly* 22(2): 197–199.

Francis. 2015. *Laudato Si'*. https://www.vatican.va/content/francesco/en/encyclicals/documents/papa-francesco_20150524_enciclica-laudato-si.html.

John Paul II. 1995. *Evangelium vitae*. https://www.vatican.va/content/john-paul-ii/en/encyclicals/documents/hf_jp-ii_enc_25031995_evangelium-vitae.html.

Martins, Alexandre A. 2019. *The Cry of the Poor: Liberation Ethics and Justice in Health Care*. Lexington, KY: Lexington Books.

Paul VI. 1968. *Humanae vitae*. https://www.vatican.va/content/paul-vi/en/encyclicals/documents/hf_p-vi_enc_25071968_humanae-vitae.html.

Romero, Miguel J., and Mary Jo Iozzio, eds. 2017. Engaging disability. *Journal of Moral Theology* 6(2): 1-254.

US Conference of Catholic Bishops (USCCB). 1993. A framework for comprehensive health care reform. http://www.usccb.org/issues-and-action/human-life-and-dignity/health-care/upload/health-care-comprehensive-care.pdf.

US Conference of Catholic Bishops (USCCB). 2023. Doctrinal note on the moral limits to technological manipulation of the human body, 20 March: https://www.usccb.org/resources/Doctrinal%20Note%202023-03-20.pdf.

Part I
Persons with Diverse Identities

Chapter 2
The Bioethics of Disability and Some Neglected Goods

Christopher Tollefsen

2.1 Introduction

This paper proposes some new directions within a bioethics that is concerned with human disability, namely, that such a bioethics should show a greater concern for the role that beauty, work, and play manifest in the lives of persons with disabilities. I call these goods "neglected" in the title of this chapter because, in one or another of several different ways, bioethics generally, and Catholic bioethics more particularly has not adequately addressed these goods in relation to human disability. The present chapter does not rectify that neglect except insofar as it proposes that more attention be paid.

2.2 Disability and the Needs of Flourishing

In an essay titled "Disability and Social Justice," published in 2010 in a Springer volume on disability, I argued, following thinkers such as Alasdair MacIntyre and Hans Reinders, that liberalism as a political theory has a difficulty in understanding certain kinds of disability in relation to classical liberalism's identification of freedom, equality, and independence as its guiding concepts in thinking about political citizenship (Tollefsen 2009; cf. MacIntyre 1999 and Reinders 2000). For a characteristic feature of disability is its rendering of certain human beings as *dependent*, sometimes profoundly so. But where liberal politics is modeled after a social contract, or some similar device, the contracting agents are typically understood as "fully cooperating members of society over the course of a complete life" (Rawls 1980, 546). Those who are profoundly disabled are thus left out of the social

C. Tollefsen (✉)
University of South Carolina, Columbia, SC, USA

J. T. Eberl (ed.), *Emerging Issues in Catholic Bioethics*, Philosophy and Medicine 153, https://doi.org/10.1007/978-3-032-08862-8_2

contract as contracting *agents*, and their status as *beneficiaries* of that contract is thrown into doubt.

By contrast, a natural law approach to politics sees the political realm as a continuation of the ethical, which is primarily concerned with questions of human flourishing. In even the best-off pre-political society of persons mutually and reciprocally pursuing human goods and caring for those who, because of some form of immaturity, infirmity, or disability, cannot care for themselves, certain needs will go unmet: needs for coordinated action, for defense against internal and external threats, and for the care of those who, through no fault of their own, in some cases cannot receive the care they need or, in other cases, cannot give the care they otherwise are obliged to provide. Political authority is thus needed just as much for human needs in which the disabled are implicated as are the abled—indeed, lack of self-sufficiency and the facts of dependence are seen, from the natural law perspective, as pervasive realities found in all human lives at some point and to some extent.

A crucial question for a perfectionist natural law approach to both ethics and politics is thus: what *are* the goods that are constitutive of human flourishing? For it is around those goods that an adequate ethics and politics will orient themselves, the former by identifying practical principles guiding the pursuit and protection of those goods, the latter by identifying the shape of a political common good towards the pursuit and protection of which political authority and law will be guided. In that earlier essay I identified life and health, knowledge, aesthetic experience, work and play, friendship, marriage, harmony with God, and the variety of forms of harmony possible within one's own complex person: harmony of choice and action; judgment and choice; judgment, choice, action, and emotion.

The attention of bioethics, including Catholic bioethics, has focused primarily, though not exclusively, on an important subset of these goods where persons with disabilities are concerned, and on a cluster of issues clearly related to those goods. Most importantly, a central question for Catholic bioethics has been making the case for the value of the lives and health of human beings with disabilities and in particular for the need for protection against the harms of abortion, infanticide, and euthanasia. Catholic bioethicists, including myself, have been focused on defending the claim that even very profoundly disabled human beings are nevertheless persons in the important metaphysical and moral senses: they are beings with a rational nature, possessed of the same natural and absolute rights against intentional attack and harm from others as are the readers and author of this essay (see, e.g., Lee and George 2008; George and Tollefsen, 2008; Jones 2009; Tollefsen 2009).

Similarly, the demands of justice require that persons with disabilities be given the same care that is expected and provided for the more fully abled: it is an injustice, for example, to refuse to feed and provide a medical intervention for a child with Down syndrome or some other disabling condition when a non-disabled child would be fed or treated (Grisez 1993, 526). So the complex good of life and health has been at the forefront of Catholic bioethics, as have the related questions of personhood and human dignity, and the moral issues of abortion, infanticide, euthanasia, artificial nutrition and hydration for those in a persistent vegetative state, genetic testing and screening, and the like. One might see, as an indication of these foci, the

"Resources" page on the National Catholic Bioethics Center's website under "Persons with Disabilities," which is primarily concerned with the issues just mentioned.[1]

Life and health is not the only good to which Catholic bioethics has attended. Friendship has been important, and Catholic thinkers have argued that the virtue of *solidarity* has a role to play in how we think of our relation to the disabled. Both Catholic and non-Catholic bioethicists have argued that "relationality" is essential to being a person and that the lives of disabled persons should be understood within the networks of care in which they are situated (Degnan 2008; Matthews 2013; Eberl 2020; Snead 2020). These concerns have been extended to a welcome emphasis on the importance of "inclusive communities" and attention to the way in which Church communities can foster the common good and belonging of the disabled and abled living together (Timpe 2018).

Further, both in Catholic thought and the wider world (as manifest in the Individuals with Disabilities Education Act [IDEA], for example) attention is given to the learning capacities of those with even significant cognitive disabilities, as prejudice against the cognitively disabled has given way to some extent to optimism and care for the cultivation of their potential for learning and their right to an education (Shakespeare 2017, ch. 6). Thus, the good of knowledge has not been neglected.

Now, one might think that if our concern is with finding learning pathways and institutions within which the cognitively disabled may thrive, then we have moved beyond the concerns of bioethics strictly speaking. Are the norms of IDEA, for example, a part of *bioethics*? It might seem not. And so, to move to the more central concerns of this paper, when we consider the pathways and institutions within which the disabled, whether cognitively, physically, or both, might flourish in regard to the goods of work, play, or the enjoyment and creation of the beautiful, it might again seem that we have moved beyond the domain of bioethics. So it is to the question of the scope of bioethics that I turn next.

2.3 The Bioethics of Disability

I want to make the case for a broad and inclusive understanding of bioethics as an initiative that, in what I will call its specific form of a *bioethics of disability*, should encompass concern for the flourishing of the disabled across all human goods. What could justify such an approach as specifically a part of *bioethics*?

I offer three reasons. First: I take it that a, and perhaps *the* central concern of bioethics, is the ethical and political norms of care and concern for those whose lives are, whether temporarily or permanently, shaped by the biological vulnerabilities of death, disease, disability, immaturity, and frailty. In this the discipline of bioethics shares its focus of attention with the practice and profession of medicine,

[1] https://www.ncbcenter.org/bioethics-resources.

the end of which is human health. But while medicine, on my view, focuses its concern in a special way upon the biological sources, and the cure and amelioration of, health-related vulnerabilities, particularly disease and disability, and is not in itself concerned with the whole of the life of the person who is biologically compromised, *bioethics* has a broader remit, viz., the all-around flourishing of the person who is, because biologically compromised, in a condition of vulnerability (Curlin and Tollefsen 2021).

This is why it is possible for bioethics to issue normative guidance on, for example, the authority of the patient as the last word regarding what forms of treatment offered by a physician are to be accepted. The physician's concern is to offer what she sees as beneficial in the domain of health to the patient; but bioethics, looking beyond the one good of human life and health to the wide range of goods constitutive of flourishing, prescribes that patients judge what is to be accepted or refused in light of the entirety of their vocation, with its personal orientation to the fuller range of human goods, specified in the patient's own personal life by their vocational commitments. Bioethics could offer no such guidance were its concern only with the single good of life and health and with health-related vulnerabilities only insofar as they had a bearing on the patient's health.

The object of bioethics thus goes beyond that of medicine to consider human flourishing more broadly. But this broader consideration is particularly needed in addressing questions concerning disability, to the extent that many or most "disabilities" are, whether from birth or by accident or other trauma or insult to the body, abiding or permanent conditions for the persons who suffer them. It is not enough, on my view, for bioethics to be concerned narrowly with the question, for example, of whether a fetal person diagnosed with Down syndrome is or is not a person, and whether such a person may, must, or must not be aborted, or must, or may not, be given life-saving treatment at birth. It is not enough, that is, for bioethics to be concerned only with threats to the life and health of such a person. For such a child's biologically-based vulnerability creates a potentially cascading series of effects for his or her complete all-around flourishing (Iozzio 2023). It is thus, on my view, part of the *bioethics of disability* that the various different ways in which these effects might follow and be recognized and combatted must be considered, evaluated, and addressed, both ethically and politically and, for a Catholic bioethics of disability, ultimately theologically.

This leads to my second reason for a broad approach to a bioethics of disability. There is a clear intersection of ways in which disability has cascading effects across all the goods of life with the dispute between the two governing paradigms of disability studies, the medical and social models. The medical model treats disability as entirely a matter of failure of biological functionality, with the "generic goal … to resolve the problem of dysfunctional and unhealthy bodies by moving them to a state of natural function and health" (Romero 2020, 290).

On the social model, by contrast,

> … to be disabled is to be excluded from community life, as that exclusion is experienced by persons who have a physical impairment or abnormality of some anatomical or psychological function. That state of exclusion occurs when arbitrary cultural preference for certain

> kinds of body function or appearances (e.g., an unexamined ethos regarding standards of beauty or what counts as unimpaired function) coalesces in the establishment of social and political limits to inclusion and participation (Romero 2020, 290–291).

A bioethics of disability must not neglect the real lessons of the medical model, and a Catholic bioethics, because of its general commitment to natural teleologies and its privation account of natural evils, cannot do so (see Colgrove 2020). Biological impairments are understood to be so from a standpoint that identifies what is an objective—even if species, sex, and age specific—health norm for an individual from which the individual has departed in a way that is a privation of the objective good of health for that individual. A Catholic bioethics of disability cannot discard the medical model except at the risk of similarly discarding much of the natural philosophy and metaphysics that is part of its intellectual heritage.

But neither can it neglect the lessons of the social model of disability, which by its nature opens the issue of disability up to the larger constellation of concerns with human flourishing overall. For the social model identifies not merely the ways in which impairments lead to health-related disabilities—as, for example, when being born blind leads to additional threats to one's life and health—but also the way in which impairments bear upon the entire range of goods due to lack of access, prejudice, ignorance, and unconsidered side-effects of otherwise well-intentioned social decisions, as when blindness leads to diminished opportunities for education, work, and the enjoyment of the fruits of culture because of ableist paradigms and structures.

The social model draws our attention to the ways in which society's response or lack thereof to impairment disables persons with respect to their ability to obtain and keep work, to pursue educational opportunities, to participate in a variety of forms of play, to have access to opportunities to enjoy and make art, and even to engage in religious worship (adults with disabilities are almost 40% more likely than others never to have attended a church, synagogue, or other place of worship according to Timpe 2018, 91). Thus, what some would call a hybrid model of disability converges with my claims about a bioethics of disability: it must be sensitive to the entire range of human goods implicated by the intersection of human biological impairment and societal barriers and discriminatory attitudes (John Paul II 2004).

Earlier, I promised three reasons why a bioethics of disability should give attention to the goods of work, play, and aesthetic experience. Here is the third. An important strain of contemporary bioethics of disability is concerned with the "expressive" argument about selective abortion of persons diagnosed *in utero* with some or other form of disability (see Asch 1988, 2000). Does the practice of selective abortion express a view that the life of a person with disability is not worth living, that it would be better if that person were not to be permitted to continue to live? The fact that very few children with Down syndrome are now born in certain countries, such as Denmark, suggests that the message is getting out (see Zhang 2020).

There are undoubtedly multiple reasons for the high rate of abortion of children diagnosed with a disability such as Down syndrome, including the recognition of the genuine difficulties of caring for such children. Yet among those reasons is the ongoing perception that disabled persons, including children and adults with Down syndrome, do not and cannot lead flourishing lives. But participation in work, enjoyment of play, and delight in the beautiful are all constitutive aspects of human flourishing in which persons with Down syndrome can and do take part in a wide variety of ways. Lack of attention to these aspects of a flourishing life perpetuates an error: that the lives of such persons do not or cannot go well for them. Attention to these goods can be an important, even if subsidiary, part of making a case against selective abortion or discarding of embryos on the basis of preimplantation genetic diagnosis.

So what, then, can be said about these goods in the context of a bioethics of disability? It is to this question that I turn next.

2.4 The Neglected Goods: Conceptual Questions

In the rest of this essay, I wish to raise a series of questions that I think have not adequately been addressed by a bioethics of disability. One could divide the terrain into questions of a conceptual, ethical, political, and theological nature; questions of all four sorts can be raised in regard to the three goods of play, work, and aesthetic experience. In this essay, I address primarily only conceptual and ethical questions, leaving political and theological questions largely to the side for reasons of space.

2.4.1 Work and Play

There is a general conceptual question concerning the nature and relationship of the goods of work and play, namely, are these one good or two. If one, why distinguish them? If two, how are they related?

I have argued elsewhere that there is a core feature of both work and play that identifies the central good-making feature of both: excellence of performance (Tollefsen 2020). It is excellence of performance that can be pursued for its own sake and that, if so pursued, is perfective of human persons and an aspect of their genuine fulfillment. What then differentiates work from play? It is that the standards of excellence in work are largely taken from the nature of the product resulting from the performance and the needs or desires that product serves, whereas the standards of excellence for play are in some sense internal to the performance itself. Thus, what it means to work well as a doctor, an architect, or a carpenter is determined in great part from the external product: the restoration of health, the creation of a building, the construction of an armoire. By contrast, what it means to play a game well is determined by the creation of an order of activity that exists simply for its own sake, for the sake of the performance, and not by any external set of needs. But the core of the good in either case lies in the performance, whether the norms of excellence for that performance are shaped externally or internally.

The significance of this conceptual issue for a bioethics of disability can be seen in a recent essay in the *Journal of Medicine and Philosophy* by Stephen Campbell, Sven Nyholm, and Jennifer Walter (CNW henceforth), "Disability and the Goods of Life" (Campbell et al. 2021). CNW divide the field of disability into four areas, including sensory, mobility, intellectual, and social disabilities. They then investigate the compatibility of such disabilities with four aspects of human well-being: happiness, knowledge, rewarding relationships, and what they call "achievement." They take "achievement to be comprised of a process and a product. The product must be competently caused, with effort, by the achiever," and they argue that, so understood, persons with each of the four sorts of disabilities are capable of the good of achievement (Campbell et al. 2021, 709).

There is a clear overlap between CNW's understanding of the good of achievement and my understanding of the good of work and play. Like CNW, I recognize that what "excellence of performance" means will be relative in certain ways. While for most non-impaired adults, for example, walking is understood as neither work nor play, for children and many persons with mobility-related disabilities, walking and learning to walk is work, and can realize the core of that good: both learning to walk, and engaging in the performance of walking as best one can is thus participation in a good that is constitutive of human well-being.

Similarly, for both CNW and myself, the good in question will often overlap with the pursuit of all other goods, since pursuit of any good – e.g., knowledge or rewarding relationships – will itself require excellence of performance and, if successful, result in achievement. Or, to take a different example, while the enjoyment of a beautiful work of art is typically not a form of work, nevertheless, coming to possess the skills necessary for that enjoyment often is work, a point I will return to later.

So there is overlap between the CNW account and my own. But focusing, as I do, on excellence of performance as the core of this good enables me to emphasize to an important extent *process*, or, better, *action* over *results*. To be engaged in the performance—the activities—aimed at achievement is itself the core of the good, and it is important not to risk evacuating such activities of meaning in the event of failure to "achieve" in a conventional sense, nor to focus exclusively, for example, on "productivity" as central to the meaning and significance of work (John Paul II 1981). Thus, it is important, as a conceptual matter, to accurately identify the nature of the good in question: as between achievement and excellence of performance, I think we should opt for the latter.

One area in which this reconceptualization might bear fruit concerns the processes of rehabilitation for those struggling to recover functionality after illness or accident. Obviously, the instrumental benefits of rehabilitation therapy seem to many important enough to warrant the directing of resources towards such therapies. Yet Tom Shakespeare notes that "[r]ehabilitation is often neglected … In healthcare, rehabilitation is low status and low priority. In disability research, rehabilitation is overlooked …" (Shakespeare 2017, 95). Shakespeare does not say why rehabilitation is neglected, but perhaps one possible reason is that it is *overly* instrumentalized. But the work of rehabilitation is just that: work; and it is of value even apart from the obvious benefits it can bring. Understanding this gives us all the more reason to put our efforts and resources into rehabilitation whenever possible.

I would like briefly to explore a possible implication of the "relativity" of excellence in performance. We should expect that in at least some contexts, work and play carried out by persons with various disabilities will look differently to us from work and play carried out by non-impaired individuals. Traditionally, such a claim would be understood in the following way: the "relativity" in question justifies a lessening of expectations and an accommodation for work that is not as good as that performed by the non-impaired.

But this is not the kind of "looking differently" that I am suggesting here. Rather than work and play "looking differently" because they fail to meet otherwise normative expectations, what I am envisaging will instead be a result of a variety of overlapping and mutually reinforcing factors: the compensatory and creative strategies adopted by persons unable to approach activities and tasks in the "standard" way; the (related) use of prosthetic devices and "hacks"; the reconceptualization of activities and tasks by persons who might see them in very different ways than the non-impaired; and in some cases, the growth and development of a culture centered around the ways that persons with a particular disability or cluster of disabilities approach work and play. So participants in a culture shaped by Deaf identity, for example, or persons with other disabilities living, working, and creating a material and moral culture together, might perform, and produce, in ways that differ from the performance and production of the non-impaired and in the process of doing so create new standards of excellence of performance. Doing this might challenge the ability of the non-impaired to recognize excellence in performance and perhaps even excellence of product in some cases, a possibility that raises ethical questions to be addressed later.

2.4.2 Aesthetic Experience

I note, in defense of my claim that the goods under discussion are neglected, that aesthetic experience warrants only a brief mention in a footnote by CNW. Once again, I believe its inclusion among the goods of a well-lived life raises essential conceptual issues even before we get to ethical questions, and I here mention two, one very general to aesthetics but relevant here, and one specific to questions of disability, but overlapping the recent discussion of work.

The general conceptual question is similar to that regarding work and play: where does the core of the good reside, so to speak? In recent work, John Finnis has tended to incorporate the good of aesthetic experience into the good of knowledge, where once he separated them (Finnis 2011). But I think this is an error. The core of aesthetic experience is the enjoyment of the beautiful, and this experience draws on different capacities, and realizes different potentialities of the person. Interestingly, some evidence for this claim comes from the realm of disability. Consider, for example, Sesha Kittay, perhaps as profoundly an intellectually disabled person as one is likely to find discussed in the philosophic literature on disability. Sesha is incapable of speech, and it is unclear how impaired she is cognitively. What is clear to her mother, Eva Feder Kittay, is that Sesha take great enjoyment in music:

> One needs nothing more than to watch this woman—who has no language, who is still incontinent, who cannot feed, dress, or in any way care for herself—listen to a Bach partita, a Beethoven symphony, one of her beloved Broadway musicals, or best of all watch her thrill to the last movement of Beethoven's Ninth, to discern the riches of her life. She listens to music as I do—at my best, that is—in a concert hall (Kittay 2019, 52).

The aesthetic enjoyment of music for Sesha seems clearly both a good, and different from the good of knowledge.

Moreover (and this will be important below), the demands of the production of aesthetic experience are likely, if it is different from the good of knowledge, to be different from the demands of the production of knowledge, just as the production of knowledge differs, e.g., from the demands of the production of health. If aesthetic experience involves the enjoyment of the beautiful, then, just as the good of work and play will be imperiled if instrumentalized in relation to product, so will this good be imperiled if instrumentalized to some other good such as knowledge.

Nor is knowledge the only good that can be mistakenly identified as the core good here. Viewing art in general as *primarily* about expression of some truth or condition of the artist, or about political protest, or the creation of a community, or about any other purpose to which a work of art might be put, can be an impediment to both the creation and appreciation of good art. This does not mean that good art cannot be expressive—it almost always is—or have for its underlying matter the materials of protest, as it often does. But subordinating or expecting subordination of aesthetic values to expressive or political values will again be an impediment to the creation and enjoyment of good art.

My view here may be contrasted with that of the Community Arts Movement, which some thinkers have incorporated into an approach to Disability Arts. Ryan Evans, for example, cites Community Arts theorist Owen Kelly in identifying "[t]wo important tenets of the community art movement—collectivism and social change … Kelly asserts that community artists should be 'concerned with collective creativity' and that they should want their art 'to be an active force for change' within their communities" (Evans 2022, 3). Evans then uses this understanding to theorize an approach to disability art that would similarly center collectivism and social change. Proponents of a "disability aesthetic" are likewise oriented towards the political uses of art by and for people with disabilities (see, e.g., Siebers 2010).

By contrast, I think we should not expect that art by or for disabled persons must primarily be a window into knowledge about disability, or expressive of what it is like to live with disability, or subordinated to the political goals or rights claims of disabled persons, even when, as it happens, it *is* such a window, or is so expressive or politically purposive (and again: it often will be). Art made or enjoyed by persons with disabilities must above all be beautiful to realize the core of this good.

So then is there no point to speaking of "disability arts" or an "aesthetic of disability"? I think an answer to this query is to be found in the earlier discussion of work. To the extent that art is created by persons with disabilities, and particularly within a subculture centered on a disability or disabilities, then a certain kind of relativity might be expected to emerge in which persons engaged in creating and developing an aesthetic culture together might create, perform, or produce, art in ways that differ from the performance, creation, and production of art by other persons, and in the process of doing so, create new standards of excellence of creation and performance. Once again, this might well challenge the ability of the non-impaired to recognize genuine excellence of performance and product in some cases. There is a parallel here to the way in which certain forms, e.g., of music in radically different cultural settings will be sufficiently alien from one another as to challenge the ability of persons from one culture to recognize beautiful music of another. I explore this possibility further below.

2.5 The Neglected Goods: Ethical Questions

The investigation of the goods of work, play, and aesthetic experience in the context of disability raises many ethical questions. Here, I briefly identify only some that seem to me of prominent importance. I divide the questions of interest to me into three "types": questions of access, questions of integration and segregation, and questions of recognition. All three kinds of questions raise additional political questions; while I do not raise those questions here, many of them will intuitively strike the reader as continuous with the ethical questions discussed.

2.5.1 Work and Play

In the domain of work and play, if we acknowledge that work and play are constitutive aspects of human well-being just as much as are health, knowledge, or friendship, and that persons with disabilities of many sorts across many different contexts can realize and benefit from participation in these goods, then it is ethically important to ensure that there is adequate access to opportunities for work and play commensurate with the—real, not imagined—capacities of the persons in question.[2]

Identifying what this means, concretely, has been challenging, historically, for a reason familiar to proponents of the social model: what persons with disability can *in fact* do has to a very great extent historically been limited by discriminatory attitudes and beliefs about what they can do. So just as diminished expectations about the capacity of persons with disabilities to learn have led to diminished educational

[2] C.f. the words of Pope John Paul II: "But it is no less important that there should be a change of heart, a conversion, on the part of every citizen and every group in society, so that they may willingly and fraternally accept the presence of handicapped persons at school, *at work and in every activity, including sport* (John Paul II, 1981, my emphasis).

opportunities, so have diminished expectations about the capacity of persons with disabilities to play or to work led to diminished opportunities to play and work.

In his discussion of inclusive communities, Kevin Timpe makes the ethical point about access to the good of play in a vivid way:

> Inclusion requires full opportunity for participation and value in one's community. Many local sports leagues, for instance, don't forbid individuals with disabilities from playing. They don't have to because they indicate in other ways that individuals with disabilities are not expected. Perhaps not wanted. Think about it for a minute: is your local soccer or lacrosse league structured in a way that makes individuals with disabilities feel welcome or valued (Timpe 2018, 85).

This suggests the need to rethink the issue of access to play for children with disabilities in a way that is not inherently segregated, an issue I return to below.

One might ask a similar question about one's work spaces: do they reveal themselves, physically and socially, to be receptive or inhospitable to persons with disabilities? In a discussion of the role of the chair in our collective work-life, Sara Hendren raises the question of what good design for people with particular needs might be like:

> In the case of disability, that would mean closely observing older adults and wheelchair users and people on the autism spectrum, say, and then abstracting from those observations, aggregating those clues for insight. What's working for these people, and what isn't working, based on what they say or do, or what they *don't* say or do? (Hendren 2020, 82)

As a side point, it is worth here making a distinction between work and employment. Employment is only one of the variety of contexts in which work is done, and it would be a mistake to identify the two. Yet the typically public nature of work done in an employment context makes it an important signifier for how seriously a society takes questions of access. As with Timpe's local soccer league, thinking about how our employment spaces are open or antagonistic to persons with disabilities is a good place to begin thinking about access to work and play more generally.

Raising the question of access very quickly leads to a related question concerning the structure of that access. Two different models are available by which to consider how one's moral obligations might be met.

For much of the first part of the twentieth century, significant forms of disability, and in particular cognitive disability, were addressed by some form of segregation. This included segregated education, institutionalized living, and a general desire to keep persons with significant disabilities out of sight. In the area of work and employment, segregation was manifest in the phenomenon of sheltered workshops, businesses that primarily employed workers with disabilities, and often remained outside the scope of labor regulations and minimum wage requirements.

Subsequent to the advances made by disability rights advocates, segregation in many of these areas has become less common, and certainly less socially expected and enforced. It is widely recognized, for example, that for education, integration is, to the extent reasonably possible, desirable both for the welfare of students with disabilities, and for the good of those who thereby learn to study, play, and spend time with a wider range of persons.

The disability rights movement has likewise resulted in gains where employment is concerned; the Americans with Disabilities Act of 1990 (ADA) makes it illegal to discriminate on the basis of disability, and as a result it is less uncommon today than in the past to see persons with disabilities at work in grocery stores, malls, and restaurants. Yet it is not *very* common, and disabled persons still remain unemployed or underemployed at higher rates than the non-impaired. This, in turn, is behind a development in the creation of employment opportunities for disabled persons that in some ways seems like a return to segregation and in other ways is clearly not. Consider, for example, the coffee shop chain Bitty and Beau's. From their website's description:

> Amy & Ben Wright are the proud parents of four children—Lillie, Emma Grace, Beau and Bitty. Lillie was born with autism, and Bitty & Beau were born with Down syndrome. After learning over 80% of people with intellectual and developmental disabilities are unemployed nationwide, the Wrights decided to create a place where people with disabilities are valued & given opportunity for meaningful work. The model? A coffee shop, with an unforgettable staff that changes the way people see other people. The original shop opened January 2016 in Wilmington, NC. It employed 19 individuals with disabilities (Bitty and Beau's 2024).

According to their website, Bitty and Beau's has gone on to open 18 shops and employ over 400 people; a Bitty and Beau's has recently opened in my own home town.

Bitty and Beau's differs from sheltered workshops in several ways: for example, there is no exemption from minimum wage laws. And because it is part of the service industry, Bitty and Beau's is public-facing in a potentially beneficial way. The Columbia South Carolina Bitty and Beau's, for example, is in the Vista area, a handful of blocks filled with trendy restaurants and hotels and frequented by young professionals. Bitty and Beau's offers an opportunity for non-disabled persons and persons with disabilities to interact in a workplace setting that is not much found elsewhere.

Yet disability advocates raise important moral questions about enterprises such as this. It is not to my purpose here to answer these concerns, but merely to sketch them. They seem to me of two basic kinds. First, there are questions to be asked about what constitutes fair and just forms of remuneration and advancement. Are employees with disabilities offered enough working hours and sufficient wage increases to increase their possibilities for independent living? Are they offered sufficient opportunities for promotion and other forms of advancement within a company that they can continue to develop their skills and excellences? The potentialities of disabled persons have historically been limited by limited expectations, and if expectations are not set higher, and more creative ways sought to meet those expectations, then workers with disabilities will not excel to the greatest extent possible.

The second kind of objection comes from the side of the business model that in one way is most promising, its public-facing quality. Bitty and Beau's, and similar ventures, offer the opportunity for the disabled and non-impaired to share space and time together; they offer a path to integration. But what motivates the non-impaired in this relationship? One journalist, reflecting on her experience at a café in Paris similar to Bitty and Beau's puts the problem like this:

> Looking back on my visit to Café Joyeux, I could see how some customers might treat it as a novelty experience rather than a legitimate place to enjoy coffee and baked goods. But others might go because they like the kind of customer service and social experience that it provides. But if both perspectives come from a place of othering, does this place increase the social isolation that people with disabilities suffer?

I identify these questions here as primarily ethical, because they reflect at root a concern with our attitudes, emotions, and personal choices: for example, how are the non-impaired best to find ways to relate to, include, and acknowledge, persons with disability in the domains of work and play in an inclusive and fully respectful way and, importantly, a way that involves equality of concern rather than pity or condescension?[3] Both in the context of employment, and in the context of the arts, to be mentioned below, disability advocates manifest a concern that persons with disabilities not be put "on display" for the edification or entertainment of others.[4] Yet inclusion does not happen without intentional effort on the part of the non-impaired. What genuine inclusion looks like in this context is thus an important question that requires further deliberation and discussion.

Having raised both the question of access and the question of employment model, I now mention a third ethical issue, one that arises directly from the conceptual discussion of the previous section. There, I suggested the following possibility: that because of a number of overlapping factors, it might not always be immediately obvious when persons with disability are doing good work (or, *mutatis mutandis*, playing well). Not only might a person with disability be working to capacity but the work of disabled persons might meet genuine standards of excellence not apparent to those around them.

This matters because of a principle of justice: good work deserves recognition. Failure to recognize such work is a form of injustice, and failure to be aware of the possibility of that failure is itself an additional failure of justice. It is thus incumbent upon those working with disabled persons, or judging their work, that they be open to the possibility that their good work will be unlike the good work of their non-disabled colleagues. Yet there is a danger here as well. Good work is at the end of the day a normative notion; its relativity is not subjectivity, nor is it relative all the way down. If it is an injustice to fail to recognize work, it is also a failure of honesty to praise all work equally, even when done poorly; and awareness that any work at all will be praised disincentivizes good work.[5] So "recognition" names both a norm of justice and implies the presence of a real standard: what ought to be recognized is work that is genuinely good.

---

[3] Harold Braswell notes the objection often found in disability rights literature that "viewed as objects of pity, disabled people are trapped in an eternally dependent state that seemingly justifies their continued marginalization," and offers a religiously grounded conception of pity, linked to the virtue of *caritas*, as a corrective (Braswell 2022).

[4] For wide ranging treatments of this phenomenon, see the essays in Thomson (1996).

[5] See the interesting discussion of the tensions between excellence and equality in Gardner (1961).

2.5.2 Aesthetic Experience

Once again I begin with the question of access. If aesthetic experience is a constitutive aspect of human well-being, then it is essential for persons with disabilities that they have access to that good. Access here must encompass both the enjoyment of works and other objects of beauty, and the creation of such works; it also must encompass opportunities to display and perform creative works of art for those on the creative side the arts.

Consider, for example, the rise of "sensory friendly" hours and programs at museums around the country. "Morning at the Museum" is a program operated by the Smithsonian Museum to offer a quieter, less crowded and less vividly lit, non-judgmental museum experience to children with, *inter alia*, autism or other sensory disabilities (Morning at the Museum 2024). Meeting such needs is distinct from meeting the perhaps better known need for structural accessibility, often mandated by the ADA, that makes possible attendance by, for example, wheelchair users, at concerts, art spaces, theatrical performances and the like, or needs for information to those incapable of sight or hearing in contexts where print or speech are the dominant modes of communication; of course, Braille programs at a concert or spoken-word tours at a museum exhibit are also important for providing access.

"Morning at the Museum" opens a window onto the important intersection of aesthetic experience, work, and education for making access possible. For art to truly be accessible to anyone requires some amount of initial labor and investment: children need to be initiated into the "language" of art, must be allowed to develop adequate familiarity with different genres and media of art, and must be encouraged to practice the forms of attention and contemplation that make appreciation and enjoyment of beauty possible. All these initial steps to making enjoyment of the beautiful possible are, I believe, increasingly neglected for all children to the extent that they are encouraged to spend an abundance of time with electronic screens and constant sensory stimulation. The challenges for creating the conditions of access in this domain are often even greater for children with disabilities, but must be embraced if this avenue of flourishing is to be made available. Had Sesha Kittay, for example, been deprived of the early opportunity to listen to Bach, and surrounded instead only with fast-moving video screens, blinking lights, and repetitive noises, perhaps because it was thought that these were all she was "capable" of attending to, her opportunities for flourishing would have been greatly and unjustly limited.

One important dimension of the good of aesthetic experience is aesthetic creation. The creation of art is, for some people, a form of play, and for others becomes work in the full sense, but creation is typically linked also to the joy of experiencing the work that is created. For the young, the standards of making art "well" are deeply internal to the performance and relative to age and capacity: everyone who has taken pride in a child's drawing knows this. Engagement in such creation is valuable and should be encouraged for all. But talent and a vocation to a life of artistic creation cut across many disabilities. The recognition of such talent can be blocked by prejudice and ignorance; and access to the forms of education and the availability of resources that make the development of talent possible is again something that should be made available to those who historically have been overlooked as candidates for skilled artistic creation.[6]

Just as there are parallels over the issue of access between work and play and aesthetic experience there are also parallels concerning integration and segregation. In an essay which has greatly influenced my thinking here, Carrie Sandahl, a disability scholar and artist at the University of Illinois at Chicago, looks at both the formation of different kinds of communities of artists and at the kinds of art made in such communities, specifically focusing on dance, but intending what she says to have application to the arts more broadly (Sandahl 2018). She identifies four models of dance that parallel in certain respects the earlier discussion of structures of work and employment.

Thus, to begin with, there is "mainstream" dance, of which ballet would be a paradigmatic instance. Such dance is "steeped in tradition and elitism, only available on a professional level to the most extraordinarily able-bodied dancer" (Sandahl 2018, 87). Implicitly, mainstream dance suggests a segregated future for dance from which disabled persons are excluded. Unsurprisingly, dance communities open to disabled persons resist such exclusion and move towards a form of integration that Sandahl terms "physically integrated dance," in which disabled and non-disabled dancers share the stage. Dancers with wheelchairs, for example, are integrated into dance involving persons lacking limbs as well as fully abled dancers in an example Sandahl gives from the dance troop Axis.

But Sandahl identifies two difficulties here that we have seen before. One is that the implicit standards setting the norm in physically integrated dance are ultimately those of mainstream dance, which "flex" to accommodate the disabled. The other is that an important source of interest for audience members in such dance is seeing how a person with this or that disability has managed to "fit in," what they have "overcome" and so on. This diversion of attention can distract from the aesthetic features of the performance. At the same time, the obvious contrast on stage between disability and non-impairment can call attention to disability and difference, again at the expense of attention to aesthetic merit, and perhaps while satisfying some non-aesthetic curiosity of the audience.

Sandahl contrast these models with two others. The first is "disability dance," dance "exclusively performed by dis-abled people (and most often composed by disabled choreographers)" (Sandahl 2018, 91), taking as her example the piece *Descent from Beauty*, performed entirely by wheelchair-users on a twisted and ramped stage designed to highlight the specificity of wheelchair motion. Disability dance here obviously does not serve the project of integration as such. But it does serve to create and establish new forms of dance that "take advantage of insider knowledge of how wheelchair users experiences space, time, partnering, and gravity," and center disabled bodies without establishing abled bodies as the comparison

---

[6]The life of Judith Scott is illustrative of many of the points I am attempting to make here; see Wikipedia, "Judith Scott (artist)".

points. And that seems helpful for considering Sandahl's fourth model, which returns to integration, but this time from the side of the disabled dancer.

One Breath is an example of "new integrated dance," choreographed and performed by the disabled dancer Lisa Bufano, who is known for her use of unusual prosthetics, such as Queen Anne table legs, when dancing. In *One Breath*, the non-disabled dancer also uses prosthetic limbs, and the as the dancers become intertwined "they become 'integrated' into a new dance form that is made possible because the choreography began with disability experience and incorporated a non-disabled dancer, rather than the other way around" (Sandahl 2018, 93). Sandahl sees this as an emerging form of both dance as an art, and of dance-and-disability culture unburdened by mainstream expectations. As in the discussion of work and play, the question of how to structure a world that includes both disabled and non-impaired persons making and enjoying art together requires further consideration of both an ethical and aesthetic nature.

This brings us to the third parallel discussion in the domain of aesthetic experience to the discussion of work and play, concerning recognition. In a recent article in *Ethics*, Rachel Fraser has argued for the existence of what she calls "aesthetic injustice," injustice committed when an agent's work is insufficiently appreciated or judged deficient, despite its merits, when that judgment is connected to broader patterns of social advantage and disadvantage (Fraser 2024). Fraser does not identify disability in her essay as among the forms of disadvantage that can generate aesthetic injustice, but clearly persons with disabilities *are* a locus of social disadvantage, as again the social model of disability has stressed.

But further, and in line with a repeated theme of this essay, art that arises out of the capacities, culture, and experience of disability is likely to be different—as can be seen vividly, for example, in the *One Breath* dance, which departs in many ways from mainstream dance. As with, for example, music from very different cultures or time periods, or twelve-tone music in Western composition, it might be difficult or impossible to immediately appreciate and enjoy an art form emerging from disabled persons or communities of disabled persons without adequate openness to and initiation into that art form. To judge that art deficient absent such an effort to engage, understand, and appreciate might well be a form of aesthetic injustice, a failure to give due recognition (as, of course, would judging it good and worthwhile independent of its actual merits).

Such failures of judgment would also be failures of human flourishing. The human capacity for fulfillment and the human community at large are benefited by new forms of realizing and appreciating the beautiful. As in the case of work and play, the possibility (indeed, I think, the reality) that art emerging from or centering disabled persons and disability communities might open up new opportunities for such flourishing requires further investigation and the obstacles to such opportunities need to be scrutinized as potential forms of injustice.

2.6 Conclusion: Reversing the Cascade

Earlier, I noted that biological impairments can create a cascading series of effects for a person's complete all around flourishing, negatively impacting access to and enjoyment of the whole range of human goods. In concluding this essay I wish to note the ways in which attending to the goods of work, play, and aesthetic experience might reverse that cascading series of effects. Consider first what is perhaps the most obvious and frequently sought benefit of creating opportunities for work and play: friendship and community.

We do things with friends, and we are friends with people with whom we do things. Creating opportunities for children with disabilities to engage in sports, and for adults to work, play recreationally, and compete in sporting events creates at the same time the opportunity for the growth of individual friendships and of communities of persons committed to pursuing the same goods. The same is true of arts communities that are either inclusive of or constituted by persons with disabilities. Indeed, for some, friendship and community might be the first reason to become involved in a sport, work, or artistic community inclusive of disabled persons; but there is no reason that multiple intrinsic goods cannot be pursued at the same time, and these neglected goods should not become overly instrumentalized.

No one can play, work, or create or enjoy art without *some* degree of knowledge, even if unarticulated and tacit. Learning a skill, recognizing opportunities in the world, and building up an awareness of available instrumentalities are all instances of knowledge-gain that can be enjoyed for their own sake as well as for the sake of the possibilities they bring. And it seems highly likely that those who work and play, compete in sports, or in a variety of forms of artistic creation enjoy health benefits from doing so. The intersection of the variety of goods is clear: one who is given no opportunity to play or work with others is unlikely to have friends, is cut off from possible sources of knowledge, and will very likely be less healthy than they could be.

And finally: to the extent that persons with disabilities work, play, and create amongst people who are Christian believers and witnesses to the Good News who treat them with dignity and respect, to that extent will Christianity be seen as a religion open to all and offering hope to all, as indeed it must be if the Church is to fulfill her mission in the world.

References

Asch, Adrienne. 1988. Reproductive technology and disability. In *Reproductive laws for the 1990s*, ed. Sherrill Cohen and Nadine Taub, 59–101. Clifton, NJ: Humana Press.

Asch, Adrienne. 2000. Why I Haven't Changed My Mind about Prenatal Diagnosis: Reflections and Refinements. In *Prenatal testing and disability rights*, ed. Eric Parens and Adrienne Asch, 234–258. Washington, D.C.: Georgetown University Press.

Bitty and Beau's. 2024. https://www.bittyandbeauscoffee.com. Accessed 23 Jul 2024.

Braswell, Harold. 2022. Take pity: What disability rights can learn from religious charity. *Journal of Medicine and Philosophy* 47:638–652.
Campbell, Stephen M., Sven Nyholm, and Jennifer K. Walter. 2021. Disability and the goods of life. *The Journal of Medicine and Philosophy* 46:704–728.
Colgrove, Nicholas. 2020. The (in)compatibility of the privation theory of evil and the mere-difference theory of disability. *National Catholic Bioethics Quarterly* 20:329–348.
Curlin, Farr, and Christopher Tollefsen. 2021. *The way of medicine: Ethics and the healing profession*. Notre Dame, IN: University of Notre Dame Press.
Degnan, Michael. 2008. Are we morally obliged to feed PVS patients until natural death? In *Artificial nutrition and hydration: The new catholic debate*, ed. Christopher Tollefsen, 39–61. Dordrecht: Springer.
Eberl, Jason. 2020. Addressing vulnerability due to cognitive impairment through Catholic social teaching. *National Catholic Bioethics Quarterly* 20:243–250.
Evans, Ryan. 2022. Taut threads, spanning ambivalences: Lessons from disability arts practitioners about how to make sense of arts-based social change. *Field: A Journal of Socially Engaged Criticism* 20. https://field-journal.com/issue-20/taut-threads-spanning-ambivalences-lessons-from-disability-arts-practitioners-about-how-to-make-sense-of-arts-based-social-change/.
Finnis, John. 2011. *Natural law and natural rights*. 2nd ed. Oxford: Oxford University Press.
Fraser, Rachel. 2024. Aesthetic injustice. *Ethics* 134:449–478.
Gardner, John W. 1961. *Excellence: Can we be equal and excellent too?* New York: Harper and Row.
George, Robert P., and Christopher Tollefsen. 2008. *Embryo: A defense of human life*. New York: Doubleday.
Grisez, Germain G. 1993. The way of the lord Jesus. In *Living a Christian life*, vol. 2. Quincy, IL: Franciscan Press.
Hendren, Sara. 2020. *What can a body do? How we meet the built world*. New York: Riverhead Books.
Iozzio, Mary Jo. 2023. *Disability ethics and preferential justice*. Washington, D.C.: Georgetown University Press.
John Paul II. 1981. Encyclical letter *Laborem exercens: On human work*. https://www.vatican.va/content/john-paul-ii/en/encyclicals/documents/hf_jp-ii_enc_14091981_laborem-exercens.html.
John Paul II. 2004. Message of John Paul II on the occasion of the international symposium on the dignity and rights of the mentally disabled person. https://www.vatican.va/content/john-paul-ii/en/speeches/2004/january/documents/hf_jp-ii_spe_20040108_handicap-mentale.html.
Jones, David A. 2009. Incapacity and personhood: Respecting the non-autonomous self. In *Incapacity and care: Controversies in healthcare and research*, ed. Helen Watt, 8–23. Oxford: The Linacre Centre.
Kittay, Eva Feder. 2019. *Learning from my daughter: The value and care of disabled minds*. New York: Oxford University Press.
Lee, Patrick, and Robert P. George. 2008. *Body-self dualism in contemporary ethics and politics*. New York: Cambridge University Press.
MacIntyre, Alasdair. 1999. *Dependent rational animals: Why human beings need the virtues*. Chicago: Open Court.
Matthews, Pia. 2013. *Pope John Paul II and the apparently 'non-acting' person*. Leominster: Gracewing.
Morning at the Museum. 2024. https://access.si.edu/program/morning-museum. Accessed 23 July 2024.
Rawls, John. (1980). Kantian constructivism in moral theory. *Journal of Philosophy 77*: 515–571.
Reinders, Hans R. 2000. *The future of the disabled in liberal society: An ethical analysis*. Notre Dame: University of Notre Dame Press.
Romero, Miguel. 2020. Disability, Catholic questions, and the quandaries of biomedicine and secular society. *National Catholic Bioethics Quarterly* 20:277–310.

Sandahl, Carrie. 2018. Disability art and culture: A model for imaginative ways to integrate the community. *Alter: European Journal of Disability Research* 12:81–95.
Shakespeare, Tom. 2017. *Disability: The basics*. New York: Routledge.
Siebers, Tobin. 2010. *Disability aesthetics*. Ann Arbor, MI: University of Michigan Press.
Snead, O. Carter. 2020. *What it means to be human: The case for the body in public bioethics*. Cambridge, MA: Harvard University Press.
Thomson, Rosemarie G., ed. 1996. *Freakery: Cultural spectacles of the extraordinary body*. New York: New York University Press.
Timpe, Kevin. 2018. *Disability and inclusive communities*. Grand Rapids, MI: Calvin College Press.
Tollefsen, Christopher. 2009. Disability and social justice. In *Philosophical perspectives on disability*, ed. Christopher Ralston and Justin Ho, 211–227. Dordrecht: Springer.
Tollefsen, Christopher. 2020. The good of play in John Finnis's *Natural law and natural rights*. *Revista Persona y Derecho* 83:571–590.
Wikipedia, Judith Scott (artist): https://en.wikipedia.org/wiki/Judith_Scott_(artist). Accessed 6 Nov 2024.
Zhang, Sarah. 2020. The last children of down syndrome. *The Atlantic Monthly*, December.

Chapter 3
The Importance of Theological Anthropology for Bioethics in the Key of Disability

Mary Jo Iozzio

3.1 Introduction

Early in my graduate studies I learned that "all theology is anthropology."[1] It is easy enough to recall the assertion, yet I wonder how to consider its insight for my musings in ethics about what is right or wrong. Those musings depend upon a variety of systems—e.g., Natural Law, Divine Command, and Virtue—that attempt to claim and conclude what is normative for human beings. If it is true that each of us has been created in the image and likeness of God, manifest expressly in Jesus of Nazareth, the crucified Incarnate God, then those systems require a reckoning with the disabled Risen Christ of Faith. Thinking in the key of disability brings the assertion on anthropology to the center of a theological bioethics that reflects the relational and dependent realities of all humankind.

In this chapter I consider how a theological bioethics based in an anthropology of Trinitarian relationality reveals a norm of radical dependence for all humankind. This anthropology recognizes the Christ of faith as Jesus risen from the dead bearing the disabling scars of the crucifixion. This norm correlates the interdependence found in philosophical understandings of relations in the Trinity with the radical dependence of relations in humankind. Such dependence is a force with which to be reckoned in the contemporary climate of absolutized autonomy and secular relativism. I ask, what does a theological anthropology of radical dependence require of a

[1] See Karl Rahner, S.J., "Current Problems in Theology," *Theological Investigations*, Vol. I: God, Christ, Mary and Grace, trans. Cornelius Ernst, OP (London: Darton, Longman and Todd, 1963): 149–200.

M. J. Iozzio (✉)
Boston College, Boston, MA, USA
e-mail: mary.jo.iozzio@bc.edu

J. T. Eberl (ed.), *Emerging Issues in Catholic Bioethics*, Philosophy and Medicine 153, https://doi.org/10.1007/978-3-032-08862-8_3

bioethics that takes its norm and imperatives from a disabled God? The essay unfolds in five sections. I (1) begin with a background primer of disability theology, (2) move to the recorded experience of (mostly) abuse, (3) present accommodations as a necessary redress, (4) offer a bioethics in the key of disability, and (5) look at *metanoia*—i.e., what needs to change for bioethics to embrace a positive approach to disability.

3.2 The Need for Disability Theology

I do not have to convince anyone of the need to respect and care for all people, regardless of race, sex, gender, religion, class, or disability status. The twentieth century taught us well, by a series of horrific offenses, that discrimination on the bases of these and similar distinctions—of language, culture, height, or girth—are unjust. Nevertheless, what is necessary today is deliberate reflection on and the need to respect and care for human diversity inclusive of people with disability and others marginalized on account of race, gender, ethnicity, and the intersectionality that often accompanies these identities. People with disability have been discriminated against and sadly are still treated shamefully by the extremes of a complete loss of their subjectivity through neglect, abuse, and murderous violence. Such treatment violates their dignity and marks perpetrators—individuals and societies alike—by sin.

This work is dedicated to *moving beyond accommodation to affirmation and advocacy for people with disabilities,* a people both marginalized and denigrated with impunity across time and place. I begin with acceptance of liberationist insights regarding God's mission to people on the margins, outside the church, and deemed less than desirable. Liberation theologies interrogate the meaning of "the poor" and find that God's preferential option applies to people with disabilities, who are often poor and otherwise marginalized on account of their vulnerability and their lack of advocates; these theologies illuminate how God is the champion of those oppressed and marginalized.[2] While some progress engaging the preferential option has been made on behalf of people with disabilities and others,[3] history reveals that many have long been refused their places in their communities of birth and of worship as well.[4] In this way of being refused, unfortunately, people with disabilities have in

[2]Among others, see Gustavo Gutierrez, *A Theology of Liberation: history, politics, and salvation* (Maryknoll: Orbis Books, 1973, 1988 rev); Leonardo Boff and Clodovis Boff, *Introducing Liberation Theology* (Orbis Books, 1986); and Jon Sobrino, *Spirituality of Liberation: toward political holiness* (Maryknoll: Orbis Books, 1988).

[3]Here I am thinking of the marginalization of women from ancient times to the present, people who have been racialized, and those who are lesbian, gay, bisexual, transgender, queer, and intersexed.

[4]On worship and inclusion, see Kevin Timpe, *Disability and Inclusive Communities* (Grand Rapids: Calvin College Press, 2018) and Erin Raffety, *From Inclusion to Justice: Disability*

the past and still typify those who are marginalized in the contemporary world. The refusal on the part of Christian communities in particular is nothing less than scandalous in a tradition that holds humankind as having been created in God's own image and likeness and with creation as a sacred trust. Justice demands that people with disabilities and others who are marginalized receive their due as persons and thereby as *imago Dei*, and that they have access to all the conditions necessary for their flourishing in the fullness of life.

Here I explore three conditions of past and present reality, each related to the *imago Dei* and each in need of reappraisal if all people are to experience the "fullness of life through Christ for all creation."[5] I argue that disadvantaging discrimination and oppression is the contour of sin, diversity the complexion of creation, and dependence the essential character of humankind. I locate my thoughts in scripture, theology, and ethics to reflect on the implications of the many mansions in God's house (Jn 14: 2), the many branches of the tree of life (cf. Jn 15:5),[6] and the many parts abiding in the one body of Christ our Lord (1 Co 12:20), proposing that human attempts through norm-making and norm-imposing to limit creative diversity confounds God's mission designs for the world and the Good News of liberation.

We have it on fairly certain grounds—from none other than Jesus and attested to by the Spirit at Pentecost (Acts 2:1–12)—that heaven/God's house is welcoming to people (and all of creation) from every tribe, race, gender, tongue, and ability in the world. I wonder about the implications of the arduous work of becoming an inclusive community. Difficulties will be encountered in meeting the challenges of accommodating, affirming, and advocating with and for people belonging to tribes, races, genders, languages, and abilities that are different from the local familiar and normative. However, to do anything less at this moment is to act contrary to God's hospitality and mission designs for the world. And I am concerned that a very great number of people in ages past and still today have been refused entry into communities of friendship and of worship, communities to which many of us here belong and to proclaim Jesus as our Lord and Savior. Many of these communities and their leaders will have much explaining to do when the Spirit of truth and love asks "why"? Or in turn when they ask, "when, Lord, did we see you hungry, thirsty, naked, sick, estranged, or imprisoned" (Mt 25: 31–46)? Our ignorance in failing them will not save us.

In order to move from accommodations for people with disabilities to affirmation and advocacy, it is necessary to uncover what has been heretofore hidden from mainstream understanding, namely, that people with disabilities are people first! Starting from a consideration of the biblical assertion that they (humankind) were

---

Ministry, and Congregational Leadership (Waco: Baylor University Press, 2022).

[5] See Council for World Mission, *Fullness of Life, through Christ, for All Creation*, CWM Strategic Priorities and Programme 2010–2019 (CWM, Ltd., 2010), http://www.cwmission.org/wp-content/uploads/2012/11/CWM-Strategic-Priorities-and-Programme-2010-2012.pdf.

[6] See Miguel J. Romero and Jason T. Eberl, "The Tree of Life: Aquinas, Disability, and Transhumanism," in Devan Stahl, ed., *Bioenhancement Technologies and the Vulnerable Body: A Theological Engagement* (Waco: Baylor University Press, 2023): Chap. 3.

made in God's image and likeness, the egregious history of treatment that people with disabilities have suffered through the ages will be exhumed and lament expressed. Accommodations are the minimum expected from this moment forward. Second, affirmation of people with disabilities will include recognition of diversity in creation inclusive of the diversity across the spectrum of abilities as well as the dependence that is the reality of all humankind upon the webs of relationships that beget, befriend, and support each and every one of us. And third, advocacy with and for people with disabilities in matters of health and well-being—bioethics—will reflect the virtues of solidarity and justice indicative of the Crucified God, Jesus of Nazareth, the Risen Christ of Faith.

3.3 From a Scandalous and Tragic Past to Accommodations, Policies, and Law

Although the tragic abuse of people with disabilities has ancient expressions, the scientific and technological revolutions of the Enlightenment period have insinuated an ideology of hegemonic normativity across the globe that confounds the readily observable reality of the diversity of creation as it belies God's fecund creative imaginary. This ideology is evident in institutions of every sort, including our churches and the missions and ministries in which we are engaged.[7] Its hegemony has effectively segregated and essentialized humankind into two groups, "the normal" and "the disabled," and distributed a power that privileges the one as it disables—by marginalizing, oppressing, and stigmatizing—the other.[8] That an ideology of segregating exclusion has been inserted into social structures and corrupted the proclamation of the Good News should itself be a cause of concern for Christians everywhere. The results of that corruption—the scandalous exclusion and oppression of people with disabilities and other non-conformists—should be remedied at once.

What we are about to do is venture into the work of rejecting the corruption of the Good News and "advance a theological perspective that recognizes differences, respects uniqueness, and regards everyone as vital to the experience of life in fullness."[9] This work considers the theological underpinnings that support the movement to include people with disabilities in all arenas of human interaction so

---

[7] Among others, see Sharon V. Betcher, *Politics and the Spirit of Disablement* (Minneapolis: Augsburg Fortress Press, 2007).

[8] On normal, see Rosemary Garland-Thomson, *Extraordinary Bodies: figuring physical disability in American culture and literature* (New York: Columbia University Press, 1997); on power, see Michel Foucault, *Power*, ed. James D. Faubion, trans. Robert Hurley, et al. (New York: The New Press, 2000).

[9] Collin I. Cowan, "Foreword: Inclusiveness and Disability," *Advocacy and Disability: Introducing the Conversation in CWM* (Singapore: Council for World Mission Ltd., 2013), 2, http://www.cwmission.org/wp-content/uploads/2013/12/Advocacy-Disability-ebook-5-May-2014.pdf.

that all may enjoy the fullness of life. Additionally, this work admits that those with the power to make and shape societies have been grossly mistaken in their judgment about the value and dignity belonging to people with disabilities and others who do not conform to the hegemonic norm. Those mistaken judgments are the bases of a history of maltreatment which people with disabilities have endured, a history that, until only recently, has been largely ignored and likely intentionally unrecorded. In effect, people with disabilities themselves and their stories of success and failure and of loves and losses have been silenced over the course of time. However, that culture of silence is no longer acceptable: unborn, newborns, infants, children, and adults have been neglected, abused, and exterminated on account of the presence of a disability.[10] Despite our unease, their history is a critical part of the history of all humankind, a history that each of us must own if the mistakes are not to be repeated, reparations made to those victimized, ideological and physical structures rehabilitated, perpetrators forgiven, and the community reconciled.

In brief: though difficult to uncover,[11] the history follows an overlapping series of categorizations beginning with attempts to understand the presence of this or that disability and to interpret the value of individuals with disabilities by, first, religious leaders and seers across the world's traditions, moving second to attempts by scientists/doctors/cognoscenti, and third to institutionalizations by those seated in the legal/sociopolitical/colonial realms of power concerning them. This series of categorizations has been characterized by silence and the absence of the voices of people with disabilities themselves, mostly as a result of the fact that history is told by the victors and those with power to dominate the narrative.[12] Fortunately, and as a result of raising consciousness among the dominant, people with disabilities are now part of the retrieval and narration of the story inclusive of their experiences.[13]

---

[10] On the ramifications of a culture of silence, see Paulo Freire, "The Adult Literacy Process as Cultural Action for Freedom" and "Cultural Action and Conscientization," *Harvard Educational Review* 68.4 (1998 [originally published in *Harvard Educational Review* 40.2–3 (1970): 205–225, 452–477]): 480–498, 499–521.

[11] Paulo Freire, Michel Foucault, Iris Marion Young, and others have written extensively on power, cultural imperialism, and marginalization. Freire's thought is instructive for those who follow: "In the culture of silence the masses are 'mute,' that is, they are prohibited from creatively taking part in the transformations of their society and therefore prohibited from being. Even if they can occasionally read and write because they were 'taught' in humanitarian—but not humanist—literacy campaigns, they are nevertheless alienated from the power responsible for their silence." Freire, "The Adult Literacy Process," 486.

[12] See Colin Barnes, "A Legacy of Oppression: a history of disability in Western culture," in *Disability Studies: Past, Present, and Future*, C. Barnes and Mike Oliver, eds (Leeds: The Disability Press, 1997), 3–24; Thomas Shakespeare, "Cultural Representation of Disabled People: dustbins for disavowal?," *Disability & Society* 9.3 (1994): 283–299; and Henri-Jacques Stiker, *A History of Disability*, trans. William Sayers (Ann Arbor: University of Michigan Press, 2000); and Herbert C. Covey, *Social Perceptions of People with Disabilities in History* (Spring Hill: Charles C. Thomas, 1998).

[13] Although the web has a host of sites started and managed by people with disabilities, one example of a "history project" lead by people with disabilities is found at the Disability Social History Project, see "Home" (2024), https://disabilityhistory.org/.

With 15 to 25% of people worldwide having one or more disability today (up to 1.75 billion of 7 billion), we can be fairly certain that people with disabilities have been among the members of the human economy from antiquity to the present. Contemporary studies offer an approach to this history through the models of disability, with models closely related to the ways that the non-disabled typified their perceptions of people with disabilities according to the social roles that the non-disabled and dominant have assigned to them. The most common models of disability are the religious/moral, the medical, and the social construction.[14] Parallel to these models are social roles, the most recognizable to which people with disabilities have been assigned are as sick, sub-human, menace, pitiable, burden, holy innocent, inspirational, amusing, and blessing.[15] Regardless of model or role, each of these interpretations include a greater or less degree of stigma, the defining mark of otherness that clears the way to marginalization and greater or lesser degrees of oppression.[16]

To the extent that those who have held power and authority record history, their resolutions concerning people with disabilities resulted in marginalization on account of the causes those powers presumed at fault. Among the causes that this history attributed to disabling conditions are divine punishment for some sin (either one's own or one's parents); consorting with evil; imbalance of humors; maternal stress during pregnancy; bestiality; menstruation; and astrology. Each of these causes encouraged perceptions that people with disabilities were more animal or otherworldly than human, they could tolerate environmental extremes and malnutrition, and they were dangerous to the societies in which they lived—a conclusion that gave license to harm them with impunity by taming, sequestering, or worse.

Scandalously, many individuals with disabilities—feared or loathed by the non-disabled—would have been exposed at or near birth or otherwise ostracized once the presence of a disability became known.[17] Infanticide by exposure was widespread and, in some state-sponsored cultic systems, the practice was mandatory. Less brutal but not less disturbing were practices that exhibited individuals with disabilities in the courts of power: as a sign of blessing, entertainment, or to pity and extend charity. Among other curiosities, the Roman gladiator games included the

[14] In addition to these three, other precisions have been offered with the tragedy/charity, expert/professional, rehabilitation, economic, and rights based, customer/empowering models. See Michigan Disability Rights Coalition, "Models of Disability" (no date; accessed June 27, 2014), http://www.copower.org/models-of-disability.html; and Carmelo Masala and Donatella Rita Petretto, "Models of Disability," in J.H. Stone and M. Blouin, eds, *International Encyclopedia of Rehabilitation* (2010), http://cirrie.buffalo.edu/encyclopedia/en/article/135/.

[15] See Wolf Wolfensberger, *A Brief Overview of Social Role Valorization: a high-order concept for addressing the plight of societally devalued people and for structuring human services* (Ontario: Valor Press, 2013 [originally published, Syracuse: The Training Institute for Human Service Planning 2004, 1998]).

[16] See Erving Goffman, *Stigma: notes on the management of spoiled identity* (New York: Simon & Schuster Inc., 1963).

[17] See Richard Slayton French, *From Homer to Helen Keller: A Social and Educational Study of the Blind* (New York: American Foundation for the Blind, 1932).

spectacle of fights between short people, the deaf, other people with varying disabilities, women, and animals. Judaism and Christianity fared a little better by prohibiting direct killing.[18] However, the biblical texts reflect the perception that impairments signal something un-Godly having been done—sin—or an evil spirit on the loose.[19] With the biblical texts, the religious model of disability takes hold and locates the presence of disabilities as punishment for some wrongdoing now in need of prayer, reparation, and expiation. Where the disability persists, the individual and their family, tribe, or clan ordinarily continues to bear the stigma of 'disabling' wrongdoing and its guilt-associated shame.[20] The early medieval period made way for the custom of caring for the sick, those with disabilities, and the poor. Outside of the support of their natal homes, people with disabilities were reduced often to poverty and they resorted to begging as a principal means of income. Wanting to follow the example of Jesus, who attended to those who were marginalized for this or that stigma, Christians began to extend compassion on the less fortunate. By the height of the Middle Ages in Europe, a "period of organized beggary" led to guilds open to people with disabilities in which leaders emerged, with rules and languages developed by the guild members.[21] The guilds represent a welcome initiative by today's standards. Yet, this same period saw the institution of "idiot cages" that kept people with disabilities confined, while the cage protected those beyond the bars. And where cages were insufficient or when the masses tired of this or that caged group—building on Plato's *Republic* and Erasmus's *Praise of Folly*—the ship of fools provided another form of distance to keep people with disabilities apart from the mainstream and exploited as members of a travelling carnival-horror-freak sideshow for port residents and visitors alike.[22]

And then there were (and some remain) institutions. Founded as a result of a system of hostels for pilgrims on their way to a holy site for both blessings and cures, hospitals for the sick and incurable became asylums for the insane and invalid. With the advent of the Enlightenment project to reject the old and quaint in favor of a rational order, new scientific ways of conceiving the individual in society and the good of society brought to the fore utopian concerns of a more perfect communion,

---

[18] See Jane Draycott, "Reconstructing the Lived Experience of Disability in Attitudes Towards Disability in Antiquity: A Case Study from Roman Egypt," *Greece and Rome* 62.2 (2015): 189–205; cf. Michel Foucault, *Madness and Civilization: A History of Insanity in the Age of Reason* (Vintage Books, 1988/Random House Inc., 1965).

[19] Consider the dialogue on theodicy in the Book of Job, his interlocutors quite certain that he or a member of his family surely sinned and brought the "Satan" on him and his children. See Job 4:7–8, 8:20, and 11:14–17.

[20] Compare the shame that many people harbor over the presence of mental illness (in particular) in the family or, similarly, the shame that surrounds many people living with HIV and AIDS and their families—the silencing that comes from shame is definitive.

[21] See Sharon Farmer, "The Beggar's Body: intersections of gender and social status," and Catherine Peyroux, "The Leper's Kiss," in Sharon Farmer and Barbara H. Rosenwein, eds, *Monks & Nuns, Saints & Outcasts* (Ithaca: Cornell University Press, 2000), 153–188.

[22] See Rosemarie Garland-Thomson, ed., *Freakery: Cultural Spectacles of the Extraordinary Body* (New York: NYU Press, 1996).

overtly including an underlying concern for the dangers lurking in the presence of people with disabilities. The isolation of the institution provided safety for the non-disabled community as well as it gave rise to better or worse care for those institutionalized, while also creating a captive population to study objectively. With concentrated access to people with a diverse array of disabilities, doctors and scientists began to investigate the causes of disabilities using medical and empirical methodologies. Some of this early science fueled the later nineteenth and early-mid twentieth centuries' eugenics movements through the subsequent sterilization of people with disabilities and other suspicious folk.[23] As long as people were institutionalized and thereby isolated from general human commerce, they were—and those who remain institutionalized remain—vulnerable to abuse, exploitation, and other dehumanizing injustices.

The confluence of social progress, science, and rational self-interest led to the systematic individualization and medicalization of all persons—normates and those "othered"—as subjects and objects. And that systematic led to an ideology of disdain for, dis-ease with, and distrust of any who did not conform to the putatively normative/ideal modern man. By the twentieth century, eugenic initiatives were set in Europe and the United States with sterilization programs and final solutions in a murderous holocaust of an untold hundreds of thousands if not millions of people with disabilities. Scandalously still, eugenics and euthanasia continue apace with neonates, children, adolescents, and adults in their prime and elderly with disabilities as the principal populations vulnerable to medical-social-scientific control.[24] While eugenics may not be institutionalized, it holds ideological power and is practiced widely in reproductive medicine and the selective abortion of fetuses. Euthanasia remains a present threat as the contemporary equivalent of exposure that is withholding life-support from a person—neonate or adult—who could thrive if given the chance. This history belies the Christian tradition's recognition that (1)

---

[23] Consider the scientific "proofs" of a biological basis for the categories of race and the subsequent discrimination against non-white peoples, especially peoples of African descent, that labeled many deviant. People from Mediterranean countries and Asia were considered to be of "questionable genetic stock" and likely to increase the number of feebleminded or criminals that became the wards of the state; it would be better to prevent them from reproducing altogether. See the discussion of "positive eugenics" in Albert Jonsen, *The Birth of Bioethics* (New York: Oxford University Press, 1998), 169–172; Paul A. Lombardo, ed., *A Century of Eugenics in America: From the Indiana Experiment to the Human Genome Era* (Bloomington: Indiana University Press, 2011); and Lombardo, *Three Generations, No Imbeciles: Eugenics, The Supreme Court, and Buck v. Bell* (Baltimore: Johns Hopkins University Press, 2008). On the biological basis of racial discrimination see, for example, Harriet A. Washington, *Medical Apartheid: the dark history of medical experimentation on Black Americans from colonial times to the present* (New York: Doubleday, 2006).

[24] See Francis, "Message to the Participants in the Symposium 'Towards a Narrative of Hope: An International Interfaith Symposium on Palliative Care'" (2024): https://www.vatican.va/content/francesco/en/messages/pont-messages/2024/documents/20240522-messaggio-palliative-care.html, and Congregation for the Doctrine of the Faith, *Samaritanus bonus* (Vatican City: Congregation for the Doctrine of the Faith, 2020): https://www.vatican.va/roman_curia/congregations/cfaith/documents/rc_con_cfaith_doc_20200714_samaritanus-bonus_en.html.

every human being is created in the image and likeness of God, (2) as such, no person is ever to be treated as an object to be discarded, to be denied access to health care equal to the abled, nor to be experimented upon without consent, and (3) diversity is to be ever more being recognized as the calling card of God's creation and our inheritance.[25]

3.4 A Disabled God and an Anthropology of Disability

In counterpose to a negative way of thinking about people with disability, I present a theological anthropology of the *imago Dei* in humankind through a Trinitarian lens of relationality and radical dependence. This anthropology rejects homogeneity in favor of diversity throughout creation inclusive of the manifold incarnations in humankind as well as the diversity revealed in the rather scandalizing disabling—through crucifixion—of the resurrected Incarnate God.[26] Today, disability identities are celebrated even as many remain vulnerable to discrimination and abuse.

People with disability and the non-disabled alike belong to a common humankind that is deigned the *imago Dei* and in which, through the *Kenosis*, God identifies with all human joys and sorrows, births and growth, disability and deaths in a continuum from the past to the present and into the future. As *imago Dei*, humankind takes its primary data from the tradition's teaching on creation, the Incarnation, and Trinitarian theology. The implications of this theology extend beyond a facile nod to the identification of God with humankind: if the Trinity is the theo-anthropological ground of being human, then the Trinitarian symbol of God must function as a template for all divine and human affairs.[27] The critical importance of a Trinitarian theological anthropology cannot be overstated. If the datum of the tradition is to be believed, then we must be earnest in engaging with God and with one another along the lines of relationship and relationality that is the manifest expression of God for Us.[28]

The Christian tradition has long discussed and defended the doctrine of the Trinity and the Trinity's action in the world as Father/Creator, Son/Redeemer, and Holy Spirit/Sanctifier. However, the Trinity carries doctrinal importance beyond the formulaic introduction to prayer: the doctrine of the Trinity holds the potential of transformative practice among believers and nonbelievers alike. If every person is

[25] See Mary Jo Iozzio, "Genetic Anomaly or Genetic Diversity: thinking in the key of disability," *Theological Studies* 66.3 (2005): 862–881; and Iozzio, "Thinking About Disabilities with Justice, Liberation, and Mercy," *Horizons* 36.1 (2009): 32–49.

[26] See Nancy Eiesland, *The Disabled God: Toward a Liberatory Theology of Disability* (Nashville, TN: Abingdon Press, 1994).

[27] See Elizabeth Johnson, *She Who Is: The Mystery of God in Feminist Theological Discourse* (New York: Crossroad Publishing Co, 1992).

[28] See Catherine Mowry LaCugna, *God for Us: The Trinity and Christian Life* (New York: HarperCollins Publishers, 1991).

created in the image and likeness of God, and as God has been revealed to us in triune relationality *in se* and *ad extra*, then relationality—not isolating autonomy or dogged self-reliance—is the form of the *imago Dei*.

As Catherine Mowry LaCugna has taught, "Trinitarian theology is the language of relationality *par excellence*,"[29] not only *in se* but *ad extra* for us and for our salvation. The relationality that subsists within the Trinity is a relationality of a willed quasi-dependence of the Three Persons in One God. We Christians do not know God in any other way than in this Tri-Personal Relationality. This relationality is as God wills it to be in salvation history, *in se* and *ad extra* dependent in form and matter though uniquely unified and subsistent in effects: "the economic Trinity is the immanent Trinity and *vice versa*."[30] It is thus *in se* wherein God enjoys/relates in God's own company and *ad extra* wherein God enjoys/relates to all that God has wrought for us—the symbol of a relational God functions with and for us.[31]

"That symbols are signs is certain: they are expressions that communicate a meaning; this meaning is declared in an intention of signifying which has speech as its vehicle."[32] Christian theology speaks of God in the terms of an I-Thou relationship. If the symbol of God functions in a vehicle of relationality, and to be in relation suggests dependence within the One and the Three, then relational dependence holds first place in the ways that we are to envision ourselves and others as the *imago Dei*.

Of course, dependence in inter- and intra-Trinitarian relationality differs exponentially from the dependence experienced by the *imago Dei*; nevertheless, dependence is in this anthropology relieved of its burdensome negative connotations. Dependence in this sense is neither a bad word nor a bad idea. It is, alternatively, liberative. I am suggesting that unity in diversity translates to the *imago Dei* as a radical dependence and, thereby, human dependent relationality: we human beings are dependent upon God and upon one another from our conceptions and births, through childhood, adolescence, adulthood, and to elder years and demise (to think otherwise is foolhardy).[33] This symbol of God for us expresses intentional relationality as the ontological ground for understanding the function of the symbol that is the *imago Dei* rightly inclusive of every manifestation of God's creative incarnate sanctifying love and contra any/every exclusion.

When we take the doctrine of the Trinity seriously it leads to us to re-think what the symbol of the *imago Dei* conveys in our lives. Thus, if we take difference or

---

[29] Catherine Mowry LaCugna, "Reconceiving the Trinity," *Scottish Journal of Theology* 38.1 (1985), 13.

[30] Karl Rahner, *The Trinity* (New York: Herder & Herder, 1970), 22.

[31] As Johnson notes, "The symbol of God functions. Neither abstract in content nor neutral in its effect, speaking about God sums up, unifies, and expresses a faith community's sense of ultimate mystery, the world view and expectation of order devolving from them, and the concomitant orientation of human life and devotion.," *She Who Is,* p. 4.

[32] Paul Ricouer, *The Symbolism of Evil* (Boston: Beacon Press, 1969), 14.

[33] See Iozzio, "Norms Matter: A Hermeneutic of Disability/A Theological Anthropology of Radical Dependence," *ET-Studies* 4.1 (2013): 89–106.

diversity as the key for theological anthropology, then our being in the world requires relationality with our human siblings, yes, and with all of the living creatures and inanimate supports upon which we depend. Thus, "a Trinitarian-inspired theological anthropology can be instructive for a Christian understanding of who (and how) God is for us, as well as who and how each of us is to be for God and for one another."[34]

Nancy Eiesland can be recognized as ushering Disability Theology in the United States (the United Kingdom was ahead by 15+ years). Eiesland gave us the image of the *Disabled God* in a sip-n-puff chair.[35] For Eiesland, this image points not only to the *imago Dei*, it points directly to the Crucified Jesus, the Risen Christ of Faith, recognized by the tortures he endured and death from which the Father raised him. Inarguably though rarely thought, that torture disabled him. Further, as Christians hold, risen from the dead he returned to the disciples to prove Mary of Magdala's proclamation that "He is Risen." This proof was confirmed for the Apostle Thomas by Jesus showing the scars *and* having Thomas put his hand into the wounds. Eiesland's work moved theology to a new way of conceiving the *imago Dei* as the disabled God. If the disabled Jesus can be recognized as God and if we mere mortals have been deigned the *imago Dei*, then surely people with disabilities count among the blessed along with the abled.

3.5 A Bioethics in the Key of Disability

Others have argued that many of the disciplines of theology are versions or a kind of sub-species of anthropology.[36] That is, whatever we humans attempt to say about God reflects what we think is true of ourselves. However, where our theologies have asserted God's perfection as the basis of intellectual, physical, and gendered norms, those theologies denied the full inheritance of the *imago Dei* to women, racialized others, and people with disabilities. The insights of these arguments about God's being and human being, though not their conclusions that have marginalized many, resonate with members of Christian communities who profess a revelation of the Incarnate God in the person of Jesus of Nazareth, the Risen Christ of Faith. And some Christians with disabilities find solace and affirmation in these insights,[37] as

---

[34] Iozzio, "Radical Bioethics and the *imago Dei*: Bioethical Implications of Access to Healthcare for People with Disabilities," *Christian Bioethics* 23.3: 243.

[35] See Nancy Eiesland, *The Disabled God: Toward a Liberatory Theology of Disability* (Nashville: Abingdon Press, 1994).

[36] Karl Rahner, SJ is arguably the first of twentieth century Catholic scholars claiming all theology must be of necessity theological anthropology. See Rahner, "Theology and Anthropology," in *Theological Investigations IX*, trans. by G. Harrison (London, UK: Darton, Longman & Todd, 1972), 28–45.

[37] See especially, Nancy L. Eiseland, *The Disabled God: toward a liberatory theology of disability* (Nashville, TN: Abingdon Press, 1994).

God Incarnate is broken, crucified, and glorified reveals the marks of disability—whose hands and feet were crippled by nails, whose side was pierced, whose face was marred beyond human semblance[38]—*without being stigmatized thereby as being disabled.* In these insights is an anthropology that affirms people who do not conform to the ideologically normative perfections that oppress; in them is a theological anthropology of the disabled God.

However, most philosophical and theological anthropologies have historically presented a limited scope of "man," an ideal that very few attain. Black, Feminist, Latino and other context-based scholars have convincingly argued that the western theo-philosophical traditions, founded in classical and Enlightenment thought, focus on things proper to white men almost singularly. Scholars in the disabilities studies movement are in the process of convincing those attentive that these same traditions focus on things proper to able-bodied, youthful, and robust men—with an even smaller number attaining the ideal. The time has come to focus instead on things proper to and good for women, people of color, and persons with disabilities, and for others to increase the scope of the ideal by including the anthropological diversity of human being-ness embodied in these heretofore invisible others. Thus, an ideal of anthropological diversity rejects the limiting notions that qualify being human and affirms that persons with disabilities and others already attain this ideal.[39]

All too often in the world of dualistic segregations, superficially identifiable differences serve to categorize and, invariably, establish hierarchies that rank individuals on the basis of their conformity to the normatively ideal. In the case of human norms the dualisms of male/female, spirit/body, white/non-white, heterosexual/homosexual, and able-bodied/disabled have relegated the second part of each pair as a defective version of the first part and to the almost inevitable oppression of the second by the first. However, when diversity is reckoned as normatively ideal, these dualisms lose their power to elevate one expression over other expressions. With diversity as the norm, dualisms no longer make sense and a search for broader measures of who counts must commence. Moreover, affirmation of diversity as a value in itself has been appropriated by liberation studies while it also confirms environmental studies that call for attention to the biodiversity of our planet and its need of care for everyone's survival and flourishing. Here too, we can recall Pope Francis's recognition of the integral ecology that supports the planet and its sentient and non-sentient populations. "Since everything is closely interrelated, and today's problems call for a vision capable of taking into account every aspect of the global crisis, I

[38] Even when the text of Isaiah is taken in a strictly Jewish context, the image of disfigurement suggests disability in a modern sense: "14. Just as there were many who were astonished at him –so marred was his appearance, beyond human semblance, and his form beyond that of mortals --… 3. He was despised and rejected by others; a man of suffering and acquainted with infirmity; and as one from whom others hide their faces he was despised, and we held him, of no account… 7. He was oppressed, and he was afflicted… 8. By a perversion of justice he was taken away" (Is 52:14, 53:3, 7, 8).

[39] Of course, people with disabilities already embody the ideal that is diversity, affirmation is the social deconstruction of hegemonic denials of their inclusion as well as the paradigmatic expression of the norm.

suggest that we … consider some elements of an *integral ecology*, one which clearly respects its human and social dimensions."[40]

Christians believe that God offers all creation the fullness of life through the workings of the Triune God. A mystery, surely, and a revelation too that the lens of theological anthropology suggests something about God in Godself and about God's creation: God is unity in diversity. This doctrine considers that in God's very being, in Godself, God is for others/God is for us. With this understanding, gathered over time from the Christian experience of God in the world, the implications of the doctrine of the Trinity reveal relationality and relationship as the metaphysical ground of God's own and all other being as one or another expression of the *imago Dei*.[41] This claim then serves as the starting-point for a new and radical paradigm in anthropology, a paradigm of relational dependence on the diversity of being in the world, even the dependence of God's being in the world. The paradigm provides the immediacy of unity in diversity that both reveals God and defines humankind, if not all of creation, as *imago Dei*—one image reflected in the kaleidoscopic diversity of being, including the diversity of disabilities present in humankind and in a planet subject to the forces of evolutionary and seismic change.

In terms of a theological anthropology, if God is three in one then the implications of unity-in-diversity/diversity-in-unity extends to the *imago Dei*. In Trinitarian doctrine, God *in se* (God in Godself) and God *ad extra* (God for Us) exposes the tri-relational Christian faith in the one God. This doctrine is not a Christian pantheism; rather, since God became incarnate in the person of Jesus of Nazareth and remains active in the world through the Holy Spirit, God's self-revelation includes a tangible manifestation of the Divinity as unity in diversity, as the tri-relational God communicating in love as Creator, Redeemer, and Sanctifier. This relationality distinguishes the persons of God as it reveals the ways in which God acts in the world. As Christians, what we know about God is known through this relational Trinitarian lens. And through that lens we come to know something about ourselves, our human being-ness, as *imago Dei*. Thus, if relationality built on a foundation of union-love characterizes God's unity in diversity then relationality, of a loving and liberatory sort, must characterize human diversity present throughout the world.

A loving and liberatory relationality belies the dualisms artificially segregating one community from another. Once those dualisms are disempowered two imperatives emerge: one that attends to the needs of those formerly ignored or oppressed—in the case of people with disabilities, accommodations—and another that attends to diversity by forming an inclusive community that affirms and advocates as it supports all people with their diversities of form and function. Rather than locate anthropological norms in superficially observable physical/cognitive/emotional characteristics, human anthropological diversity locates the norms of human expression in the relationships that affirm human flourishing by supporting the functional

[40] Francis, *Laudato Si'* (Vatican City: Libreria Editrice Vaticana, 2015), §137: https://www.vatican.va/content/francesco/en/encyclicals/documents/papa-francesco_20150524_enciclica-laudato-si.html.

[41] See Mowry LaCugna, *God for Us*.

capabilities deigned suitable by and for each.[42] A move beyond a normatively static and rather boring ideal—of the white heterosexual male—to functioning capabilities proper to diversity opens the dialogue on human being-ness to include the diversity of human expressions far and wide and moves from that putatively normative and idealized man to all those deemed previously defective, less than, or disabled.

By way of affirming the wide expression of humankind, the anthropological diversity presented by persons with disability is uniquely indicative of human being-ness. Further, that diversity exposes what is most true of human beings *qua imago Dei*: in order to be and to thrive, human beings must accept and affirm their relational dependence upon one another. Relational dependence, which I have identified elsewhere as radical dependence, is human unity in diversity; as such, dependence holds normative force for humankind.[43] As a dynamic force, this norm requires nothing less than affirmation of every person regardless of their race, gender, or ability and, to the extent that they have been denied, the explicit, intentional, and robust affirmation of people with disabilities of every kind. Practically, that affirmation of dependence will require explicit, intentional, and robust access to the exercise of their functioning capabilities inclusive of: a natural lifespan; physical health and opportunities for sexual satisfaction and reproductive choice; access to pain relief; use of creative, intellectual, and spiritual potential; relationships with others; self-determination; economic-social-political participation and reciprocity; ecological balance; recreation; non-interference; and freedom.[44]

---

[42] See Martha Nussbaum, "Human Capabilities, Female Human Beings," in *Women, Culture, and Development: A Study of Human Capabilities*, ed. Martha Nussbaum and Jonathan Glover (Oxford, UK: Clarendon Press, 1995).

[43] See Iozzio, "Norms Matter: a hermeneutic of disability/a theological anthropology of radical dependence," *ET-Studies* 4.1 (2013): 89–106.

[44] As Nussbaum notes, "Being able to live to the end of a human life of normal length (not dying prematurely); being able to have good health (to have adequate nourishment and shelter, having opportunities for sexual satisfaction and choice in matters of reproduction, being able to move from place to place); being able to avoid unnecessary and non-beneficial pain and to have pleasurable experiences; being able to use the senses (to imagine, to think, and to reason and to do these things aided by adequate education, and to enjoy and produce aesthetic and religious materials); being able to have attachments to things and persons outside ourselves (to love those who love and care for us, to grieve at their absence, to experience longing and gratitude); being able to form a conception of the good and to engage in critical reflection about the planning of one's own life (to seek employment outside of one's home and to participate in political life); being able to live for and to others, to recognize and show concern for other human beings, to engage in various forms of social interaction (being able to imagine the situation of another and have compassion for that situation, to have the capability for both justice and friendship, protection of freedom of assembly and freedom of speech); being able to live with concern for and in relation to animals, plants, and the world of nature; being able to laugh, to play, to enjoy recreational activities; being able to live one's own life and nobody else's (having certain guarantees of noninterference with certain choices that are especially personal and definitive of selfhood such as choices regarding marriage, child-bearing, sexual expression, speech and employment); being able to live one's own life in one's own surroundings and context (this requires freedom of association and freedom from unwarranted search and seizure, a certain sort of guarantee of the integrity of personal property)." Nussbaum, "Human Capabilities, Female Human Beings," in *Women, Culture, and Development*, 83–85.

These functional capabilities point to activities that contribute to human flourishing and lead to the recognition that capabilities are thereby measures of the human good: wherever they are provided, there a good life can ensue regardless of the presence or absence of disability. As they pertain to persons with disabilities, these capabilities refocus and build upon the relations in need of identification to determine the demands of justice as well as the expectations of an affirming, inclusive community of faith. Moreover, as relational dependence epitomizes the human condition even as it characterizes the *imago Dei*, identifying and affirming the needs that people with disabilities may have today and that others will acquire tomorrow can lead to provisions of those capabilities for all God's people for the fullness of life.

Finally, if diversity is God's own calling card—both in Trinitarian doctrine and in creation—then an affirmation of diversity, and particularly the diversity in humankind inclusive of people with disabilities, is an affirmation of God in the created order and a cause for thanks and praise. To the extent that people with disabilities represent a significantly large cohort of the human population and to the extent that their normative diversity is uniquely indicative of being human—disabilities are both widespread and non-discriminatory—to these extents, diversity signals God's fecund imaginary … as it was in the beginning, is now, and ever will be. What more are we to do if we are to be communities worthy of God's image and likeness in moving beyond accommodations to welcoming, affirming, and integrating people with disabilities in our assemblies? What revisions to our existing physical and philosophical structures and their systemic-institutionalized support are required? And what force does a theological anthropology of radical dependence require of a people who fashion their norms in and imperatives from a disabled God?

What more? If we are to be worthy communities, we must do nothing less than accommodate and affirm. Political will is an important though insufficient step; many hearts and minds are yet in need of *metanoia* if systemic-institutionalized ableism is to lose its power to oppress. And intentional relationships across the spectrum of being human in a world of fecund asymmetry and a creative imaginary are to be celebrated according to the mission designs of the Holy Trinity and God for us. Nothing less than friendship will do.

3.6 *Metanoia*: What Needs to Change for Bioethics to Embrace Disability as a Positive Experience?

Theological anthropology forms the bases of most of our musings about ourselves, our relationships, and our appreciation or lack thereof for the persons we encounter each day. Philosopher and lawyer Max Ehrmann wrote the poem "Desiderata" in 1927, published posthumously in 1948. Its popularity may have waned of late but we human beings continue to burn with desires, with *desiderata* of many kinds, some mundane, others profound. Whatever the desires of people with disability, their family members, friends, those who care for and about them, and those who do not, I think we can take Ehrmann's words to head and heart in our strivings for a better tomorrow:

> Go placidly amid the noise and haste, and remember what peace there may be in silence.
> As far as possible, without surrender, be on good terms with all persons.
> Speak your truth quietly and clearly;
> And listen to others, even to the dull and ignorant; they too have their story.
> ...
> Beyond a wholesome discipline, be gentle with yourself.
> You are a child of the universe, no less than the trees and the stars;
> [And] you [with disability and you the nondisabled] have a right to be here.
> ...

It bears repeating Ehrmann's words: You have a right to be here!

As I follow the work of Rosemarie Garland-Thomson, who argues that her "proposition here, then, is that the field of bioethics is an appropriate arena of knowledge-building practice into which disability studies can be brought to bear."[45] For Garland-Thomson, disability bioethics offers a particular glimpse into the experience of bioethics writ large. Here disability bioethics engages not only the medical sciences' approach to disability, it engages the interdisciplinary critical disability studies with healthcare, medical research, and biomedical research. Thus, disability bioethics must be attentive to sociocultural factors affecting the person living with this or that/these or those disabilities. It is critical that the field not focus on impairment, limitations, or functioning capacities but a *de rigueur* focus on the persons seeking care and/or understanding of the conditions they experience.

The point of disability bioethics is to become aware of and attentive to human flourishing in the healthcare arena *and* in the commons of cultural, political, and environmental space. Further, disability bioethics is necessarily interpersonal and diverse of persons in need of critical care as well as care-providers' cultural competence to respond appropriately to their needs. Garland-Thomson offers this competence to address the theologically normative/normate anthropology of the mainstream.

> Informed by Alicia Ouellette's "disability principlism" and Scully's "disability epistemology,"[46] Garland-Thomson notes that disability cultural competence goes beyond sensitivity to language and adjustments in activities of daily living; it is developing competencies for using the world effectively, maintaining our dignity, exercising self determination, cultivating resilience, recognizing and requesting accommodations, using accessible technology, finding community, maintaining successful relationships—all as persons living with disabilities. Thus, disability cultural competence is for people with disabilities and about living with a disability at the same time. It brings disability culture to people currently

---

[45] Rosemarie Garland-Thomson, "Disability Bioethics: From Theory to Practice," *Kennedy Institute of Ethics Journal* 27.2 (2017): 325.

[46] See Alicia Ouellette, *Bioethics and Disability: Toward a Disability Conscious Bioethics* (Cambridge University Press, 2011) and Jackie Leach Scully, "Epistemic Exclusion, Injustice, and Disability," in Adam Cureton and David Wasserman, eds, *The Oxford Handbook of Philosophy and Disability* (New York: Oxford University Press, 2018): 296–309.

> identified as disabled and their families and caregivers as well as people who may in the future identify as disabled.[47]

Disability illiteracy is no longer acceptable. Of course, attention has been increasing over the decades since the 2006 United Nations' Convention on the Rights of Persons with Disabilities and the United States' 1990 Americans with Disabilities Act. However, the theological and bioethics academies have yet to embrace the liberating force of these instruments and to apply them in their deliberations on care for people with disabilities.[48] Surely, people with disability are, as the saying goes, people first. As such, what concerns the mainstream concerns also those on the peripheries: women, gender minorities, and people with disability.

Capacity-based anthropologies ultimately fail the majority of our neighbors near and far.[49] Further, our capacities will wane with age or disablement. Denial is not an option. How then ought we to proceed in recognizing and embracing the dynamics of change in a body and in a world subject to the promise of eternity?[50]

We must proceed with recognition "that the *telos* of those [even] with severe cognitive disabilities is defined in relation to God, we can affirm an anthropology that recognizes all human beings as created by God in their particularity and made for intimate relationship with God."[51] Further, as "human beings are the sorts of beings we are because we are cared for by other human beings, then the human being's ontological status and corresponding moral status need to be acknowledged by the larger society that makes possible the work of those who do the caring required to sustain us. That is what we each require if we are some mother's child, and we are all some mother's child."[52] Here we can envision a "theological anthropology, overseen by the hermeneutics of grace, [that] helps to prevent bioethics from becoming mere external behavioral reform and social activism."[53]

Finally, even as many have gained cultural competence in matters related to race, gender, and ethnicity, just as many have not learned the basics of the cultural competence concerns of people with disability. Time is critical in bringing this competence to the mainstream. As Garland-Thomson reminds us: "All people are either

[47] Garland-Thomson, "Disability Bioethics," *Kennedy Institute of Ethics Journal* 27.2: 331.

[48] See Devan Stahl, *Disability's Challenge to Theology: Genes, Eugenics, and the Metaphysics of Modern Medicine* (Notre Dame: University of Notre Dame Press, 2022).

[49] See Martha Nussbaum and Saul Levmore, *Aging Thoughtfully: Conversations about Retirement, Romance, Wrinkles, and Regret* (New York: Oxford University Press, 2020).

[50] See Alasdair MacIntyre, *Dependent Rational Animals: Why Human Beings Need the Virtues* (Peru, IL: Open Court Publishing, 2001).

[51] Joshua Cockayne, "Disability, Anthropology, and Flourishing with God: A Kierkegardian Account," *Religions* 11.4 (2020): 199.

[52] Eva Feder Kittay, "The Personal is Philosophical is Political: A Philosopher and Mother of a Cognitively Disabled
Person Sends Notes from the Battlefield," *Metaphilosophy* 40.3–4 (2009): 265.

[53] Scott Stiegemeyer "Theological Anthropology for Bioethics," *Dignitas* 30.1: 13.

active patients or patients in waiting. Developing disability cultural competence will prepare all of us to navigate that status and experience effectively."[54]

References

Betcher, Sharon V. 2007. *Politics and the Spirit of Disablement*. Minneapolis: Augsburg Fortress Press.

Boff, Leonardo, and Clodovis Boff. 1986. *Introducing Liberation Theology*. Maryknoll: Orbis Books.

Cockayne, Joshua. 2020. Disability, Anthropology, and Flourishing with God: A Kierkegardian Account. *Religions* 11 (4): 189–204.

Council for World Mission. 2010. *Fullness of Life, Through Christ, for All Creation*. CWM Strategic Priorities and Programme 2010–2019. Singapore: CWM, Ltd. http://www.cwmission.org/wp-content/uploads/2012/11/CWM-Strategic-Priorities-and-Programme-2010-2012.pdf

Feder Kittay, Eva. 2009. The Personal is Philosophical is Political: A Philosopher and Mother of a Cognitively Disabled Person Sends Notes from the Battlefield. *Metaphilosophy* 40 (3–4): 606–627.

Foucault, Michel. 2000. *Power*, ed. James D. Faubion. Trans. Robert Hurley, et al. New York: The New Press.

Francis. 2015. *Laudato Si': On Care for Our Common Home*. Vatican City: Libreria Editrice Vaticana. https://www.vatican.va/content/francesco/en/encyclicals/documents/papa-francesco_20150524_enciclica-laudato-si.html.

Freire, Paulo. 1988. "The Adult Literacy Process as Cultural Action for Freedom" and "Cultural Action and Conscientization". *Harvard Educational Review* 68 (4): 480-498, 499-521.

Garland-Thomson, Rosemarie. 1997. *Extraordinary Bodies: Figuring Physical Disability in American Culture and Literature*. New York: Columbia University Press.

Garland-Thomson, Rosemarie. 2017. Disability Bioethics. *Kennedy Institute of Ethics Journal* 27 (2): 323–339.

Gutierrez, Gustavo. 1988. A Theology of Liberation: History, Politics, and Salvation. Maryknoll: Orbis Books, 1973, 1988 rev.

Iozzio, Mary Jo. 2017. Radical Bioethics and the *imago Dei*: Bioethical Implications of Access to Healthcare for People with Disabilities. *Christian Bioethics* 23 (3): 234–260.

Mowry LaCugna, Catharine. 1985. Reconceiving the Trinity. *Scottish Journal of Theology* 38 (1): 1–23.

Nussbaum, Martha. 1995. Human Capabilities, Female Human Beings. In *Women, Culture, and Development: A Study of Human Capabilities*, ed. Martha Nussbaum and Jonathan Glover. Oxford: Clarendon Press.

Raffety, Erin. 2022. *From Inclusion to Justice: Disability Ministry, and Congregational Leadership*. Waco: Baylor University Press.

Rahner, S. J. Karl. 1963. Current Problems in Theology. In *Theological Investigations, Vol. I: God, Christ, Mary and Grace*. Trans. Cornelius Ernst, OP, 149–200. London: Darton, Longman and Todd.

Rahner, S. J. Karl. 1970. *The Trinity*. New York: Herder & Herder.

Ricouer, Paul. 1969. *The Symbolism of Evil*. Boston: Beacon Press.

Romero, Miguel J., and Jason T. Eberl. 2023. The Tree of Life: Aquinas, Disability, and Transhumanism (chap. 3). In *Bioenhancement Technologies and the Vulnerable Body: A Theological Engagement*, ed. Devan Stahl. Waco: Baylor University Press.

[54] Garland-Thomson, "Disability Bioethics," *Kennedy Institute of Ethics Journal* 27.2: 336.

Sobrino, Jon. 1988. *Spirituality of Liberation: Toward Political Holiness*. Maryknoll: Orbis Books.
Stahl, Devan. 2022. *Disability's Challenge to Theology: Genes, Eugenics, and the Metaphysics of Modern Medicine*. Notre Dame: University of Notre Dame Press.
Stiegemeyer, Scott. 2023. Theological Anthropology for Bioethics. *Dig* 30 (1): 13–21.
Timpe, Kevin. 2018. *Disability and Inclusive Communities*. Grand Rapids: Calvin College Press.

Chapter 4
Accompanying the Transgender Patient in Catholic Health Care: Morally Permissible Interventions

Peter J. Cataldo

4.1 Introduction

In his *motu proprio* updating the statutes of the Pontifical Academy of Theology, Pope Francis provides an explanation of theological reflection that is also an appropriate standard for the theological and ethical evaluation of questions in Catholic health care:

> … theological reflection is urged to *develop with an inductive method,* which starts from the different contexts and concrete situations in which peoples are inserted, allowing itself to be seriously challenged by reality, to become discernment of the "signs of the times" in the proclamation of the salvific event of the God-agape, communicated in Jesus Christ (Pope Francis 2023b, n. 8)[1]

In light of this standard for theological reflection, the analysis proffered here is informed by the challenge of reality posed by the care of individuals with gender dysphoria in Catholic health care. Specifically, the question examined in this chapter is whether medical and surgical interventions on secondary sex characteristics for the treatment of adults diagnosed with gender dysphoria in Catholic health care are morally consistent with Catholic teaching and with the Catholic moral tradition on the creation of the human person as either female or male. Consideration of this question will include: an overview of the suffering of persons with gender dysphoria and the status of medical and surgical interventions for the treatment of gender

[1] The English translation of this text is from Google Translate. For the Latin original see https://www.vatican.va/content/francesco/la/motu_proprio/documents/20231101-motu-proprio-ad-theologiam-promovendam.html

P. J. Cataldo (✉)
Ethicist, Cataldo Consulting Services, LLC, Hampton, NH, USA

J. T. Eberl (ed.), *Emerging Issues in Catholic Bioethics*, Philosophy and Medicine 153, https://doi.org/10.1007/978-3-032-08862-8_4

dysphoria; an explanation of Catholic teaching on the creation of the human person as a composite unity of body and soul and other central, ontological components of this teaching; an explanation of the status of sex in the body/soul unity; an application of certain principles in the Catholic moral tradition, such as the principle of totality; and an evaluation of certain recent ecclesial documents relating to the issue of hormonal and surgical interventions for gender dysphoria.

4.2 Treating the Suffering of Those with Gender Dysphoria with Hormonal and Surgical Interventions[2]

Studies and reviews have consistently shown that individuals with gender dysphoria experience significantly higher rates of suicidal ideation and suicide than general populations, even though the ranges among studies vary. A meta-synthesis of both gray and peer-reviewed health literature showed that

> … an average of 56% of participants experienced lifetime suicidal ideation and 29% had *ever* attempted suicide. Given the rate of suicidality in the general population, 0.5% and 3.7%, respectively, it is clear that transgender individuals are disproportionately impacted by this phenomenon. Indeed, even over *the past year*, where 51% of participants ideated and 11% attempted, suicidality is approximately 14 and 22 times higher than in the general public. (Adams, Hitomi, and Moody 2017, 70–71)

The 2015 U.S. Transgender Survey summarized its findings on suicide ideation and suicide attempts among transgender individuals in this way:

> … transgender adults have a prevalence of past-year suicide ideation that is nearly twelve times higher, and a prevalence of past-year suicide attempts that is about eighteen times higher, than the U.S. general population. The 2015 U.S. Transgender Survey (USTS), which is the largest survey of transgender people in the U.S. to date, found that 81.7 percent of respondents reported ever seriously thinking about suicide in their lifetimes, while 48.3 percent had done so in the past year. In regard to suicide attempts, 40.4 percent reported attempting suicide at some point in their lifetimes, and 7.3 percent reported attempting suicide in the past year. (Herman Jody et al. 2019, 1)

Findings from the 2015 USTS show that 60.5% of transgender individuals from ages 18 to 25 reported *suicide thoughts* in the past year compared to 8.3% of individuals surveyed in the in the 2015 National Survey on Drug Use and Health (NSDUH); (USTS vs NSDUH respectively) 41.5% to 4.1% ages 26 to 49; and 24.6% to 2.6% ages 50 and above. In a comparison with results from the 2015 NSDUH regarding *suicide attempts* in the past year, the USTS showed respectively 10.2% to 1.6% ages 18 to 25; 5.6% to 0.5% ages 26 to 49; and 2.1% to 0.3% ages 50 and above (Herman Jody et al. 2019, 30). A recent study found that *lifetime suicidal ideation* for the population (ages 16–84) of people who identify their gender identity with their biological sex (cisgender) was 12.5% compared to 36.1% among

[2] For a more comprehensive citation of scientific literature on the suffering of gender dysphoria and the effects of hormonal and surgical treatment by the author see Cataldo (2019).

transgender individuals; respectively, for *suicidal ideation in the past 12 months* 3.2% compared to 13.3%; and for *suicide attempts* in the past 12 months 0.4% compared to 1.9% (Bränström et al. 2022, 407).

Individuals with gender dysphoria also experience higher levels of mental health disorders. For example, a systematic review of the literature regarding lifetime Axis I mental disorders (formerly classified as mental health and substance use disorders) among individuals with gender dysphoria found "a high lifetime frequency of 53.2% of Axis I mental disorders in GD [gender dysphoria] individuals. Among the large groups of mental disorders, those of mood and anxiety appear to be the most frequent in the studied GD population, with frequencies of 2.1% and 26.8%, respectively" (de Freitas et al. 2020, 101; Substance Abuse and Mental Health Services Administration 2016, 5). Another study concluded that "untreated transgender people had higher prevalence rates of possible or probable depressive disorder than controls and trans people already on cross-sex hormones" (Witcomb et al. 2018, 313). The World Professional Association for Transgender Health in its "Standards of Care for the Health of Transgender and Gender Diverse People, Version 8," affirmed that "some studies have shown a higher prevalence of depression ... and suicidality ... among TGD [transgender and gender diverse] people ... than in the general population" (Coleman et al. 2022, S171).

Studies and reviews of the effects of hormonal therapy and surgery on mental health symptoms for transgender people, including those with gender dysphoria, generally show positive results (van Leerdam et al. 2023; Baker et al. 2021; Nguyen et al. 2018).[3] For example, a prospective longitudinal study on the reduction of anxiety and depression symptoms in transgender people after 18 months of gender-affirming hormonal therapy (GAHT) found that "a significant reduction in symptoms of depression in transgender individuals after 18 months of starting GAHT, with a more than one-fifth decrease in the number of participants whose scores reflected a possible or probable depressive disorder" (Aldridge et al. 2021, 1813). Studies and reviews also indicate in general that surgical interventions can improve the well-being of individuals with gender dysphoria (Javier et al. 2022; Park et al. 2022). For example, a systematic review of literature on this question found that "overall, most of the studies comparing pre-and post-operative data on quality of life, body image/satisfaction, and overall psychological functioning among individuals with gender dysphoria suggest that GAS [gender-affirming surgery] leads to multiple, significant psychological benefits" (Wernick et al. 2019, 482). Given the growing evidence of the positive effects of hormonal and surgical treatment on the psychological well-being of transgender people and those with gender dysphoria, are any such treatments compatible with Catholic teaching on the creation of the person as either female or male? The first step in answering this question is to provide a brief outline of the teaching.

[3] The reporting of general positive results is compatible with the fact that studies on medical and surgical interventions differ in degree with respect to quality of evidence and other shortcomings.

4.3 Catholic Teaching and Tradition on the Creation of the Human Person as Female or Male and the Body/Soul Unity

Since a goal of this analysis is to demonstrate how medical and surgical interventions for the treatment of adult patients with gender dysphoria in Catholic health care is consistent with Catholic teaching on the creation of the human person, it is important to present the central components of this teaching as they are found in the magisterium. The fact that the biological dimension of this teaching as it was historically considered in its origins was wrong and lacking the benefit of the true account of sex determination found in contemporary science is not relevant for this analysis and will not be considered. The presentation of this teaching will focus on its ontological dimension which, as historically considered, does not depend on a biologically accurate account of the characteristics of human beings as female or male, but only on the fact that the teaching recognized a generally valid biological distinction between male and female. An ontological consideration of the teaching focuses on the metaphysical principles or sources for existing as a female or male person, on how humans exist as either male or female *qua* being, or, in other words, precisely insofar as they have being. The fact that the ontological dimension of Catholic teaching on the creation of the human person as either female or male was historically developed in the context of an incorrect biology does not invalidate the metaphysics of the teaching.[4] Although the orders of being represented in biology and metaphysics are related, they are also distinct with respect to their principles and designata, and this distinction allows for the latter to remain valid with respect to its order even if there is a defective understanding of the former.

Catholic teaching has constantly maintained that God creates the human person as a unity of body and soul. For example, the Council of Ephesus, in 431, stated that "We do not say, in fact, that the nature of the Word underwent a transformation and became flesh or that it was changed into a complete man composed of soul and body" (Denzinger 2012, 94). The Fourth Lateran Council, in 1215, taught that God creates "at once out of nothing both orders of creatures, the spiritual and the corporeal, that is, the angelic and the earthly, and then the human creature, who, as it were, shares in both orders, being composed of spirit and body" (Denzinger 2012, 266). Later, the Council describes Jesus Christ as being made "true man, composed of a rational soul and a human body" (Denzinger 2012, 266). This teaching was later reaffirmed by the First Vatican Council in 1870 (Denzinger 2012, 601). The Council of Vienne, in 1312, declared that "We define that from now on whoever presumes to assert, defend, or obstinately hold that the rational and intellectual soul is not of itself and essentially the form of the human body is to be censured as heretic"

[4] For the view that the antiquated biology of Aristotle and Thomas Aquinas invalidates the metaphysics of Catholic teaching on the creation of the human person as female or male see, for example, Horan (2019, 2, 14, 149–151). For a historical review of Aquinas's biology of the sexes and his theological concept of woman, see Allen (1997) and Allen (2002).

(Denzinger 2012, 290). *Gaudium et spes* declared that "Though made of body and soul, man is one. Through his bodily composition he gathers to himself the elements of the material world; thus they reach their crown through him, and through him raise their voice in free praise of the Creator" (Second Vatican Council, 1965, n. 14). In his examination of moral agency and the natural moral law in *Veritatis splendor*, St. John Paul II writes that the unity of body and soul is:

> ... *the unity of the human person,* whose rational soul is *per se et essentialiter* the form of his body. The spiritual and immortal soul is the principle of unity of the human being, whereby it exists as a whole—*corpore et anima unus*—as a person. (John Paul II 1993, n. 48)

Pope Benedict XVI, in discussing love as *agape* and *eros*, shows that human dignity is rooted in the unity of body and soul: "... man is a being made up of body and soul. Man is truly himself when his body and soul are intimately united ..." (Benedict XVI 2005, n. 5). In 2021, Pope Francis spoke to the Fifth International Conference sponsored by the Pontifical Council for Culture and the Cura Foundation on its theme of Exploring the Mind, Body & Soul in which he acknowledges the teaching on the unity of body and soul:

> ... the person ... [is] an inseparable unity of body and soul, the latter being endowed with intellect and will.... We are not pure spirits; for each of us, everything starts with our body, but not only: from conception to death, we do not simply *have* a body; we *are* a body. (Francis 2021)

The *Catechism of the Catholic Church* summarizes this constant teaching: "The unity of soul and body is so profound that one has to consider the soul to be the 'form' of the body: i.e., it is because of its spiritual soul that the body made of matter becomes a living, human body; spirit and matter, in man, are not two natures united, but rather their union forms a single nature" (Catechism of the Catholic Church 1997, n. 365).

As considered in Catholic teaching and tradition, body and soul are two principles or sources of the being of an individual person; they are not themselves existing beings that somehow exist parallel to each other. The soul has traditionally been considered the form of the human body, and, precisely as form, it is the principle through which an individual human *exists as* the *kind* of being it is. The term "body," understood as a principle of being, does not denote a particular physical body but the potentiality (the "matter") to become an actual human body, and the term "form" denotes the actualization of matter to exist in a human way as this individual human being. For example, St. Thomas Aquinas in *De ente et essentia* explains that "... through form, which actualizes matter, matter becomes an actual being and this particular thing" (Aquinas 1968, ch. 2, 2). The soul, as form of the body, communicates being to the potentiality for human bodily existence to actually be a body:

> ... the soul is united to the body as its proper form.... That by which a thing, from being potentially, is made an actual being, is its form and act. Now the body is made by the soul an actual being from existing potentially, since to live is the being of a living thing. (Aquinas, *Summa contra gentiles*, bk. II, ch. 57)

As has been shown, Catholic teaching affirms that the human person exists as a composite unity of body and soul. The soul as form alone does not exist and the body as matter alone does not exist, but when these two principles of being compenetrate each other, the resulting unified composite exists through its own single act of existence. As Aquinas explains, "… the being that a composite substance has is not the being of the form alone nor of the matter alone but of the composite, and it is essence according to which a thing is said to be" (Aquinas 1968, ch. 2, 3). Just as an existing individual human being is composed of the intrinsic unity of body and soul, so too is the human species a composite: "For as it belongs to the notion of this particular man [any individual with a human essence] to be composed of this soul, of this flesh, and of these bones; so it belongs to the notion of man to be composed of soul, flesh, and bones; for whatever belongs in common to the substance of all the individuals contained under a given species, must belong to the substance of the species" (Aquinas 1947a, I, q. 75, a. 4). The "essence of anything whatever," Aquinas argues, "is that which is signified by its definition. But the definition of man signifies not form alone but also matter, since matter must be comprised in the definition of material things" (Aquinas, *Compendium theologiae*, ch. 154). Thus, the essence of an individual human being is that by which the individual exists as a human, and, as such, this human essence is composed of both the matter of the individual ("body") and the form ("soul") of this matter.

The material principle of an actual individual human being enables the soul in its universal character as human form to be individualized in this or that particular human being. Understood in this way, matter serves as the principle of individuation, for Aquinas:

> … matter is the principle of individuation … the matter which is the principle of individuation is not just any matter, but only designated matter. By designated matter I mean that which is considered under determined dimensions. This kind of matter is not part of the definition of man as man, but it would enter into the definition of Socrates if Socrates could be defined. The definition of man, on the contrary, does include undesignated matter. In this definition we do not put this particular bone and this particular flesh, but bone and flesh absolutely, which are the undesignated matter of man. (Aquinas 1968, ch. 2)

The multiplication of universal human nature in many individuals is made possible through matter as an ontological principle. Because this type of matter first has the capacity for dimensions in general and is subject to dimensions, it allows for the particular dimensions of this or that individual. The potential for bone and flesh in general (undesignated mater) in correlation with human nature allows for human nature to be individualized with this or that particular flesh and bone (designated matter) of a specific, actually existing person. Thus, the potential for existing with measurable bodily dimensions such as body mass index (BMI) in general, or bone, or flesh as such, and other similar general dimensions, underlies the potential of an individual to exist with a specific BMI or with these specific bones and tissue. "And by these indeterminate dimensions," Aquinas argues, "matter is made to be this signate matter [designated matter], and thus gives individuality to a form, and thus also by matter there is caused the numerical diversity of things in the same species"

(Aquinas, *Super Boethium de Trinitate*, q. 4, a. 2).[5] In sum, when the human soul as form unites with matter that is properly disposed to receive the human form, an actual human individual comes into being.

This actually existing composite unit of body and soul exists as a substance. As a substance, the individual human cannot be predicated of another entity and exists in and through itself, not in another (Aquinas, *Commentary on the Metaphysics*, bk. 5, ch. 1, lect. 10; 1947a, I, q. 2, a. 1). In the *Summa theologiae*, Aquinas argues that the "word substance signifies not only what exists of itself—for existence cannot of itself be a genus … but, it also signifies an essence that has the property of existing in this way—namely, of existing of itself; this existence, however, is not its essence" (Aquinas 1947a, I, q. 3, a. 5 *ad* 1). The meaning of substance for Aquinas, as John Wippel has shown, "is that it have a quiddity to which it belongs to exist not in something else" (Wippel 2000, 232–233; Aquinas 1947a, III, q. 77, a.1 *ad* 2). As there are beings whose existence is in themselves and not in something else, so there are things that have being in a qualified sense "to which it belongs to be in something else" (Wippel 2000, 234; Aquinas, *Commentary on the Metaphysics*, bk. 12, ch. 1). Such things are called "accidents" and the "something else" in which such things exist are substances. The former do not possess substantial being but accidental being and exist only insofar as they inhere in something that is a substance. Thus, for example, the height, weight, hair, or position of an individual human being as accidents exist only if the individual exists as a substance and can change without the individual changing as a substance. The individual as an existing human remains even if such characteristics change. Etienne Gilson summarizes well the interrelationship of substance and accidents: "… within our experience, there are no substances without accidents any more than there are accidents without substance, still it is the accidents which belong to the substance, not the substance to the accidents" (Gilson 1956, 30).

In Catholic teaching, sex and gender are not constitutive of the human being *qua* human and *qua* substance.[6] Sexual identity is constitutive of actual, individual human beings in Catholic teaching, but is not included in the definition of a human substance (Second Vatican Council, 1965, nn. 12 and 14; *Catechism of the Catholic Church*, nn. 355–368). For Aquinas, sexual identity as female or male are accidents of the individual as a human substance, and like any accident sexual identity is not part of the definition of the human essence (see Aquinas, *Quaestiones disputatae de anima*, q. 12 *ad* 7). In particular, sexual identity is what he terms an "inseparable accident." An inseparable accident is caused by "the principles of the individual" as opposed to "the principles of the species" and has a "permanent cause" in the subject in which the accident inheres (Aquinas, *Quaestiones disputatae de anima*, q. 12 *ad* 7; Aquinas, *Commentary on the Sentences*, d. 17, q. 1, a. 2 *ad* 2.). "Principles of the individual" here may be interpreted to refer to the designated matter of the

---

[5] See also *Commentary on the Sentences*, d. 23, q. 1, a. 1; d. 25, q. 1, a. 1 *ad* 3; d. 12, q. 1, a. 1 *ad* 3; *Summa contra Gentiles*, bk. 4, ch. 65; Eberl (2020, 31–34); Wippel (2000, 353–375); Owens (1988).

[6] For a different view see Ashton (2024, 79–84), Bedford and Eberl (2016), and Finley (2015).

individual as an animal, which disposes specific individuals to have these bones and this flesh as either males or females (Aquinas 1968, ch. 6; Aquinas, *Commentary on the Metaphysics*, bk. 10, ch. 1, lect. 9). For Aquinas, then, the human individual cannot come into being and exist without being either a female or male, but this does not entail that being male or female is part of the human essence.

Catholic teaching affirms this ontology of sexual identity as it relates to the creation and existence of the individual human being, reflected in the teaching of Genesis that "God created man in his image; in the divine image he created him; male and female he created them" (Gen 1:27):

> Man and woman have been created, which is to say, willed by God: on the one hand, in perfect equality as human persons; on the other, in their respective beings as man and woman. "Being man" or "being woman" is a reality which is good and willed by God: man and woman possess an inalienable dignity which comes to them immediately from God their Creator. [Gen 2:7, 22] Man and woman are both with one and the same dignity "in the image of God". In their "being-man" and "being-woman", they reflect the Creator's wisdom and goodness. (Catechism of the Catholic Church 1997, n. 369)

As an inseparable accident, sexual identity is inextricably tied to a human individual's act of existence ("being-woman," "being-man"), which commences at the point when body and soul are unified by God's creative act of the individual. Ontologically, this means that the sexual identity of an individual person as female or male remains unchanged so long as the person's act of existence remains. Due to the ontological contingency of individual existence, the ontological dimension of Catholic teaching on the creation of the human person as female or male is compatible with the fact that, for example, certain individuals have disorders of sex development, or that trans people experience themselves as other than their biological sex, or that medicine may lack certitude about a particular person's sex (Cataldo 2019; Cataldo 2024; Hull 2006, 94–97, 102–103, 110). Hence, these facts need not be explained as God' deliberate plan in order to make sense of them (Bouchard 2024).

4.4 Morally Permissible Interventions to Treat Gender Dysphoria in Adults

Given the metaphysical dimension of Catholic teaching on the creation of the human individual as either female or male, is there a basis in the teaching that would morally allow medical or surgical interventions for the treatment of gender dysphoria in adults? Are there interventions that would both respect and preserve the sexual identity of a person within the body/soul unity and treat the serious effects of gender dysphoria in adults? The biological distinction between primary and secondary sex characteristics in humans has metaphysical implications that provide a basis for the moral permissibility of certain interventions consistent with Catholic teaching.

Primary sex characteristics are physical properties that are directly related to reproduction. Primary sex characteristics include the sex chromosomes XX and XY, gamete cells, and the reproductive organs. Secondary sex characteristics include physical properties that are to one degree or another indirectly related to reproduction; these include mammary glands, body hair, size of hips, size of larynx, bone structure, and muscle mass, all of which develop differently in females and males (Federman 2006, 1507–14; Wizemann and Pardue 2001, 65, 67; Fairbairn 2013, 161, 163–164; Schiebinger et al. 2011–2020). Because sex characteristics are in part effects caused by sex hormones such as progestogens, androgens, and estrogens, sex hormones are not themselves sex characteristics.

Both the primary and secondary sex characteristics comprise the biological sex of an individual human being, but what is relevant for the question of this analysis is whether only the primary sex characteristics are both necessary and sufficient for God's creative act of the human individual to be either female or male. Elsewhere, I have described this ontological status of the primary sex characteristics as "integral" for the creation of an individual person as male or female (Cataldo 2019; Cataldo 2024). The primary sex characteristics are ontologically both necessary and sufficient for, or integral to, an individual being created as female or male, because they constitute the natural potentiality that allows the individual *to exist* as either female or male.

This ontological role of the primary sex characteristics is reflected in Catholic teaching on the communion of men and women in their complementarity. Regarding the notion of complementarity itself, St. John Paul II, in his teaching on the first chapters of Genesis, explained how the complementarity of male and female is an inextricable part of being-female or being-male:

> … masculinity and femininity. These are, as it were, two "incarnations" of the same metaphysical solitude before God and the world. They are two ways of "being a body" and at the same time a man, which complete each other. They are two complementary dimensions of self-consciousness and self-determination and, at the same time, two complementary ways of being conscious of the meaning of the body. (John Paul II, *General Audience*, 21 November, 1979a)

For the teaching of St. John Paul II, femininity and masculinity, as modes of being, and as conceived by God their creator, comprise what it means to be a human individual. They are mutually integral components of what it means to be an individual human substance, and each are not intelligible without the other as they exist in an actual, human individual. In the real individual, each property complements the other in the person's self-consciousness, self-determination, and consciousness of being a body. This metaphysical view of the complementarity of the sexes in Catholic teaching is also supported by Thomistic metaphysics. For Aquinas, real being is being in act; in fact, even potency is unintelligible without its ordering to actual being. Norris Clarke, S.J. eloquently argued that "*relationality* is a primordial dimension of every real being, inseparable from its substantiality, just as action is from existence," and that, "[*t*]*o be* fully is to be *substance-in-relation*" (Clarke and Norris 1993, 13–14; Eberl 2020, 116–117; Ratzinger 1990, 542; Hull 2006, 113–115). "Thus," Clarke concludes in the case of a human person, "a personalized being must obey the basic dyadic ontological structure of all being, that is, *presence in itself and presence to others*" (Clarke and Norris 1993, 71). As a logical extension, the complementarity of the sexes may legitimately be regarded as an ontological form of the fundamental relationality of being itself.

Central to the complementarity of men and women is an ordination to the generation of new life. This is observed in the way that the *Catechism* connects the complementarity of masculine and feminine with the Genesis texts on procreation:

> Man and woman were made "for each other"–not that God left them half-made and incomplete: he created them to be a communion of persons, in which each can be "helpmate" to the other, for they are equal as persons ("bone of my bones ...") and complementary as masculine and feminine. In marriage God unites them in such a way that, by forming "one flesh", [Gn. 2:24] they can transmit human life: "Be fruitful and multiply, and fill the earth." [Gn. 1: 28] By transmitting human life to their descendants, man and woman as spouses and parents co-operate in a unique way in the Creator's work [cf. Second Vatican Council, 1965, n. 50]. (Catechism of the Catholic Church 1997, n. 372)

In his teaching on Genesis, St. John Paul II articulated a fundamental theological and metaphysical understanding of the connection between the complementarity of the sexes and the communion of persons (*communio personarum*) with procreation:

> In the mystery of creation–on the basis of the original and constituent "solitude" of his being–man was endowed with a deep unity between what is, humanly and through the body, male in him and what is, equally humanly and through the body, female in him. On all this, right from the beginning, the blessing of fertility descended, linked with human procreation (cf. Gn 1:28). (John Paul II 1979b, 1980, 1982)[7]

If we overlay Catholic teaching on the creation of the individual human person as a unity of body and soul participating in a complementarity of the sexes on the one hand, with the distinction between the primary and secondary sex characteristics on the other, a case can be made that it is the primary sex characteristics, not the secondary sex characteristics, which are essential for the creation of the individual human person as female or male. The secondary sex characteristics certainly contribute to what it means to be male or female and are necessary for being-male or being-female, but insofar as they are indirectly related and to, and are not direct causes of, the coming-to-be of new life as female or male, they lack the kind of ontological potentiality that is necessary for an individual to be created and exist as an individual human substance with the inseparable accidents of female or male. The direct role that the primary sex characteristics have in procreation gives them an ontologically primary role in determining the coming-to-be of an individual as female or male in God's creative act. Given that the secondary sex characteristics are necessary, but not integral or both necessary and sufficient, for the creation of the individual human person as female or male, their loss or alteration as a treatment for gender dysphoria does not disrupt the fundamental ontological order of the

---

[7] See Ashley (1995, 530, 544–545) for the helpful explanation of "complementary equality" as "God's original intention for the relation of man and woman" in which "completing each other" refers to women and men enabling each other to "equally attain the perfection of human personhood."

human body *precisely as it exists as male or female* and does not detract from God's creation of the individual as a female or male.

Given what is argued here regarding the ontological nature is of the secondary sex characteristics, the simple fact that they are sex characteristics is not a reason to prohibit interventions on them for the treatment of gender dysphoria in adults. Thus, the ethical evaluation of treatment for gender dysphoria in Catholic health care ought not lump all interventions in the same category of fundamental alteration of the body/soul unity. As this analysis is intended to show, not all interventions are "sex-reassignments" in the sense that they change the fundamental order of the human body in the body/soul unity. To reduce or conflate all interventions to treat gender dysphoria in adults as an attempt at fundamentally transforming a person's existence begs the question by assuming the very thing in need of proof; namely, that there is no ontologically significant difference between the primary and secondary sex characteristics for the creation of the individual human person as male or female.

Although it can be shown that a moral prohibition against medical and surgical interventions on the secondary sex characteristics to treat adults with gender dysphoria based simply on the fact that these are sex characteristics is not valid, the moral justification of such interventions needs additional support. This support comes from the moral principles of totality, double effect, cooperation in wrongdoing, and from a consideration of theological scandal. I have addressed the application of these principles to the question of this chapter in more detail elsewhere (Cataldo 2019, 9–17; Cataldo 2024, 105–110, 112–114). Here, it will be sufficient to provide the broad outlines of how these principles may be applied to morally justify the use of medical and surgical interventions on secondary sex characteristics to treat adults with gender dysphoria within Catholic health care.

The principle of totality is a moral guide for evaluating the removal or alteration of a part of the human body. It presumes that the individual human body is itself an integral part of a whole which is the unity of body and soul, and that the good of any bodily part is subordinated to the good of the whole. The principle has developed to include three basic criteria for morally acceptable interventions on bodily parts that have serious consequences: (1) serious harm to the well-being of the patient is threatened; (2) the intervention is likely to be effective; and (3) the resulting good for the whole is proportionate to any loss (Pius XII 1960d). Beginning with the antecedent to the principle in the work of St. Thomas Aquinas (1947a, II-II, q. 65), and continuing through the Catholic moral tradition, the received interpretation of the principle has reduced the meaning of "totality" to physical totality or integrity and has restricted the "good of the whole" about which the principle is concerned to the physical good of the human body as a whole. This physicalistic and reductionistic interpretation of the principle is not warranted. The subject of the principle is not merely a physical being, but a unity of body and soul with integral parts that are physical and immaterial; therefore, the good of the whole evaluated by the principle must be inclusive of both physical and immaterial or spiritual dimensions of the person as they function within the unity of body and soul. Moreover, as will be shown below, Pope Pius XII's interpretation of the principle, which is often cited as authoritative, supports a use of the principle that allows for a notion of totality and the good of the whole that is inclusive of the spiritual dimension of the human person.

I have argued for this thesis elsewhere (Cataldo 2019, 10–13; Cataldo 2024, 105–110), but it is important to highlight here some central factors regarding Pius XII's interpretation of the principle. First, in his texts that either explicitly or implicitly refer to the principle of totality *qua* principle, there is nothing that necessarily reduces the notion of totality or restricts the principle itself to the physical body. Second, Pope Pius employed three meanings of "organism," only one of which referred to the physical organism of the human body. Third, Pope Pius does apply the principle of totality to the body and its organs, but in each instance, this is an *application* in which the notion of totality for the particular case is appropriately identified with the physical body. However, this fact alone does not entail that Pope Pius is reducing the *notion of totality and the principle itself* in a physicalistic way. A particular application of the principle of totality ought not be conflated with the principle itself. Fourth, in certain texts Pope Pius distinguishes between the physical unity of a subsisting individual person and the community as a whole. As with the previous point, what he states about the physical totality of the person in these texts ought not be confused with the principle itself, but more importantly, he underscores the essential contradistinction in these cases between the intrinsic unity of the person and the extrinsic unity of the community or the state (Pius XII 1958, 1960a, 1960b, 1960d, 1960e). This intrinsic unity is not reducible to the physical body but is true of the composite unity of the person, body and soul. Fifth, there are several texts from Pope Pius where an integrated view of the whole person, body and soul, is explained either directly or indirectly in relation to the principle of totality (Pius XII 1960e, 1960c, 1960a, 1960f, 1960g).[8] As a result of this perspective of the principle, it can be legitimately argued that medical and surgical interventions on the secondary sex characteristics to treat adults with gender dysphoria meet the criteria of the traditional principle of totality.

This same treatment may be validly justified by an application of the principle of double effect to the Catholic health care institution. The traditional principle was developed as a moral guide to preserve the integral goodness of an act's object, intention, and circumstances when the act has both good and bad foreseen effects (Cataldo 2022). The double-effect act in this case is a medical or surgical intervention which has the good effect of preventing suicide and the bad effect of indirectly contributing to *an individual's attempt* to alter God's creative act of the individual as either female or male. It may be shown that medical and surgical interventions on the secondary sex characteristics to treat adults with gender dysphoria can meet the four conditions of the traditional principle: (1) the moral object of such acts is the amelioration of a serious health condition and the alteration of a healthy secondary sex characteristic is not *per se* immoral; (2) the intention of the institutional

[8] See Kelly (1955, 379): "It might be objected that the popes have referred the principle of totality only to the whole body or the whole physical organism. This is true of the formulation of Pius XI; but Pius XII has spoken not only of the body and the physical organism, but also of the good of the "being as a whole" [1960b]—which, it seems, is best interpreted as the person."

provider is the amelioration of a serious health condition and the prevention of grave harm; (3) the bad effect of contributing to the patient's attempt to alter God's action of being created as either female or male is not the means by which the good effect is achieved (the good effect is directly achieved by altering bodily parts that *per se* are healthy and ought not be altered, but *per accidens* are contributing to a dysphoria with mortality risks); and (4) the good effect is morally proportionate to the bad effect, especially considering that the sex characteristics affected by the intervention are not primary sex characteristics.[9]

Since medical and surgical interventions on secondary sex characteristics for the treatment of adults with gender dysphoria can be morally justified through an application of the principle of double effect, such interventions can also be justified in a similar way by the principle of cooperation in wrongdoing.[10] The principal act of wrongdoing could be identified as the patient's attempt to alter God's creative act of the individual as either female or male. The cooperator is the Catholic health care institution that allows and facilitates the medical and surgical interventions to ameliorate a serious health condition and to prevent suicide. These actions are *per se* good and are what the institution intends. The patient appropriates these morally good acts by the Catholic health care institution for the wrongful purpose of attempting to alter God's creative act. The likelihood that the patient's condition will be ameliorated and suicide prevented are proportionate to the indirect contribution that the institution makes to the patient's attempt to alter how God created her or him. Importantly, foreseeing that the patient will appropriate the good actions of the institution and that these actions will indirectly contribute to the patient's attempt is not necessarily to intend or approve of what the patient attempts. Foreseeing a bad consequence as a result of one's actions does not by itself constitute an intention of or willingness for the same. Under the conditions described here, any cooperation by a Catholic health care institution in the principal act of a patient would be justified, remote or proximate, mediate material cooperation.

The potential for a Catholic health care institution to occasion insurmountable theological scandal by allowing and facilitating the limited interventions described here is significantly reduced by the fact that any cooperation in wrongdoing is morally justified.[11] Communications and education have been essential in the Catholic moral tradition for the mitigation of theological scandal associated with morally justifiable cooperation in wrongdoing (Aquinas 1947a, II-II, q. 43, a. 7; a. 8 *ad* 5; McHugh and Callan 1958, 601; Davis 1935, 335). Communications and education for health care staff, parishes, and communities that include the moral justification for interventions on secondary sex characteristics to treat gender dysphoria in certain adult cases, that are appropriately adjusted for various audiences, and that speak

[9] For an elucidation of the principle's criteria see Porter (2015) and Mangan (1949).

[10] The principle of cooperation in wrongdoing is essentially the principle of double effect applied to a cooperator and acts of cooperation.

[11] For an explanation of the traditional distinctions in the principle of cooperation in wrongdoing, see McHugh and Callan (1958, 615–627). For an explanation of the traditional notion of theological scandal, see Prümmer (1956, 102–103).

to the various clinical, theological, and ethical factors involved, are important measures for mitigating the risk of insurmountable scandal. Such efforts should include, as appropriate, an explanation why support for the interventions in question does not equate to support for what has been termed "gender ideology." In this regard, and with respect to mitigating the risk of insurmountable theological scandal generally on the question, it is important to make use of helpful analogies. These analogies engage examples of how the Catholic Church has supported certain kinds of actions that involve association with activities that are contrary to Catholic teaching, but where the risk of insurmountable theological scandal has been avoided or mitigated. Such examples include support of vaccines derived from cell lines linked to aborted fetal tissue, support for abortion legislation that reduces but does not eliminate abortion, and millennia of caring for individuals whose persistent behavior or lifestyle is contrary to Catholic teaching (Cataldo 2019; Cataldo 2024).[12] The development and issuing of clear clinical guidelines for staff regarding interventions on secondary sex characteristics for the treatment of adults diagnosed with gender dysphoria is another helpful communications and educational measure that mitigates the risk of insurmountable theological scandal.

4.5 An Evaluation of Recent Ecclesial Teaching and Documents

4.5.1 Dignitas Infinita (On Human Dignity)

Dignitas infinita, from the Dicastery for the Doctrine of the Faith, is a declaration that reflects on the meaning of human dignity and identifies grave and current violations of it. Paragraph 60 addresses the role of human dignity with respect to the question of "sex change" and begins by articulating the foundation of human dignity in the unity of body and soul:

> … humans are inseparably composed of both body and soul. In this, the body serves as the living context in which the interiority of the soul unfolds and manifests itself, as it does also through the network of human relationships. Constituting the person's being, the soul and the body both participate in the dignity that characterizes every human. (Dicastery for the Doctrine of the Faith 2024, n. 60)

The declaration then concludes with the following assessment of what it calls "sex change intervention":

> It follows that any sex-change intervention, as a rule, risks threatening the unique dignity the person has received from the moment of conception. This is not to exclude the possibility that a person with genital abnormalities that are already evident at birth or that develop

[12] As with all analogies, the analogates in these analogies differ in various ways, but the analogon, or the item shared among the analogates, is the same; namely, the lack of theological scandal over actions associated with wrongdoing.

> later may choose to receive the assistance of healthcare professionals to resolve these abnormalities. However, in this case, such a medical procedure would not constitute a sex change in the sense intended here. (Dicastery for the Doctrine of the Faith 2024, n. 60)

There are a number of reasons why *Dignitas infinita* need not be interpreted as necessarily prohibiting medical and surgical interventions on secondary sex characteristics to treat adults with gender dysphoria. First, the declaration does not explicitly prohibit hormonal therapy as an intervention. Indeed, the term for "intervention" in the medical context of the original Italian text, "*intervento*," connotes surgery. Second, the prohibition against surgical interventions (or any intervention) is an "in principle" prohibition, which as such does not preclude any and all interventions. Third, in the Catholic moral tradition an "in principle" prohibition against removing a bodily organ derives from the fact that the organ or body part has a *per se* usefulness; that is, considered in itself the organ or part is ordered to the good of the whole body. Thus, on the level of its *per se* usefulness there is a corresponding universal obligation not to remove the organ or part. However, due to particular circumstances an organ or part might, *per accidens*, not be useful or be harmful. In such cases, there might not be a prohibition against removing the part or organ corresponding to the fact that incidentally considered (*per accidens*) they are harmful (Cataldo 2004; Cataldo 2008). Thus, *per se*, and generally, interventions on the secondary sex characteristics to treat adults with gender dysphoria are not permissible, but *per accidens*, some might be allowed because even though the external actions might be the same as a "sex change" intervention, the intervention considered in itself need not be not the same morally due to the circumstances.

For example, organs or body parts such as breasts or the Adam's apple are *per se* useful to the good of the whole body, but when, *per accidens*, they contribute to the serious adverse effects of gender dysphoria, surgery to treat these effects in particular cases, and where the intervention does not threaten the unique dignity of the human person, can be permissible even though they involve the same external acts as those used in an intervention whose purpose is "sex change." The circumstances of such cases demonstrate that the moral object of the Catholic health care institution is to prevent adverse effects by means that do not attempt to alter the fundamental order of the patient's body or God's creative act. This difference between *per accidens* allowing an intervention that "in principle" or *per se* is prohibited is also analogously exhibited by the "in principle" prohibition against foregoing or withdrawing medically administered nutrition and hydration. Because food and water are *per se* are useful, there is a universal obligation to use them for the conservation of life. However, this obligation is not in force if under certain circumstances and *per accidens* food and water are no longer useful to the whole body or are harmful (Congregation for the Doctrine of the Faith 2020, 2007; USCCB 2018, n. 58). Fourth, the permissibility of interventions to treat "genital abnormalities" is *one* example of the "in principle" status of the prohibition. There is nothing in the text of *Dignitas infinita* that indicates that all other possible examples are excluded, such as interventions on secondary sex characteristics to treat the serious adverse effects of gender dysphoria in certain adult cases.

4.5.2 *Doctrinal Note on the Moral Limits to Technological Manipulation of the Human Body*

On 20 March, 2023, the Committee on Doctrine of the United States Conference of Catholic Bishops issued a *Doctrinal Note on The Moral Limits to Technological Manipulation of the Human Body* (hereafter, "*Note*"). For the most part, the Committee correctly and helpfully presents an overview of Catholic teaching and tradition on the created order in general and the created, natural order of the human person in particular. The general explanations of the body/soul unity of the human person, the notion of individuation of human nature in individuals, and the relation of human sexual differentiation in this unity are accurate as far as they go. However, the *Note* contains a number of flaws. This analysis will briefly focus on two central problems: (1) its interpretation and application of the principle of totality, and (2) its moral and ontological conflation of any and all hormonal and surgical interventions to treat adults with gender dysphoria as attempts to change the fundamental order and finality of the body and the human person as an intrinsic unity of body and soul.

In its presentation of the principle of totality (without mentioning it by name), the *Note* focuses on the teaching of Pope Pius XII and reduces the principle and Pius's understanding of it to physicalistic meanings (USCCB 2023, n. 15). This reductive interpretation by the *Note* can be legitimately concluded because there is no attempt to present a broader interpretation of totality and the principle that is inclusive of the non-physical aspects of the body/soul unity, nor does the *Note* address an alternative application of the principle with this more integrative understanding. It is on the basis of this reductive interpretation of the principle and Pope Pius's teaching of it to physical concepts and meanings that the *Note* then prohibits any and all chemical or surgical interventions to treat gender dysphoria:

> Catholic health care services must not perform interventions, whether surgical or chemical, that aim to transform the sexual characteristics of a human body into those of the opposite sex or take part in the development of such procedures. (USCCB 2023, n. 18)

However, as I have argued above and elsewhere, Pius did not interpret and apply the principle and its concepts in a physicalistic way, but rather he had an integrated view of totality as it functions in the principle that incorporates the body/soul unity of the person inclusive of the mental, physical, and spiritual dimensions the person (Cataldo 2019; Gremmels 2016a; Gremmels 2016b; Austriaco and Giorgio 2022). The *Note's* entire critique of interventions to treat gender dysphoria rests on an erroneous, reductive view of Pope Pius's understanding and application of the principle of totality.

A second central problem with the *Note* is that it does not differentiate between primary and secondary sex characteristics in any of its references to sex characteristics. As a result, the *Note* never considers whether the distinction has significant metaphysical and moral consequences for the issue of treating adult patients with gender dysphoria in Catholic health care. In its sweeping conflation of, and prohibition against, any and all hormonal and surgical interventions it erroneously assumes that the primary and secondary sex characteristics are morally equal without showing why this is valid. The *Note* assumes the very thing in need of proof and begs the question. Because the *Note* assumes that there is no moral or metaphysical distinction between the primary and secondary sex characteristics that has significance for God's creative act of the human person, it erroneously assumes that the only ultimate intention ("aim") for any and all gender dysphoria interventions is "to alter the fundamental order and finality of the body and to replace it with something else" (USCCB 2023, n. 16). However, that the secondary sex characteristics represent the "fundamental order and finality of the body" is the very thing in question. The *Note* provides no explanation as to why the primary and secondary sex characteristics are metaphysically and morally equal, or conversely, why their difference makes no moral or metaphysical difference.

4.6 Conclusion

In an address to the Alphonsian Academy, Pope Francis reminded us that moral theologians

> … are required to enter into a living relationship with the People of God, engaging in particular with the cry of the least, to understand their real difficulties, to look at existence from their perspective, and to offer them answers that reflect the light of the eternal love of the Father. Faithful to Alphonsian tradition, you try to offer a model of Christian life that, with respect for the demands of theological reflection, is not however a cold morality, theoretical morality (Francis 2023a).

The analysis of this chapter has been an attempt to understand the real difficulties of people with gender dysphoria who wish to be treated in Catholic health care, and how Catholic health care can respond to their suffering in a way that is consistent with Catholic teaching on the creation of the human person as female or male. It is an attempt to address this need through a morally legitimate *via media* between jettisoning or altering fundamental Catholic doctrine and the complete prohibition against all hormonal and surgical treatment. The ontological and moral significance of the distinction between the primary and secondary sex characteristics represents the fulcrum upon which balance between faithfulness to Catholic teaching and meeting the needs of these suffering patients is achieved.

References

Adams, Noah, Maaya Hitomi, and Cherie Moody. 2017. Varied reports of adult transgender suicidality: Synthesizing and describing the peer-reviewed and gray literature. *Transgender Health* 2 (1): 60–75.

Aldridge, Zoë, et al. 2021. Long-term effect of gender-affirming hormone treatment on depression and anxiety symptoms in transgender people: A prospective cohort. *Andrology* 2021 (6): 1808–1816.

Allen, Prudence. 1997. *The concept of woman: The Aristotelian revolution 750 bc–ad 1250*. Cambridge: Eeerdmans Publishing.

Allen, Prudence. 2002. *The concept of woman: The early humanist reformation 1250–1500 volume II part 1*. Eeerdmans Publishing: Cambridge.

Aquinas, Thomas. 1947a. *Summa theologica*. Trans. Fathers of the English Dominican Province. https://www.ccel.org/a/aquinas/summa/home.html.

Aquinas, Thomas. 1947b. *Compendium theologiae*. https://aquinas.cc/la/en/~CT.BookI.C154.5.

Aquinas, Thomas. 1947c. *Summa contra gentiles*. https://aquinas.cc/la/en/~SCG2.C57.12.

Aquinas, Thomas. 1949. *Quaestiones disputatae de anima*. https://aquinas.cc/la/en/~QDeAn.

Aquinas, Thomas. 1953. *Super boethium de trinitate*. Thomas Aquinas: On Boethius' De Trinitate: English (isidore.co).

Aquinas, Thomas. 1961. *Commentary on the metaphysics*. https://aquinas.cc/la/en/~Metaph.

Aquinas, Thomas. 1968. *On being and essence*. Trans. Armand Maurer, C.S.B. Toronto: The Pontifical Institute of Mediaeval Studies.

Aquinas, Thomas. 2025. *Commentary on the sentences*. https://aquinas.cc/la/en/~Sent.I.

Ashley, Benedict, O. P. 1995. *Theologies of the body: Humanist and christian*. Braintree, Massachusetts: The Pope John Center.

Ashton, Sam. 2024. *Beyond male and female? A theological account of intersex embodiment*. London: T & T Clark.

Austriaco, Nicanor Pier Giorgio. 2022. Is self-amputation for survival morally justifiable? *The Linacre Quarterly* 89 (2): 206–211.

Baker, Kellan E., et al. 2021. Hormone therapy, mental health, and quality of life among transgender people: A systematic review. *Journal of the Endocrine Society 2021* 5 (4): 1–16.

Bedford, Elliott Louis, and Jason T. Eberl. 2016. Is the soul sexed? Anthropology, transgenderism, and disorders of sex development. *Health care Ethics USA* 24:18–33.

Benedict XVI. 2005. *Deus caritas est*. https://www.vatican.va/content/benedict-xvi/en/encyclicals/documents/hf_ben-xvi_enc_20051225_deus-caritas-est.html.

Bouchard, Charles. 2024. Catholic priest and ethicist: why the church needs LGBTQ people outreach: A LGBTQ Catholic Resource. https://outreach.faith/2024/02/catholic-priest-and-ethicist-why-the-church-needs-lgbtq-people/.

Bränström, Richard, et al. 2022. Transgender-based disparities in suicidality: A population-based study of key predictions from four theoretical models. *Suicide Life-Threatening Behavior* 52 (3): 401–412.

Cataldo, Peter J. 2004. Pope John Paul II on nutrition and hydration: A change of Catholic teaching? *The National Catholic Bioethics Quarterly* 4 (3): 513–536.

Cataldo, Peter J. 2008. The ethics of pope John Paul's allocution on care of the PVS patient: A response to J.L.A. Garcia. In *Artificial nutrition and hydration: The new catholic debate*, ed. Christopher Tollefsen, 141–161. Dordrecht: Springer.

Cataldo, Peter J. 2019. Catholic teaching on the human person and gender dysphoria. *Health Care Ethics USA* 27 (3): 9–17.

Cataldo, Peter J. 2022. The principle of the double effect as preserving integral goodness: A brief historical overview. *Health Care Ethics USA* 30 (2): 3–11.

Cataldo, Peter J. 2024. Treating persons with gender dysphoria in Catholic health care: Catholic teaching interpreted and applied. In *Trans life and the Catholic church today*, ed. Nicolete Burbach and Lisa Sowle Cahill, 105–114. London: T&T Clark.

Catechism of the Catholic Church. 1997. Second Edition. Vatican City: Libreria Editrice Vatican.

Clarke, W., and S. J. Norris. 1993. *Person and being*. Milwaukee, WI: Marquette University Press.

Coleman, E., et al. 2022. Standards of care for the health of transgender and gender diverse people, version 8. *International Journal of Transgender Health* 23 (Suppl 1): S1–S259.

United States Conference of Catholic Bishops (USCCB). 2023. *Doctrinal note on the moral limits to technological manipulation of the human body*. https://www.usccb.org/resources/Doctrinal%20Note%202023-03-20_0.pdf.

Congregation for the Doctrine of the Faith. 2007. *Responses to certain questions of the United States conference of catholic bishops concerning artificial nutrition and hydration: Commentary*. https://www.vatican.va/roman_curia/congregations/cfaith/documents/rc_con_cfaith_doc_20070629_commento-responsa_en.html.

Congregation for the Doctrine of the Faith. 2020. *Samaritanus bonus: On the care of persons in the critical and terminal phases of life*. https://www.vatican.va/roman_curia/congregations/cfaith/documents/rc_con_cfaith_doc_20200714_samaritanus-bonus_en.html.

Davis, Henry. 1935. *Moral and pastoral theology. volume one*. New York: Sheed and Ward.

de Freitas, Larissa Dias, et al. 2020. Psychiatric disorders in individuals diagnosed with gender dysphoria: A systematic review. *Psychiatry and Clinical Neurosciences* 74 (2): 99–104.

Denzinger, Heinrich. 2012. *Enchiridion symbolorum*. 43rd ed. San Franscisco: Ignatius Press.

Dicastery for the Doctrine of the Faith. 2024. *Dignitas infinita*. https://www.vatican.va/roman_curia/congregations/cfaith/documents/rc_ddf_doc_20240402_dignitas-infinita_en.html.

Eberl, Jason T. 2020. *The nature of human persons: Metaphysics and bioethics*. Notre Dame, Indiana: University of Notre Dame Press.

Fairbairn, Daphne J. 2013. *Odd couples: Extraordinary differences between the sexes in the animal kingdom*. Princeton, New Jersey: Princeton University Press.

Federman, Daniel D. 2006. The biology of human sex differences. *The New England Journal of Medicine* 354 (14): 1507–1514.

Finley, John. 2015. The metaphysics of gender: A Thomistic approach. *The Thomist* 79:585–614.

Francis. 2021. Message to participants in the fifth international conference entitled "Exploring the mind, body & soul: How innovation and novel delivery systems improve human health." https://www.vatican.va/content/francesco/en/messages/pont-messages/2021/documents/papa-francesco_20210508_videomessaggio-mindbodysoul.html.

Francis. 2023a. Address to participants in the conference promoted by the Alphonsian academy. https://www.vatican.va/content/francesco/en/speeches/2023/march/documents/20230323-accademia-alfonsiana.html.

Francis. 2023b. *Ad theologiam promovendam*. https://www.vatican.va/content/francesco/la/motu_proprio/documents/20231101-motu-proprio-ad-theologiam-promovendam.html.

Gilson, Etienne. 1956. *The christian philosophy of St. Thomas Aquinas*. Trans. L.K. Shook, C.S.B. New York: Random House.

Gremmels, Becket. 2016a. Sex reassignment surgery and the catholic moral tradition: Insight from pope Pius XII on the principle of totality. *Health Care Ethics USA* 24:6–10.

Gremmels, Becket. 2016b. More insight from Pius XII, a reply to Brugger and Brehany, and a clarification. *Health Care Ethics USA* 24:7–17.

Herman Jody L, Taylor NT Brown, and Ann P. Haas. 2019. *Suicide thoughts and attempts among transgender adults: Findings from the 2015 U.S. transgender survey*. University of California School of Law Williams Institute. https://williamsinstitute.law.ucla.edu/wp-content/uploads/Suicidality-Transgender-Sep-2019.pdf.

Horan, Daniel P., O. F. M. 2019. *Catholicity & emerging personhood: A contemporary theological anthropology*. Maryknoll, NY: Orbis Books.

Hull, Carrie. 2006. *The ontology of sex: A critical inquiry into the deconstruction and reconstruction of categories*. New York: Routledge.

Javier, Christienne, Charlie R. Crimston, and Fiona Kate Barlow. 2022. Surgical satisfaction and quality of life outcomes reported by transgender men and women at least one year post gender-affirming surgery: A systematic literature review. *International Journal of Transgender Health* 23 (3): 255–273.

John Paul II. 1979a. *General Audience*. https://www.vatican.va/content/john-paul-ii/en/audiences/1979/documents/hf_jp-ii_aud_19791114.html.
John Paul II. 1979b. *General audience*. https://www.vatican.va/content/john-paul-ii/en/audiences/1979/documents/hf_jp-ii_aud_19791121.html.
John Paul II. 1980. *General audience*. https://www.vatican.va/content/john-paul-ii/en/audiences/1980/documents/hf_jp-ii_aud_19800109.html.
John Paul II. 1982. *General audience*.
John Paul II. 1993. *Veritatis splendor*. https://www.vatican.va/content/john-paul-ii/en/encyclicals/documents/hf_jp-ii_enc_06081993_veritatis-splendor.html.
Kelly, Gerald, S. J. 1955. Pope Pius and the principle of totality. *Theological Studies 16* (3): 373–396.
Mangan, Joseph T. 1949. An historical analysis of the principle of double effect. *Theological Studies* 10 (1): 41–61.
McHugh, John A., and Charles J. Callan. 1958. *Moral theology: A complete course. Volume I*. New York: Joseph F. Wagner, Inc.
Nguyen, Hillary B., et al. 2018. Gender-affirming hormone use in transgender individuals: Impact on behavioral health and cognition. *Current Psychiatry Reports* 20 (12): 110. https://www.ncbi.nlm.nih.gov/pmc/articles/PMC6354936/pdf/nihms-1007763.pdf.
Owens, Joseph. 1988. Thomas Aquinas: Dimensive quantity as individuating principle. *Mediaeval Studies* 50:279–310.
Park, Rachel H., et al. 2022. Long-term outcomes after gender-affirming surgery 40-year follow-up study. *Annals of Plastic Surgery* 89 (4): 431–436.
Pius XII. 1958. Address to the participants in the tenth national congress of the Italian society of plastic surgery, Retrieved Oct 4, 1958, from https://www.vatican.va/content/pius-xii/it/speeches/1958/documents/hf_p-xii_spe_19581004_chirurgia-plastica.html.
Pius XII. 1960a. Address to the Italian medical-biological union of St. Luke. In *The Human Body: Papal Teachings*, Monks of Solesmes, ed., 55–56. Boston, Massachusetts: St. Paul Editions.
Pius XII. 1960b. Address to the first international congress of histopathology. In *The Human Body: Papal Teachings*, Monks of Solesmes, ed., 194–208. Boston, Massachusetts: St. Paul Editions.
Pius XII. 1960c. Address to a group of country doctors. In *The Human Body: Papal Teachings*, Monks of Solesmes, ed., 276. Boston, Massachusetts: St. Paul Editions.
Pius XII. 1960d. Address to the delegates at the 26th congress of urology. In *The human body: Papal teachings*, Monks of Solesmes, ed., 277–279. Boston, Massachusetts: St. Paul Editions.
Pius XII. 1960e. Address to military surgeons. In *The human body: Papal teachings*, Monks of Solesmes, ed., 281–282. Boston, Massachusetts: St. Paul Editions.
Pius XII. 1960f. Address to the congress of psychotherapy and clinical psychology. In *The human body: Papal teachings*, Monks of Solesmes, ed., 227–237. Boston, Massachusetts: St. Paul Editions.
Pius XII. 1960g. Address to a group of eye specialists. In *The human body: Papal teachings*, Monks of Solesmes, ed., 375–377. Boston, Massachusetts: St. Paul Editions.
Porter, Jean. 2015. Choice, causality, and relation: Aquinas's analysis of the moral act and the doctrine of double effect. *American Catholic Philosophical Quarterly* 89 (3): 479–504.
Priimmer, Dominic M., O. P. 1956. Trans. Gerald W. Shelton. *Handbook of moral theology*. Cork: The Mercier Press.
Ratzinger, Joseph Cardinal. 1990. Concerning the notion of person in theology. *Communio* 17:439–454.
Schiebinger, L. et al. 2011–2020. G*endered innovations in science, health & medicine, engineering, and environment*. https://genderedinnovations.stanford.edu/terms/sex.html.
Second Vatican Council. 1965. *Gaudium et spes*. https://www.vatican.va/archive/hist_councils/ii_vatican_council/documents/vat-ii_const_19651207_gaudium-et-spes_en.html.
Substance Abuse and Mental Health Services Administration. 2016. *DSM-5 changes: implications for child serious emotional disturbance*. Rockville, MD: Substance Abuse and Mental Health Services Administration. https://www.ncbi.nlm.nih.gov/books/NBK519708/pdf/Bookshelf_NBK519708.pdf.

United States Conference of Catholic Bishops (USCCB). 2018. *Ethical and religious directives for Catholic health care services*. 6th ed. Washington, D.C.: United States Conference of Catholic Bishops.

van Leerdam, Taylah R., Jeffrey D. Zajac, and Ada S. Cheung. 2023. The effect of gender-affirming hormones on gender dysphoria, quality of life, and psychological functioning in transgender individuals: A systematic review. *Transgender Health* 8 (1): 6–21.

Wernick, Jeremy A., et al. 2019. A systematic review of the psychological benefits of gender-affirming surgery. *Urologic Clinics of North America* 46 (4): 475–486.

Wippel, John F. 2000. *The metaphysical thought of Thomas Aquinas*. Washington, D.C.: The Catholic University of America Press.

Witcomb, Gemma L., et al. 2018. Levels of depression in transgender people and its predictors: Results of a large matched control study with transgender people accessing clinical services. *Journal of Affective Disorders* 235:308–315.

Wizemann, T. M., and M. L. Pardue. 2001. Institute of medicine, committee on understanding the biology of sex and gender differences. In *Exploring the biological contributions to human health: Does sex matter?* Washington, D.C.: National Academies Press.

Chapter 5
Epistemic Humility, Intuitive Cognition, and the Care of Transgender Persons

Daniel P. Horan

5.1 Introduction

Among the most pressing and controversial of the emerging topics in bioethics in recent years has been how to understand and respond to transgender persons seeking care within Catholic health care systems. Several ecclesially authoritative bodies have sought to establish prohibitory guidelines, which call for the denial of care for gender variant patients. The United States Conference of Catholic Bishops' Committee on Doctrine's publication of "Doctrinal Note on the Moral Limits to Technological Manipulation of the Human Body" (USCCB 2023) and the Vatican's Dicastery for the Doctrine of the Faith's publication of *Dignitas infinita* ["On Human Dignity"] (DDF 2024, n. 60) have provided support for those Catholic health care institutions wishing to limit or outright deny gender-affirming treatment for transgender persons.[1] But even prior to the publication of these texts, certain Catholic health care providers and ecclesiastical authorities—notably, local ordinaries of dioceses in the United States—had begun issuing local guidelines that advocated for the denial of care for transgender persons, especially hormone treatment for pre-pubescent children and young adults and surgical intervention for adults.[2] Among the most common reasons given for refusal of care is the explicit or tacit

[1] I am prescinding from analyzing these two documents in this chapter due to limits in space and scope. For a sampling of my previous assessment of these texts, see Horan 2023a, 2024a and 2024b.

[2] This is something that David Palmieri, a Catholic school teacher in Massachusetts, has been tracking for several years. Palmieri hosts an open-access Google folder that contains all available diocesan policies on so-called "gender policies": https://drive.google.com/drive/folders/1yVCuDjFJKdwTZIsHK0rWfZmuISr3sKgF

D. P. Horan (✉)
Saint Mary's College, Notre Dame, IN, USA
e-mail: dhoran@saintmarys.edu

J. T. Eberl (ed.), *Emerging Issues in Catholic Bioethics*, Philosophy and Medicine 153, https://doi.org/10.1007/978-3-032-08862-8_5

rejection of the reality of transgender identities, sometimes going so far as to reject the notion of "Gender Dysphoria," an official Diagnostics and Statistical Manual (DSM-V) categorization. Several of these diocesan policies frame the call to deny care for trans* persons within the context of "truth," claiming that the experience that trans* patients have of themselves regarding gender identity is mistaken, errant, delusional, or even sinful.[3] At the heart of this dynamic is a refusal to believe not only the experiences of trans* people, but also their own knowledge of self. In other words, there is at once a refusal on the part of ecclesiastical authorities and Catholic health care institutions to believe or trust what is known and communicated by trans* patients while also claiming and asserting an almost apodictic certainty and epistemological authority over against the trans* person themselves. I argue in this chapter that such a dynamic constitutes a form of what philosophers call "epistemic injustice." While I am not the first to make such an assessment of the denial of care or treatment of trans* persons (Fricker and Jenkins 2017; Argyriou 2021; and Bullock 2023), I do so with an eye toward identifying resources within the Catholic philosophical and theological tradition that advances a counterargument in favor of care for trans* patients within the context of Catholic health care institutions.

The modest aim of this chapter is to argue that one among other cases for Catholic health care institutions to provide care for trans* patients can be made on the grounds of epistemology and self-knowledge. The current trend among Catholic ecclesiastical leaders and many Catholic health care institutions is to restrict or deny care altogether on grounds that are philosophically problematic, rooted as they are in epistemic hubris and the denial of first-person authority of patients that ought to be privileged over the interpretative judgement of the patients' identities or experiences by ministers or practitioners. There are other arguments than can and ought to be made in favor of broadening access to care for trans* patients, including from psychological and endocrinological perspectives, but those efforts will be left for specialists in those areas to argue.

This chapter is organized into three parts. First, drawing on some recent philosophical work, I show that the denial of care for transgender patients is a form of both testimonial and hermeneutical epistemic injustice, which harms both trans* patients as well as the broader social community. Second, I make a case for trusting the knowledge and testimony of trans* patients over the remote knowledge or judgment of others using the philosophical argument from first-person authority. I then draw on the work of medieval philosopher and theologian John Duns Scotus to argue for the primacy of knowledge arising from intuitive cognition over that which

[3] I generally use the terms "transgender" and "trans*" interchangeably in this chapter. "Trans*" has been widely adopted in recent years as providing a more inclusive term that encompasses a range of gender identities both within and outside the binary. For more information, I refer readers to the *Oxford English Dictionary*, which added the term in 2018 and defines it as "originally used to include explicitly both transsexual and transgender, or (now usually) to indicate the inclusion of gender identities such as gender-fluid, agender, etc., alongside transsexual and transgender." (See "Trans*, *Adj.*" *Oxford English Dictionary*, Oxford University Press, December 2023, doi:10.1093/OED/1137339111).

result from an act of abstractive cognition, which provides a framework for believing trans* patients and affirming their knowledge of themselves. Finally, I propose that Catholic health care professionals cultivate the virtue of epistemic humility, which originates within the theological context of interreligious dialogue but is appropriated here to provide a constructive resource in the work toward epistemic justice. I close by making a case for transgender patient care within Catholic health care systems that is more capacious than is suggested by current ecclesiastical guidance.

5.2 Denial of Care of Transgender Patients as Epistemic Injustice

For at least half a century, philosophers have explored the phenomenon that not all speakers are treated with the credibility or received with the credulity that is commensurate with their knowledge.[4] Some people are, as a result of some or multiple forms of prejudicial or stereotypical judgement, dismissed as irrational, ignorant, incompetent, untrustworthy, unintelligent, "crazy", or worse, while still others are granted undue authority and inflated credibility without objective warrant. This dynamic of asymmetrical ascription of credibility and authority to individuals on the basis of gender, race, ability, sexual orientation, native language, or some other dimension of social location has elicited a number of substantive analyses over the years. For example, the poststructuralist philosopher Michel Foucault lectured in the 1970s about what he termed "subjugated knowledge," referring in part to "a whole series of knowledges that have been disqualified as nonconceptual knowledges, as insufficiently elaborated knowledges: naïve knowledges, hierarchically inferior knowledges, knowledges that are below the required level of erudition or scientificity" (Foucault 2003, 7). The postcolonial theorist Gayatri Chakravorty Spivak, writing about the capacity of subaltern persons to act as testimonial agents of their own experience, named the denial of this credibility a form of "epistemic violence" (Spivak 1988). And more recently, feminist philosopher Miranda Fricker has popularized the concept of "epistemic injustice" to help theorize this phenomenon across a range of contexts and disciplines (Fricker 2007), which has catalyzed a robust and interdisciplinary field of study.[5]

---

[4] While some scholars can trace such observations of epistemic asymmetry to the nineteenth century as in the case of Sojourner Truth's 1867 acknowledgement of the denial of Black women's experiential and intellectual authority (May 2014), I am focusing here on the more narrowly demarcated scholarly movement arising from the work of mid-twentieth-century poststructuralist and feminist philosophers onward. Certainly, the *experience* of epistemic injustice predates this terminology and the theoretical framing found in contemporary scholarship and, as a phenomenon, it has been present throughout human history.

[5] Although some philosophers have sought to expand the categories of epistemic injustice beyond those outlined by Fricker (for example, Pohlhaus 2017 and Coady 2010), I am following Fricker's

What is uncovered through philosophical reflection on the phenomenon of epistemic injustice is the way in which power functions to advantage some speakers and hearers, while disadvantaging others. This results in an asymmetry of power that unfairly harms someone (or group of people) precisely as a knower. Fricker understands power in this case as "social power," by which she means, "power is a socially situated capacity to control others' actions" and this power can operate both actively and passively (Fricker 2007, 4 and 9–14).[6] One subset of social power is "identity power," which is the means by which individuals and groups are conceptualized, labeled, imagined, and viewed. Within a social context, some people experience "identity prejudice" as a result of the deleterious and discriminatory views predicated of them by others on account of their social location or type. In the case of epistemic injustice, this is made manifest in the exchange between a speaker and hearer wherein the hearer discredits or deflates the credibility of the speaker on account of identity prejudice that informs the hearer's perception of the speaker (Fricker 2007). In summarizing this phenomenon, Ian James Kidd, José Medina, and Gaile Pohlhaus, Jr., explain:

> *Epistemic injustice* refers to those forms of unfair treatment that relate to issues of knowledge, understanding, and participation in communicative practices. These issues include a wide range of topics concerning wrongful treatment and unjust structures in meaning-making and knowledge producing practices, such as the following: exclusion and silencing; invisibility and inaudibility (or distorted presence or representation); having one's meanings or contributions systematically distorted, misheard, or misrepresented; having diminished status or standing in communicative practices; unfair differentials in authority and/or epistemic agency; being unfairly distrusted; receiving no or minimal uptake; being coopted or instrumentalized; being marginalized as a result of dysfunctional dynamics; etc. (Kidd et al. 2017, 1).

Fricker (2007) argues that there are two fundamental yet distinct types of epistemic injustice: testimonial and hermeneutical.[7]

Testimonial injustice occurs when a person's statement, perspective, or knowledge is given more or less credibility by a hearer on account of prejudices predicated of the speaker. As Morten Byskov explains, "Testimonial injustices wrong someone in their capacity as a speaker or knower because the increased or decreased credibility accorded to their testimony is based not on any relevant concerns, but on prejudices that have nothing to do with whether the speaker or knower should be granted credibility" (Byskov 2021, 117). The dismissal or diminishment of credibility of a person's testimony—in this case, any propositional claim and not merely juridical attestation—on account of identity prejudice causes harm because it results in what Fricker calls a "credibility deficit" that effectively erases or silences the

---

framework, which has proven the contemporary standard referent for such analysis.

[6] Fricker provides an explicit and more technical working definition of this concept of social power: "a practically socially situated capacity to control others' actions where this capacity may be exercised (actively or passively) by particular social agents, or alternatively, it may operate purely structurally" (Fricker 2007, 13).

[7] Subsequently, scholars have sought to expand these respective categories; see Wanderer 2017, Medina 2017, and Fricker 2017.

experience, perspective, or knowledge of the speaker (Fricker 2007, 17–29). Conversely, testimonial injustice may result in the unfair inflation of credibility for others who either benefit from positive identity prejudices or stereotypes, or are otherwise not burdened by negative ones. Fricker uses Harper Lee's 1960 novel *To Kill a Mockingbird* to illustrate this point through the character of Tom Robinson, a Black man falsely accused of a crime, whose (literal) testimony is deflated and ultimately ignored on account of the white jury's and the broader community's racist identity prejudice (Fricker 2007, 21–29). He is not believed, and one might argue that he *cannot be believed* given the epistemic injustice he experiences within such a white supremacist context. Meanwhile, the credibility of the white members of the same community is often unfairly inflated, illustrating the other side of the epistemic injustice coin.

Unlike testimonial injustice, which describes the harm caused to an individual knower or speaker on account of their credibility deficit arising from the prejudice of a hearer, hermeneutical injustice refers to a larger social context in which the meaning-making capacity and understanding of a marginalized group is disadvantaged or absent altogether. Fricker explains,

> This sort of marginalization can mean that our collective forms of understanding are rendered structurally prejudicial in respect of content and/or style: the social experiences of members of hermeneutically marginalized groups are left inadequately conceptualized and so ill-understood, perhaps even by the subjects themselves; and/or attempts at communication made by such groups, where they do have an adequate grip on the content of what they aim to convey, are not heard as rational owing to their expressive style being inadequately understood (Fricker 2007, 6–7).

Fricker notes that testimonial injustice can lead to forms of hermeneutical injustice (Fricker 2007, 162) when, as Byskov explains, "testimonial injustices structurally affect what is included in a collective pool of knowledge it leads to an underrepresentation of the experiences of marginalized individuals and groups, in turn affecting their ability to make sense of their experiences" (Byskov 2021, 117). One can see how the marginalization of certain populations' knowledge, experience, wisdom, and perspective can result in their not being understood or seen as credible by the powerful and others whose mode of interpretation already precludes the experience and knowledge of the marginalized group. The consequent lack of shared social knowledge or understanding also limits the ways marginalized individuals and groups can make sense or name their own identities and the realities they embody.

Both categories of epistemic injustice name harm that is done to those people who are often already in precarious or otherwise marginalized social locations. We can witness the manifold ways this dynamic plays out in various contexts and according to a range of identity characteristics, such as race, sex, gender, ability, education, and nationality, among others. One context within which epistemic injustice has been examined in recent years is that of health care institutions (Carel and Kidd 2014; Blease et al. 2016; Kidd and Carel 2016; Carel and Kidd 2017; Kidd and Carel 2018). While health care settings have been the focus of some scholars in general, we are starting to see increasing attention given to the way trans* patients specifically have experienced epistemic injustice in health care settings (Nordmarken and Kelly 2014; Fricker and Jenkins 2017; Bullock 2023). Building on this initial work exploring the ways trans* patients have experienced epistemic injustice in health care settings, I will briefly highlight how the denial of care for trans* patients can be understood as both a form of testimonial injustice (2.1) and hermeneutical injustice (2.2).

5.2.1 Denial of Care as Testimonial Injustice

The denial of gender-affirming care, inclusive of a range of therapies from mental health treatment to hormone treatment to surgical interventions, on grounds that deny the reality of a trans* person's experience of themselves and knowledge of their gender identity is a form of testimonial injustice. The condition of the possibility for this denial of care often stems from a cis-normative setting in which those who govern health care institution policies, determine the assessment criteria for care, and make clinical decisions on individual cases benefit from and contribute to "social systems and practices that maintain a cis-normative culture, which privileges and normalizes cisgender experiences" (Nordmarken and Kelly 2014, 143). There is a pervasive, if not always conscious, presumption that cisgender providers and administrators—as well as ecclesiastical leaders influencing Catholic health care institutions—know better than trans* patients themselves about the patients' respective gender identities. Even in settings where transgender care may be available, often the gatekeepers to that care are situated in a position where they benefit from credibility excess, where their opinions, interpretations, and judgements are granted more authority and credibility than they ought to (Fricker 2007, 17). Such health care providers become the arbiters of whether or not a given trans* patient "meets the criteria" for care, and those determined not to meet the given set of criteria would then be denied care. As Fricker and Jenkins note, "Such a person would suffer a testimonial injustice: their testimony concerning their conviction that they were trans and had a genuine need to access transition-related medical procedures was subject to an unjustified credibility deficit stemming from identity prejudice concerning trans identities" (2017, 272). Health care providers could determine that they did not believe this particular trans* person's testimony and cause harm by refusing treatment. This dynamic is often further compounded by the persistence of mental health stigma, which pathologizes trans* experiences and makes "them vulnerable to having their reports of their own experience dismissed on the spurious grounds that mental health problems made them unreliable or even deceptive" (Fricker and Jenkins 2017, 272).

Another form of testimonial injustice for trans* patients in health care settings is the frequency with which they encounter harmful "microaggressions" (Sue 2010), which can dehumanize, objectify, and traumatize trans* people seeking support and assistance. Scholars have developed typologies to identify and categorize the range of harmful microaggressions trans* people experience, including misgendering, universalizing of trans* experience, exoticization, signals of discomfort/

disapproval, endorsement of gender normativity, denial of cultural transphobia, denial of bodily privacy, and denial of the provider's own implicit or overt transphobia (Nadal et al. 2012; Nadal 2013, 80–107). As Nordmarken and Kelly observe, "These often unintentional slights, snubs, or insults are embedded within a larger stream of communication. They can be confusing to receive and difficult to notice, pinpoint, or recognize, leaving them particularly challenging to address, which can amplify the stress they cause" (2014, 145–146). Furthermore, "the ways that health care providers often interact with trans patients in the midst of treatment create inadequate and inequitable services" (Nordmarken and Kelly 2014, 146). Given the way such transphobic microaggressions function within the framework of communication, which is essential to adequate and ethical health care treatment, trans* patients suffer from epistemic injustice as the credibility of their testimony is further diminished through such discursive actions.

Fricker identifies another iteration of testimonial injustice that is commonly experienced by trans* people, especially in health care settings: "pre-emptive testimonial injustice," an *a priori* judgement of an individual's trustworthiness as a knower and narrator (Fricker 2007, 130). The consequence of this sort of injustice is the wholesale refusal to solicit information from an individual, an act of omission informed by a hearer's presumption that, in this case, a trans* patient is simply wrong or cannot be trusted. Fricker notes that the result of this modality of testimonial injustice is the silencing of the one harmed. Collectively, this plays a contributing factor in the broader phenomenon of hermeneutical injustice, while at the individual level it results in further marginalization, dehumanization, and erasure. It does not take much effort to see how preemptive testimonial injustice leads to distorted policy decisions and treatment plans because those most directly affected by the policies and treatment options are never consulted, nor are their views solicited in the first place.

The other way trans* patients might experience testimonial injustice is in the total refusal of a health care institution or system to offer any form of transgender treatment. This is increasingly the case within the context of Catholic health care as diocesan guidance, national episcopal conference statements, and Vatican dicastery documents often explicitly call for the refusal of transgender care in Catholic health care. Such instructions and guidance are predicated on a form of testimonial injustice. This arises from cis-normative assumptions about gender, experience, and health care practices, which are framed by ecclesiastical authorities and health care providers who wrongly benefit from credibility excess within this context while simultaneously reinscribing credibility deficits for trans* patients. This dynamic further harms trans* patients as they seek gender-affirming care and are told that no such options are available to them because, according to these ecclesiastical and ethical guidelines, their experience is untrue and perhaps even framed as sinful. In cases that moralize the experience of trans* patients, what is certainly a form of testimonial injustice is further compounded by pastoral malpractice.

5.2.2 Denial of Care as Hermeneutical Injustice

Within the context of the Catholic Church in the United States generally, and Catholic health care institutions specifically, the phenomenon of testimonial injustice experienced by trans* people exists as both a cause and, at times, also a result of hermeneutical injustice. According to Fricker, hermeneutical injustice is "the injustice of having some significant area of one's social experience obscured from collective understanding owing to a structural identity prejudice in the collective hermeneutical resource" (Fricker 2007, 155). In the case of trans* people, the denial of the veracity of their experience and existence by those in socially powerful positions not only causes individual harm as we see in the case of testimonial injustice, but it also impacts a larger, collective context of knowing. In recent years the increasing visibility of the trans* community and the broader sharing of personal narratives have been met by some with accusations that this phenomenon is a "novelty" or a "fad," often in an effort to discredit and deflate the credibility of trans* people and their allies.[8] This dynamic serves to perpetuate hermeneutical injustice in two ways. First, the lack of common, social referents, language, and terminology around trans* people and transgender experiences is, in part, the result of the hermeneutical marginalization the trans* community has experienced for generations. The refusal to recognize individual cases (as a result of pervasive testimonial injustice) and the broader suppression of the existence of gender variant individuals outside the binary has resulted in a sclerotic sense that cis-normative experience and frameworks are the only ones that are true and extant. This has deprived whole communities of accessing epistemological resources to understand and relate to the reality and experience of trans* people.

Second, and perhaps more insidiously, this same social dynamic also deprives trans* people themselves from understanding their own experiences and identities. Referring to an example of a cisgender woman who is the victim of what today we would readily identify as "sexual harassment," Fricker notes how hermeneutical injustice functions when there is no commonly held understanding or language to name this harm. "The cognitive disablement prevents her from understanding a significant patch of her own experience: that is, a patch of experience which it is strongly in her interests to understand, for without that understanding she is left deeply troubled, confused, and isolated, not to mention vulnerable to continued harassment." Fricker adds, "Her hermeneutical disadvantage renders her unable to make sense of her ongoing mistreatment, and this in turn prevents her from protesting it, let alone securing effective measures to stop it" (Fricker 2007, 151). Within the context of transgender care, trans* people suffer from the effects of hermeneutical injustice twice over: from its effects in reifying cis-normative and binary assumptions about gender that render health care providers and others from

[8]Allyship is a complicated positionality that has received increased critical analysis by philosophers in recent years, especially as it relates to the trans* community; see McKinnon 2017, 167–174.

understanding or naming their experience, which restricts the resources available to them; and they are positioned to misunderstand their own experience and selves as they are also at an epistemological disadvantage as a member of a social community that has not welcomed trans* testimony to inform the collective understanding of the world.

The Catholic Church generally and Catholic health care institutions specifically can be said to represent just such a social community that experiences and perpetuates this kind of hermeneutical injustice. One of the most common criticisms of ecclesiastical documents that seek to regulate the availability of and access to transgender care (as well as use of gender-specific restrooms, pronouns and preferred names, and so on) at Catholic institutions is that the construction of such texts almost never involve consultation with actual members of the trans* community (Horan 2019b; Horan 2020; Horan 2021; Horan 2022; Scott 2023; Lysaught 2023; Horan 2023a; Horan 2023b; Lysaught 2024; and Horan 2024a). Furthermore, the most common justification for this lack of consultation of actual trans* people is the hubristic assumption (that is, pre-emptive testimonial injustice) that a narrow reading of ancient and medieval theological sources establishes an immutable binary gender framework (for example, see USCCB 2023 and Finley 2015). This is often framed in terms of a singular "Catholic theological anthropology" (Horan 2019a, 156–166). Given the overt privileging of one kind of knowledge—usually some version of a neo-Thomist interpretation of sex and gender—over against, and often to the total exclusion of, any other kinds of knowledge is a textbook illustration of hermeneutical injustice.[9] The harm perpetrated against trans* patients in this case is manifested in the presumptive discrediting of their own knowledge of themselves as valid, which then reinforces the willful ignorance of the powerful in this social community who deny access to transgender care. There is also harm present for the powerful within the same social community because they are deprived of the expansion of individual and collective knowledge and broader experience of the world as it actually is, thereby reinforcing the social community's vincible ignorance.

What I have aimed to illustrate in this section is the way in which the refusal to believe or trust the experience and knowledge of trans* people, particularly as it relates to their own gender identity, has not only contributed to the denial of transgender care in particular and systemic ways, but it also constitutes a form of epistemic injustice. This pattern of refusing to believe or ignoring entirely the knowledge of trans* patients ought to be suspended and measures to overcome the testimonial and hermeneutical injustices that arises from such treatment of trans* patients ought to be pursued. I will argue later in this chapter that one approach to remedying these injustices within Catholic health care is for practitioners (and, by extension, health care administrators and ecclesiastical leaders) to embrace the virtue of "epistemic

---

[9] There are several illustrations of this recourse to a kind of Thomism to justify the rejection of trans* testimony about their own gender and identity, such as Schumacher 2016, DiCamillo 2017, and Kersten 2016. It is not uncommon to see an apodictic sense of authority and truth in the claims that trans* experiences, and therefore persons, do not exist. Such a claim is seen in, among others, Corradi 2015.

humility." But before making that constructive suggestion, I want to turn to insights from both contemporary philosophy and medieval theology related to the primacy of intuitive cognition over abstractive intellection as a philosophical and theological justification for why believing trans* patients and restoring due credibility to their testimony within health care settings is not antithetical to the Catholic intellectual tradition but is actually supported by approaches within the tradition.

5.3 First-Person Authority and the Primacy of "Intuitive Cognition"

For centuries, philosophers have been exploring questions related to self-knowledge and the apparent "asymmetry between our knowledge of our own minds and our knowledge of other minds" (Child 2013, 533).[10] In recent decades, the result of this topic has been a focused epistemological inquiry into "first person authority," a concept that gained significant traction thanks to the work of Donald Davidson in the early 1980s (see Davidson 2001). Davidson explains succinctly that, "When a speaker avers that he has a belief, hope, desire or intention, there is a presumption that he is not mistaken, a presumption that does not attach to his ascriptions of similar mental states to others" (Davidson 2001, 3). The primary issue at hand is to understand why (and in some instances, how) this is the case. The argument centers on a distinctive kind of authority that rests with a knower's knowledge of themselves which, the argument from first person authority goes, ought to supersede or be privileged over against another person's propositional claim about the knowledge of the self-avowing speaker. The way that Davidson and his followers explain this prioritization of self-knowledge over the knowledge claim of another person about the same subject has to do with the necessity of interpretation. While there is certainly a chance for self-delusion, confusion, or some other error in the experience of self-knowledge, which means it is not infallible, nevertheless we may assert that there is a *presumption* of authority predicated of the knower about their own knowledge over a second or third person's:

> The difference follows, of course, from the fact that the assumption that I know what I mean necessarily gives me, but not you, knowledge of what belief I expressed by my utterance. It remains to show why there must be a presumption that speakers, but not their interpreters, are not wrong about what their words mean. The presumption is essential to the nature of interpretation—the process by which we understand the utterances of a speaker. This process cannot be the same for the utterer and for his hearers (Davidson 2001, 12).

[10] For the sake of brevity and focus on the question at hand, I will prescind from providing a survey of the breadth of analyses of self-knowledge and its relationship to the knowledge and minds of others. Needless to say, from the Ancients to Descartes to Wittgenstein and beyond, this question has and remains an ongoing and robust topic of debate with a range of approaches and theories. Given its applicability to the case of trans* persons, their self-avowal of gender, and the question of epistemic injustice in health care, I narrow the inquiry in this section to advocates of first-person authority and related medieval Catholic philosophical resources.

In the case of Davidson's account of first-person authority, which of course has received some criticism from some philosophers,[11] the explanation of the seeming *a priori* assumption of the privileging of self-knowledge rests on the necessity that the second or third person's knowledge requires evidence in need of interpretation. Relatedly, Davidson notes, "There is a presumption—an unavoidable presumption built into the nature of interpretation—that the speaker usually knows what he means. So, there is a presumption that if he knows that he holds a sentence true, he knows what he believes" (Davidson 2001, 14; cf. Child 2013).

Framed within the context of debates about first-person authority, we can see how instances of epistemic injustice in the denial of care for trans* persons involve a rejection of the presumption that a knower/speaker who expresses a self-avowal about their gender is in a privileged position to affirm what they say to be true. In such cases, it is the hearer/interpreter who assumes for themselves a privileged epistemic positionality, claiming that the knower/speaker is obviously wrong, confused, mentally ill, or otherwise mistaken. In addition to the epistemological harm arising from such epistemic hubris on the part of the hearer/interpreter, there is an existential and ethical harm that follows. In a well-known essay on trans* identities and first-person authority, Talia Mae Bettcher takes up the issue of first-person avowals of gender identity and the authority that such statements of self-knowledge bear (Bettcher 2009). She argues that one ought to think about such self-avowals in terms of ethical authority rather than in strictly epistemological terms (Bettcher 2009, 100–103). In addition to the harm that the refusal to afford trans* persons appropriate credibility occasions, Bettcher believes that there are ethical violations in refusing to recognize or accept a trans* person's "existential identity," which she describes as determined by "*who* rather than *what* one is" (Bettcher 2009, 99). As a result, moral harm is perpetrated in the process of refusing both the first-person authority of an individual and their right to privacy and autonomy (Bettcher 2009, 103–115).

While Bettcher downplays the epistemic norm of first-person authority in favor of an ethical norm, Gus Turyn (2023) has more recently made a compelling case for recognizing both an epistemological and an ethical set of norms when it comes to thinking about first-person authority, especially as it relates to gender identity: "How we treat others is obviously enormously significant, but what we believe about others is also enormously significant." Turyn continues:

> What we believe about others shapes how we treat others, both to the extent that we are likelier to treat others in the right way if we believe the right things about them and to the extent that we cannot fully treat others in the right way if we do not believe the right things about them. Our beliefs, just like our actions, can be sources of wrongdoing and harm, and an adequate set of norms of FPA [first-person authority] must take this relationship into account (Turyn 2023, 17).

I am inclined to agree with Turyn's argument, particularly as he eventually makes the case that, in response to the twofold epistemic and ethical understandings of

[11] For a survey of the critiques and a defense of Davidson's proposal, see Child 2007.

first-person authority that ought to prioritize the knower's knowledge over an interpreter's, the hearer/interpreter ought to actually believe the knower/speaker and not merely tolerate their views or "play along" with others about something they do not believe (Turyn 2023). How a hearer/interpreter might arrive at that point advocated for by Turyn varies by person and cannot be predicated universally of all people. However, it is perhaps possible to identify an initial normative response. And it is this that I take up in the last section of the chapter when I make the case for Catholic health care practitioners specifically and, by extension, Catholic health care institutions generally embracing the virtue of "epistemic humility."

While not articulated in precisely the same way as contemporary philosophical discourse presents it, the concept at the heart of the principle of first-person authority can be found centuries earlier in the epistemology of the medieval Franciscan philosopher and theologian Bl. John Duns Scotus (d. 1308). Framed as the primacy of what he calls intuitive intellection over abstractive intellection, Scotus offers a theologically significant philosophical argument that aligns with what advocates of first-person authority assert today. This is of particular interest to us because it may provide an intellectual framework that is at once both theologically orthodox from within the Christian intellectual tradition and in general alignment with many contemporary proposals for the primacy of first-person authority in contexts related to gender identity.

Scotus affirms the broadly held Aristotelian concept of abstractive cognition which, under the auspices of the "agent intellect," takes empirical data that has been acquired through sense perception and generates a mental image (the so-called *phantasm*). As Mary Beth Ingham explains, "Through various stages of abstraction, mental processes and imagination, the activity of human knowing moves toward greater and greater immateriality, generality, and conceptual clarity. This process is called *mediated* (taking place through various steps) and provides the bridge between a material world and an immaterial knower" (Ingham 2017, 35). This admittedly abbreviated and simplified summary gestures to Aristotle's explanation—and the medieval appropriation of his thought—for how the human mind is able to know anything of the world.[12] For Scotus, this general understanding of human cognition is "the most obvious and common way that we know the world around us" (Ingham 2017, 36). But Scotus believes there is another form of cognition, one that is both more perfect and immediate. He calls this "intuitive cognition." In the *Lectura*, his first commentary on the *Sentences* of Peter Lombard, Scotus writes about the distinction between abstractive and intuitive cognition: "We should know that there can be two kinds of cognition and intellection in the intellect: one intellection can be in the intellect inasmuch as it *abstracts* from all existence; the

[12] Space limitations prevent a fuller presentation of the medieval psychological assumptions and metaphysical context undergirding Scotus's innovative developments here. For a brief summary, see Cross 2014, 3–17; Ingham and Dreyer 2004, 26–28; and Cross 1998.

other intellection can be of a thing insofar as it is present in its existence" (*Lectura* II, d. 3, pars 2, q. 2, n. 285; trans. in Pasnau 2003, 296).[13]

Scotus's epistemological argument centers on the assertion that we can have "some kind of direct, perceptual acquaintance with reality" (Pasnau 2003, 298). The idea here is that normal modes of intellection, rooted in an Aristotelian understanding of sense perception and reduction to the phantasm, requires several degrees of mediation. In contrast to this normal abstractive cognition, intuitive intellection does not require mediation or sense knowledge in the same way, but instead the knower knows in the most immediate way. Allan Wolter explains, "intuitive intellection has the thing itself, rather than a substitute species, interacting with the mind or soul to cause actual knowledge of itself" (Wolter 1990, 101). This is especially true as it pertains to self-knowledge. In this regard, Scotus argued that this form of intuitive cognition and the resulting knowledge acquired—what I will call "intuitive knowledge"—are essential to everyday human existence and knowing. Scotus explains, "If we were not to have intuitive cognition of anything, we would not know whether our own acts were present to us, or at least would not know about those acts with any certainty. But this is false, therefore etc." (*Ordinatio* IV, d. 49, q. 8 trans. in Pasnau 2003, 298). While scholars debate what actually constitutes the final position or view Scotus held about intuitive cognition given the development of his thought throughout his written corpus (see Wolter 1990 and Dumont 1989), and Scotus quite clearly asserts the capacity for intuitive cognition in the afterlife as the condition of the possibility for the beatific vision, there appears to be sufficient scholarly support to stipulate that Scotus also held that humans have the real

[13] Richard Cross suggests that the most substantive account Scotus gives of the distinction between abstractive and intuitive cognition appears in his *Quodlibetal* questions: "Speaking of simple apprehension, or of the intellection of a simple object, one [act] can be indifferent in relation to an existent or non-existent object, and also indifferent in relation to an object not really present, just as to one really present. And we often experience this in ourselves, because we understand universals or the quiddities of things equally whether they have real extramental existence (*ex natura rei esse extra*) in some *suppositum* or not; and it is the same for their presence or absence … There is another act of cognizing, which, although we do not experience so evidently in ourselves, is possible. It is precisely of a present object, as present, and of an existent object, as existent. This is proved, because everything that is a perfection of cognition absolutely, and which can pertain to a sensitive cognitive power, can eminently pertain to an intellective cognitive power. But it is a perfection in an act of cognizing, in so far as it is a cognition, to perfectly attain the first thing cognized. But this is not perfectly attained unless it is attained in itself, and not merely in some diminished likeness, derived from [the object]. But a sense power has this perfection in its cognition, because it can attain its object in itself, as existent and as present in real existence, not merely attaining it in a diminished way, in some kind of diminished perfection. Therefore this perfection pertains to an intellective [power] in its cognizing. But it could not pertain to it unless it cognized an existent thing, as present in its proper existence" (*Quodlibet* 6, nn. 7–8; trans. in Cross 2014, 43).

Robert Pasnau suggests that, for as important and innovative a concept as intuitive cognition/intellection is for the Subtle Doctor, Scotus "devotes relatively little space to the topic" (Pasnau 2003, 297). Nevertheless, there are several instances within Scotus's *oeuvre* in which he develops his thinking over time. Examples of such locations include: *Ordinatio* I, d. 1, pars 1, q. 2, nn. 34–36; *Ordinatio* II, d. 3, pars 2, q. 2, n. 321; *Quodlibet* q. 6, nn. 18–19 (Scotus 1975); *Quaestiones super libros Metaphysicorum Aristotelis* 7, q. 15, n. 18; and *De primo principio* 4.89 (Scotus 1966).

capacity for intuitive cognition *pro statu isto*. The act of intuitive cognition is predicated on the immediacy and existential reality of that which is known which, as Ingham asserts, "contains something that ensures that I am not being deceived." This is because, as Ingham also states, "something about this activity called intuition depends upon the actual existence of the experience in all its dimensions" (Ingham 2017, 38).

For our purposes, it is worth noting that Scotus's privileging of intuitive knowledge over abstractive knowledge says something about the significance of both proximity and in establishing verity. This is notable for us in two ways. First, it suggests that we ought to privilege the experiential, intuitive knowledge (arising from the act of intuitive cognition) of trans* persons about their own gender identity over contestations by other parties with mere reference to abstractive knowledge, which is subject not only to mediation but also necessary interpretation. Second, it subordinates any claim to knowledge of a universal essence (or "common nature") to the particularity of a given individual's identity (see Horan 2019a, 123–142). This second point raises challenging questions about our longstanding reliance on Aristotelian-Thomistic anthropological frameworks like the ones commonly deployed in Catholic health care debates in order to silence, dismiss, or deny the reality of trans* persons and their experiences, which results in epistemic injustice as previously examined. There is an ethical, pastoral question for cisgender persons that arises from this admittedly dense philosophical reflection on epistemology. Who are we to claim to know better than the particular individual his, her, or their true identity with recourse only to our remote and abstractive knowledge of a given person?

A trans* person's knowledge of themselves and their gender, whether framed within the context of first-person authority or as something understood arising from intuitive cognition, ought to be granted the highest degree of respect and should be afforded due credibility and presumed veracity within Catholic health care. We turn now in the last section to consider how the embrace and cultivation of the virtue of "epistemic humility" may allow for individual practitioners and institutions to begin working toward greater equity and justice when it comes to care of transgender persons.

5.4 Cultivating the Virtue of "Epistemic Humility"

So far, we have seen how common reasoning used to justify the denial of care to trans* patients within Catholic health care institutions—which can be summarized as the denial, dismissal, or erasure by practitioners of trans* patients' self-knowledge of their gender—constitutes a form of epistemic injustice that causes harm to both the trans* patients themselves and the social community more broadly. This form of epistemic injustice can be characterized as both testimonial and hermeneutical. In an effort to lay a philosophical foundation for privileging the self-knowledge of trans* patients regarding gender identity, we looked at first-person authority arguments and the medieval Catholic case for the primacy of knowledge arising from the act of intuitive cognition over that of abstractive cognition, especially as informed by the influential work of John Duns Scotus. Returning now to the context of contemporary Catholic health care, we might ask ourselves: how ought health care practitioners and institutions respond to trans* patients in light of these points? My aim in this last section is an admittedly brief and humble one. I believe that what is required of those involved in Catholic health care decision-making as it relates to transgender care is an embrace of "epistemic humility." This suggestion does not resolve all the problems, concerns, or questions that will continue to arise in this admittedly dynamic area of medical and psychological treatment for trans* patients. However, it does offer a starting-point, and one that is significantly more pastoral and humane than what has been proposed to date by a range of dioceses, bishops' conferences, Vatican dicasteries, and some Catholic health care institutions.

Within theological circles, the term epistemic humility has appeared most frequently within the context of interreligious dialogue (see Kraft 2006 and Borchert 2013). Framed as a response to religious triumphalism and epistemic hubris, epistemic humility is a virtuous practice that one cultivates in order to, minimally, arrive at some degree of religious tolerance in a pluralistic context. The description of epistemic humility as outlined by Donald Marvin Borchert invites adoption beyond responses to religious intolerance and triumphalism, suggesting a way of thinking and acting that bears potency for our purposes here:

> Advocates of a worldview who practice epistemic humility, while holding their worldview in high regard, are open to the *possibilities* that

(a) their worldview might not contain the whole truth and nothing but the truth,
(b) alternative worldviews might contain important valid insights, and
(c) their worldview might need enrichment and even modification because of new insights either generated by their fellow adherents or borrowed from advocates of alternative worldviews (Borchert 2013, xiii).

Because the fundamental issue examined in this chapter is that of belief—who is believed, who is refused credulity, on what grounds, etc.—there appears to be a strong resonance between religious belief and belief about gender. Both subjects are complex and the arguments about these subjects are notably fraught, at times even violent. This makes Borchert's characterization of differing "worldviews" particularly helpful in thinking about gender identity and responses to gender variant patients within Catholic health care. We might think, for instance, of the presumed gender binary latent in Catholic directives and pastoral guidance as one kind of worldview. It is certainly held in "high regard," as Borchert puts it, by many within the Catholic community. And there can be a kind of gender-binary triumphalism within these debates and contexts that echo that of extreme religious sectarianism.

But what if those people who sincerely believe that there is a divinely established gender binary that governs human identity, both psychologically and physiologically, were to admit to the *possibility* that other gender worldviews, informed by other experiences and knowledge, might be true and salutary? What if, instead of a triumphalist attitude informed by one's abstractive cognition about another's identity or experience in the world, the hearer/interpreter (in this case, those involved in Catholic health care) sought to overcome epistemic injustice by affording the knower/speaker the credibility they rightly deserve? According to Borchert's schema, those previously in a position of epistemic hubris might actually learn, grow, and find themselves adjusting their own knowledge and worldview.

In health care settings marred by the historical and present reality of epistemic injustice, the embrace and cultivation of epistemic humility takes the form of virtue. It welcomes curiosity, new knowledge, and renewed understanding about the world and those who move through it on pathways and in contexts not necessarily like one's own. Cultivating the virtue of epistemic humility can also serve as the condition of the possibility of working toward epistemic justice and avoiding the real harms that previous stances of epistemic hubris have wrought. In many ways, cultivating the virtue of epistemic humility with regard to transgender care and approaching trans* patients can be viewed as a spiritual practice. It requires empathy and openness, patience and active listening skills, and it challenges ministers and practitioners to consider experiences and worldviews that are sometimes radically unlike their own.

On the personal level, for ecclesial ministers and health care practitioners alike, this invites a form of ongoing conversion, which is at the heart of the Christian life. On the institutional level—recognizing the harms of epistemic injustice in the denial of care to trans* patients and affirming the primacy of first-person authority in these cases—ought to challenge individuals and organizations not to rush to make broad and sweeping policy decisions that can cause further harm to trans* patients.

5.5 Conclusion

In closing, this chapter presented a case for providing care for trans* patients at Catholic health care institutions grounded in epistemological and self-knowledge concerns. I argued that the denial of care can be understood as a form of epistemic injustice, especially when the foundation for the denial centers on the refusal to believe trans* patients and grant them due credibility. According to the framework developed by Miranda Fricker, this form of epistemic injustice can be understood in terms of both testimonial and hermeneutical injustice. One way forward in working toward epistemic *justice* is to believe trans* patients in their self-avowal about gender, while affirming their first-person authority and proximate knowledge arising from intuitive cognition. Relativizing one's own abstractive knowledge in relationship to the self-knowledge of a trans* patient is a concrete illustration of what epistemic humility can look like in practice. The key element here is a commitment to ongoing intellectual and moral conversion, learning along the way and remaining open to necessary shifts in one's worldview. While much of the current ecclesiastical guidance discouraging or outright prohibiting medical care for trans* patients at Catholic health care institutions is founded on epistemically unjust presumptions, this is not the only way forward. If more ecclesiastical leaders and health care practitioners embraced epistemic humility when approaching questions of transgender care in Catholic health care institutions, perhaps a more hospitable, safe, and supportive environment would be formed in which the best practices of epistemic justice meet the best practices of clinical treatment.

References

Argyriou, Konstantinos. 2021. Misgendering as epistemic injustice: A queer STS approach. *Las Torres de Lucca: Revista Internacional de Filosofía Política* 10:71–82.

Bettcher, Talia Mae. 2009. Trans identities and first-person authority. In *"You've changed": Sex reassignment and personal identity*, ed. Laurie J. Shrage, 98–120. New York: Oxford University Press.

Blease, Charlotte, Havi Carel, and Keith Geraghty. 2016. Epistemic injustice in healthcare encounters: Evidence from chronic fatigue syndrome. *Journal of Medical Ethics* 43:549–557.

Borchert, Donald Marvin. 2013. *Embracing epistemic humility: Confronting triumphalism in three Abrahamic religions*. Lanham: Lexington Books.

Bullock, Carla G. 2023. Experiencing 'epistemic injustice' within transgender healthcare. *Sexuality, Gender & Policy* 6:47–50.

Byskov, Morten Fibieger. 2021. What makes epistemic injustice an 'injustice?'. *Journal of Social Philosophy* 52:116–133.

Carel, Harvi, and Ian James Kidd. 2014. Epistemic injustice in healthcare: A philosophical analysis. *Medicine, Healthcare and Philosophy* 17:529–540.

Carel, Harvi, and Ian James Kidd. 2017. Epistemic injustice in medicine and healthcare. In *The Routledge handbook of epistemic injustice*, 336–246.

Child, William. 2007. Davidson on first person authority and knowledge of meaning. *Nous* 41:157–177.

Child, William. 2013. First-person authority. In *A companion to Donald Davidson*, ed. Ernie Lepore and Kirk Ludwig, 533–549. New York: John Wiley & Sons.

Coady, David. 2010. Two concepts of epistemic injustice. *Episteme* 7:101–113.

Corradi, Richard. 2015. Transgender Delusion. *First things* (October): 17–19.

Cross, Richard. 1998. *The physics of duns Scotus: The scientific context of a theological vision*. Oxford: Clarendon Press.

Cross, Richard. 2014. *Duns Scotus's theory of cognition*. New York: Oxford University Press.

Davidson, Donald. 2001. *Subjective, intersubjective, objective: Philosophical essays volume 3*. New York: Oxford University Press.

DiCamillo, John A. 2017. Gender transitioning and Catholic health care. *National Catholic Bioethics Quarterly* 17:213–223.

Dicastery for the Doctrine of the Faith (DDF). 2024. Declaration dignitas infinita on human dignity. Vatican City: Liberia Editrice Vaticana.

Dumont, Stephen D. 1989. Theology as a science and duns Scotus's distinction between intuitive and abstractive cognition. *Speculum* 64:579–599.

Finley, John. 2015. The metaphysics of gender: A Thomistic approach. *The Thomist* 79:585–614.

Foucault, Michel. 2003. *Society must be defended: Lectures at the collége de France, 1975–76*. Trans. David Macey. New York: Picador.

Fricker, Miranda. 2007. *Epistemic injustice: Power & the ethics of knowing*. New York: Oxford University Press.

Fricker, Miranda. 2017. Evolving concepts of epistemic injustice. *In The Routledge Handbook of Epistemic Injustice,* 53–60.

Fricker, Miranda, and Katharine Jenkins. 2017. Epistemic injustice, ignorance, and trans experiences. In *Routledge companion to feminist philosophy*, ed. Ann Garry, Serene J. Khader, and Alison Stone, 268–278. London: Routledge.
Horan, Daniel P. 2019a. *Catholicity and emerging personhood: A contemporary theological anthropology*. Maryknoll, NY: Orbis Books.
Horan, Daniel P. 2019b. What's the path to true dialogue about gender in the church? National Catholic Reporter.
Horan, Daniel P. 2020. The truth about so-called 'gender ideology. National Catholic Reporter.
Horan, Daniel P. 2021. Recent transphobic statements from bishops make truth claims without facts. National Catholic Reporter.
Horan, Daniel P. 2022. Church's anti-LGBTQ policies drive people away—and the policies are sinful, too. National Catholic Reporter.
Horan, Daniel P. 2023a. US Bishops' document against transgender health care is a disaster. National Catholic Reporter.
Horan, Daniel P. 2023b. Let's hear transgender people tell their own stories. National Catholic Reporter.
Horan, Daniel P. 2024a. The strawman of 'Gender Theory' in the Vatican's new document. New Ways Ministry Bondings 2.0. https://www.newwaysministry.org/2024/04/08/the-strawman-of-gender-theory-in-the-vaticans-new-document/. Accessed 16 Oct 2024.
Horan, Daniel P. 2024b. In dignitas infinita, the Vatican quarrels with a theory of its own making. Outreach: An LGBTQ Catholic Resource. https://outreach.faith/2024/04/in-dignitas-infinita-the-vatican-quarrels-with-a-theory-of-its-own-making/. Accessed 16 Oct 2024.
Ingham, Mary Beth. 2017. *Understanding john duns scotus: Of reality the rarest-veined unraveller.* St. Bonaventure, NY: Franciscan Institute Publications.
Ingham, Mary Beth, and Mechthild Dreyer. 2004. *The philosophical vision of John duns Scotus*. Washington, DC: Catholic University of America Press.
Kersten, Katherine. 2016. Transgender Conformity. *First things* (December): 25–31.
Kidd, Ian James, and Havi Carel. 2016. Epistemic injustice and illness. *Journal of Applied Philosophy* 33:172–190.
Kidd, Ian James, and Havi Carel. 2018. Healthcare practice, epistemic injustice, and naturalism. *Royal Institute of Philosophy Supplement* 84:211–233.
Kidd, Ian James, José Medina, and Gaile Pohlhaus Jr. 2017. *The Routledge handbook of epistemic injustice*. London: Routledge.
Kraft, James. 2006. Religious tolerance through religious diversity and epistemic humility. *Sophia* 45:101–116.
Lysaught, M. Therese. 2023. US Bishops' transgender document fails morally, theologically—and is poorly argued. National Catholic Reporter.
Lysaught, M. Therese. 2024. Vatican's own gender ideology makes 'Dignitas Infinita' incoherent. National Catholic Reporter.
May, Vivian M. 2014. Speaking into the void? Intersectionality critiques and epistemic backlash. *Hypatia* 29:94–112.
Medina, José. 2017. Varieties of Hermenutical injustice. In *The Routledge handbook of epistemic injustice*, 41–52. Routledge.
Nadal, Kevin L. 2013. *That's so gay! Microaggressions and the lesbian, gay, bisexual, and transgender community*. Washington, DC: American Psychological Association.
Nadal, Kevin L., Amy Sklonik, and Yinglee Wong. 2012. Interpersonal and systemic microaggressions toward transgender people: Implications for Counseling. *Journal of LGBT Issues in Counseling* 6:55–82.
Nordmarken, Sonny, and Resse Kelly. 2014. Limiting transgender health: Administrative violence and microaggressions in health care systems. In *Health care disparities and the LGBT population*, ed. Vickie L. Harvey and Teresa Heinz Housel, 143–166. Lanham: Lexington Books.
Pasnau, Robert. 2003. Cognition. In *The Cambridge companion to duns Scotus*, ed. Thomas Williams, 285–311. New York: Cambridge University Press.

Polhaus, Gaile, Jr. 2017. Varieties of epistemic injustice. In *The Routledge handbook of epistemic injustice*, 13–26. Routledge.

Schumacher, Michelle M. 2016. Gender ideology and the ‘artistic’ fabrication of human sex: Nature as norm or the remaking of the human? *The Thomist* 80:363–423.

Scott, Katie Collins. 2023. An inside look at how the Boston Archdiocese is crafting gender-identity guidelines. National Catholic Reporter.

Scotus, John Duns. 1966. *De primo principio*. Trans. Allan B. Wolter. Chicago: Franciscan Herald Press.

Scotus, John Duns. 1975. *God and creatures: The quodlibetal questions*. Trans.Frank Alluntis and Allan B. Wolter. Princeton: Princeton University Press.

Spivak, Gayatri Chakravorty. 1988. Can the subaltern speak? In *Marxism and the interpretation of culture*, ed. Cary Nelson and Lawrence Grossberg, 271–315. Champaign: University of Illinois Press.

Sue, Derald Wing. 2010. *Microaggressions in everyday life: Race, gender, and sexual orientation*. New York: John Wiley and Sons.

Turyn, Gus. 2023. Gender and first-person authority. *Synthese* 201 (122): 1–19.

United States Conference of Catholic Bishops (USCCB) Committee on Doctrine. 2023. Doctrinal note on the moral limits to technological manipulation of the human body. Washington, DC: USCCB Publications.

Wanderer, Jeremy. 2017. Varieties of testimonial injustice. In *The Routledge handbook of epistemic injustice*, 27–40. Routledge.

Wolter, Allan B. 1990. Duns Scotus on intuition, memory, and our knowledge of individuals. In *The philosophical theology of John duns Scotus*, ed. Marilyn Mc Cord Adams, 98–124. Ithaca, NY: Cornell University Press.

Lee, Harper. 1960. *To Kill a Mockingbird*. Philadelphia: J. B. Lippincott & Co. Publishing

McKinnon, Rachel. 2017. Allies behaving badly: gaslighting as epistemic injustice. In *The Routledge Handbook of Epistemic Injustice*, 167-174.

Part II
Public and Global Health

Chapter 6
More than the Greatest Good for the Greatest Number: Developing a Catholic Public Health Ethic

Mary E. Homan

6.1 Introduction

Public health ethics is fundamentally shaped by how we define and understand public health, especially in terms of our shared and individual definitions of health and the ethics of multiple publics. An inherent ethical tension between the good of individuals and the good of the State exists. Unfortunately, in practice, public health ethics is often reduced to a utilitarian ethic that justifies interventions or imposes behavioral modifications at the infringement of personal rights based on the greater good. In this chapter, I will briefly explore the history of public health as it relates to improving the health of communities, current approaches to public health ethics, and the resultant challenges of moral relativism and utilitarian-directed outcomes. I will suggest a new conceptual apparatus to support public health's mandate to promote health and prevent disease, disability, and injury in the population grounded in a Catholic virtue approach to human flourishing.

6.2 Ethical Theories and Public Health Practice

In beginning any discussion about public health ethics, we must clarify what encapsulates public health. The American Public Health Association describes public health as promoting and protecting the health of all peoples and their communities (American Public Health Association n.d.-b). Public health involves preventive

M. E. Homan (✉)
Department of Theology & Ethics, CommonSpirit Health, Chicago, IL, USA
e-mail: mary.homan@commonspirit.org

J. T. Eberl (ed.), *Emerging Issues in Catholic Bioethics*, Philosophy and Medicine 153, https://doi.org/10.1007/978-3-032-08862-8_6

action and screening before someone gets sick. The three core functions of public health are assessment, policy development, and assurance. There are subsequently 10 Essential Services in Public Health which fall into these categories to "actively promote policies, systems, and overall community conditions that enable optimal health for all and seek to remove systemic and structural barriers that have resulted in health inequities" (Centers for Disease Control and Prevention 2020). The 10 Essential Services (hereafter, ES) center around equity, which is defined as "a fair and just opportunity for all to achieve good health and well-being."

One of the challenges to the success of public health interventions resides within our varied understandings of health, both on the individual level and in the judgement of another. Tensions can arise between the negative and positive conceptions of health where someone is healthy because of the absence of disease versus being healthy because of having well-being (Holland 2015, 112). In the negative conception, one might ask, does a person have a healthy life if there is the presence of disease, regardless of how that disease affects their daily activities? Using a positive conception of health, one might ask, if a person does not have emotional or social well-being, are they healthy? For those less familiar with health promotion activities, the Social Ecological Model explains that relationships, community, and society impact an individual's health and health behavior (McLeroy et al. 1988). For example, if I have a friend who likes to take walks, we have complete sidewalks in our neighborhood, and it is safe to walk in our neighborhood, I am more likely to take walks and therefore keep healthy according to the positive conception of health, and yet, I may still have diabetes so I am considered unhealthy. A contrary example would be that I work a very difficult job and, during my workday, smokers are given the opportunity for an extra break. I want to join my colleagues in a break from our difficult work and I smoke because the organization does not have a prohibition on doing so nor is there a local law prohibiting smoking at a worksite and I build relationships with my colleagues. I therefore choose an unhealthy behavior, but I might have emotional well-being. What happens then when a municipality implements a smoke-free ordinance and now I cannot smoke at my very stressful job? Am I healthier? Do I have well-being?

I remember early in my public health career interviewing parents of children with severe asthma symptoms, who often utilized emergency department services, or were subsequently hospitalized. One of our questionnaires expressly looked at the burden of disease on the family. Often the parents underreported the severity or burden of their child's asthma on the rest of the family (Kurz et al. 2007) such that even if a child had recently visited the emergency department or had been admitted, the parent would report that at this exact moment in time this child was healthier than they have ever been or that the child's asthma was better controlled because of the great care they received at an acute care facility. An insurer would disagree because the cost of an emergency department visit or hospitalization significantly outweighs the cost of an inhaler or nebulizer treatment. A public health practitioner similarly might be concerned about the need for a more timely preventive service rather than tertiary response so that a person's quality of life is improved.

Not only do public health practitioners determine which definition of health they design interventions around, they must also determine which public is being served and at whose bequest. Accordingly, Faden et al. (2019) suggest there are four characteristics of public health interventions as they relate to public health ethics: 1. it is a public or collective good; 2. its promotion involves a particular focus on prevention; 3. promotion often entails government action, and 4. it involves an intrinsic outcome orientation (Faden et al. 2019).

On a daily basis public health professionals struggle to attend to the needs and values of multiple publics while operating in a complex social arrangement where the action, or inaction, of government entities influence what essential services are considered as such. There remains an inherent tension between the individual and the State and the usual ethical theories employed suggest very different understandings of what the obligation or role of the State is to an individual or population. Table 6.1 shows the common understandings of deontology, libertarianism, utilitarianism, and virtue ethics as they relate to public health.

Moving the focus of an intervention from an individual to a population also challenges the applicability of certain ethical theories, in particular many of the resources we find in clinical ethics, and as such, public health ethics differs from clinical or biomedical ethics (Ortmann et al. 2016). In Table 6.2, we can see the comparisons Ortmann et al. draw from the language of familiar ethical theories: "harm avoidance" as a form of consequentialism and "professional duty" as a form of deontology.

To practice this application of public health ethics to public health action, when teaching graduate public health students (many of whom were already public health professionals or medical students), I challenged them to choose from the four major ethical theories (deontology, consequentialism, utilitarianism, and virtue)[1] to justify a particular public health action based on that week's course materials and to articulate the role of the individual and the State. What followed was a typical application of utilitarianism with a few clinicians using principlism[2] in lieu of deontology and a few brave souls trying to apply virtue ethics. Similarly, in training workshops with local health departments, for which ethics education is required for accreditation (Bernheim et al. 2013),[3] we discussed that individuals often rely on the ethical

[1] While utilitarianism and consequentialism are often conflated with utilitarianism being a type of consequentialism, Holland (2015, 18) argues that consequentialism is not of much use unless we can explain why one set of consequences is better or worse. Hence the need for utilitarianism which "says that the right action is that which brings about the best consequences, and the best consequences are those that maximize well-being, welfare or benefit."

[2] Holland (2015, 37) writes "It is questionable as to how useful these four principles of biomedical ethics are in public health." Even as a clinical ethicist, I find the four principles maddening because there is no way to figure out which principle eclipses another.

[3] PHAB Measure 11.1.2 A: Ethical issues identified and ethical decisions made.

Table 6.1 Types of common ethics theories and operative values for public health application

Ethics Theory	Operative Values	Understanding of the Good/ Public Health Application
Deontology (Kant, Ross, Rawls)	• Right action in terms of duties and moral rules focus on the rightness of an act and not on what results from the act • The rights of individuals are taken seriously while providing minimal protection for the poor and marginalized • Words that are suggestive of a deontological approach are "duty" or "everybody must....," as well as "right", "obligation", "always", "universal", "autonomy", "self-determination," "freedom", "respect," "commandments", "code of ethics".	• Obedience to the rule is the good means that will lead to the good end • Equal distribution of social goods unless unequal distribution is to everyone's advantage because citizens are essentially equal
Libertarianism (Locke, Nozick)	• The function of the state is to protect individual liberty • Rights of individual take priority over the good of the state • Rejects "welfare state" which redistributes wealth or creates particular opportunities for the poor and marginalized	• Task of public health is to minimize harm to the individual in the least intrusive way possible Would reject or challenge mandatory public health initiatives like smoke-free initiatives
Utilitarianism (Bentham, Mill)	• Greatest happiness of the greatest number (Bentham) • Duties, rights, and virtues to be means to the end of increased social utility (Mill) • Rights are protected so long as they preserve the greater good	• What is in the State's interest is the best thing for the populace • Social cost to groups could be justified when there is a perception that a greater good could come about Mandatory vaccination Social distancing/quarantine
Virtue (Aristotle)	• Human actions are not isolated events without a context, performed apart from the character and history of the moral agent • Stresses the social character of persons while simultaneously affirming the inviolable dignity of individual persons. • Asks "Who am I?" "Who ought I become?" "How ought I get there?"	• Task of state is to be involved to promote health and well-being by engaging the community • Challenge to political, socio-cultural, and economic systems that tolerate and/or exploit disparities. Dismantling of unjust systems; intersectionality Community engaged research

Mary E. Homan, DrPH, MA, MSHCE

Revised October 2013, January 2015, January 2018, July 2025

Table 6.2 Comparison of areas of focus/tendency in clinical ethics and public health ethics

Clinical Ethics Focus/Tendency	Public Health Ethics Focus/Tendency
Treatment of disease and injury	Prevention of disease and injury
Medical interventions by clinical professionals	Range of interventions by various professionals
Individual benefit seeking and harm avoidance based on health care provider's fiduciary relation to a patient	Social, community, or population benefit seeking and harm avoidance based on collective action
Respect for individual patients	Relational autonomy of interdependent citizens
Professional duty to place the interests of the patient over that of provider	Duty to the community to address health concerns that individuals cannot solve and that require collective action
Authority based on the prestige and trustworthiness of the physician and the medical profession as a whole	Authority based on law, which is a principal tool of public health policy for creating health regulations
Informed consent sought from an individual patient for a specific medical interventions	Community consent and building a social consensus through ongoing dialogue and collaboration with the public
Justice concerns largely limited to treating patients equally and ensuring universal access to health care	Central concern with social justice regarding health and achieving health equity

From: Public Health Ethics: Global Cases, Practices, and Context

theory that they use the most in their daily life (O'Toole 1998). Therefore, when considering a public health action, local health departments might/should/could find ways to reason with those operating from a different ethical theory.

Public health ethics continues to develop as a field distinct from bioethics or medical ethics. The current codes of public health ethics (American Public Health Association 2019) and recent writings from leaders in the field of public health ethics (Callahan and Jennings 2002; Gostin and Powers 2006; Jennings 2016; Lee 2012) bring together a constellation of ethical theories as applicable to the work of public health. I observe two major problems with that approach. One, certain ethical theories just do not fit public health ethics well. Two, a risk of ethical relativism manifests when, in an effort to be pluralistic and inclusive, we deny that universal norms exist. The problem with ethical relativism according to Jennings et al. (2003, 87) "is that it seems to condone every permutation of human behavior, if the surrounding cultural context would also condone it. So neither internal critique of a malfunctioning culture nor outside critique on the basis of universal human standards—such as human rights—are possible if the theory of ethical relativism is accepted."

We see the consequence of ethical relativism in the way certain communities treat certain kinds of public health issues such as banning safe needle exchanges (Boyce 2024) or breaking up tent communities (Olson and Pauly 2021) whereas other communities readily supply fentanyl testing strips (McCollum et al. 2024) or provide safe spaces for pregnant persons (Mkandawire-Valhmu et al. 2018). And yet, for a discipline so focused on equity, the default reliance on harm reduction or the 'greatest good for the greatest number' threatens at-risk, historically, and chronically disenfranchised populations and fails to bring about the flourishing of individuals and communities. Public health interventions become reduced to supporting the good of the many, at the detriment of the most vulnerable, or adhering to compliance standards that satisfy few (Carman and Timsina 2015). Similarly, professionals adhere to codes of ethics thinking they are being ethical when really such guidelines are designed to eliminate discernment and establish the lowest standard of acceptable conduct. As the trio of Dominican moralists remind us, "Proposing ethical guidelines in a manner that is acceptable to the majority of the public is dangerous because it may justify whatever is popular at the expense of what is ethical" (Ashley et al. 2006, 226).

6.3 Reframing a Public Health Ethic

Those who worked diligently on such an important document like the *Public Health Code of Ethics* explain that "in terms of ethical theory, the task force did not adopt a deductive approach based on covering principles of either a utility (i.e. health) maximizing sort or a rights-based orientation. We attempted to articulate the notion that justice pertains both to the outcomes of public health policies and practices and to the processes and procedures used to achieve those outcomes" (Jennings 2020, 190). Often those who teach public health ethics, whether for graduate MPH students or in providing education for local health departments, do not have formal ethics training. Alternatively, theological ethicists opine about how public health should conduct itself when not engaged in the work of improving the public's health. What follows now includes a larger discussion about justice and the common good and, while shaped by Catholic moral theology and the social teachings of the Church, remains accessible for even the most secular of public health professionals.

Lisa Sowle Cahill (2005, 8) helps ground this topic with her opinion that "[t]heologians informed by the common good tradition of Catholic social teaching are among those who most consistently try to present justice concerns as integral to bioethics." Byron (1998, 10) describes the common good as "the social conditions that allow people to reach their full human potential and to realize their human dignity." The common good can only manifest when there is relationship between individuals and community, a reciprocity where the community enables the flourishing of the individual and individual behavior affects the flourishing of the community (Lebacqz 1986). It is "the good of each human person … intimately related to the good of the whole community" (Center of Concern 2008). These descriptions of the common good look, feel, and sound more like the ideal of public health in part because of the attention to relationship and flourishing, where we must put forward the needs of some in a way that may feel unbalanced or uncomfortable to others. Gustavo Gutiérrez would call this the preferential option for the poor and diagnose that our discomfit arises from the need to ask *why* the poor are poor (Gutiérrez 2009).

I begin here by suggesting that the "job of justice in its most pressing role looks first to the conditions of the most profound disadvantaged" (Powers and Faden 2006, 95). This first examination is a "fundamental option for the marginalized" (Cochran 1999, 45). Producing complex statistical analyses about the health disparities of a particular community for a peer-reviewed journal is neither sufficient in tackling health inequity (Robeznieks 2021) nor in alignment with ES #1.[4] We must acknowledge that persons are "made vulnerable by racial, cultural, social, and institutional attitudes and practices that privilege one group at the expense of another" (Lumas 2011, 75). Vulnerable populations not only are less able to protect their own interests but also have less opportunity to do so, and may also be at high risk to experience different health outcomes due to disparities, barriers, or special needs (Danis and Patrick 2002, 312). However, some have suggested that the word "vulnerable" is problematic as well and is used "to avoid the discomfort associated with naming structural racism as the true root cause of many public health problems. Sometimes referring to those who are clinically vulnerable to disease because of their age is appropriate, but generally calling an entire population (or sub-population) vulnerable is not" (Avila 2020).

We cannot ignore how racism, systemic bias, distrust of the medical system, and unfair treatment by clinicians have negatively impacted many in our communities. Accordingly, the common good challenges other understandings of social justice,[5] such as distributive and structural justice, which are often used in the public health literature. Distributive justice requires that the goods of society be distributed in an equitable manner to all members of society (Tieman and Hamel 2008, 7). It is a sense of justice where a minimum of material resources are considered an absolute necessity for human life. A distributive justice lens regards fairness or equity as the starting point. Distributive justice could consider that fair allocation of health care resources includes provision of needed counseling as well as other social services. However, if there is only a focus on the distribution of material goods and resources, the scope of justice is restricted because it fails to see the importance of structural and institutional contexts (Young 1990, 20). Fabrice Jotterand et al. (2023, 104–105) readily admit that the virtue of justice is the most demanding as it requires concern both at the individual level and the institutional level.

---

[4] ES #1: Using data and information to determine the root causes of health disparities and inequities (Centers for Disease Control and Prevention 2020)

[5] "Social justice is the view that everyone deserves equal rights and opportunities" (American Public Health Association n.d.-a)

While distributive justice and structural justice fill the public health equity literature, a shared experience of human flourishing remains just out of reach. We must look beyond fair allocation of resources and examine the relative deprivation vulnerable persons experience. Relative deprivation is the "lack of sources of self-respect that are deemed essential for full participation in society" (Daniels et al. 2002, 25). If someone has significant health disparities, reduced access to health care or social services, or are historically disenfranchised, some ethicists argue such persons should receive preferential treatment as we saw with the various resource allocation schemas during the height of the COVID-19 pandemic (Emanuel et al. 2020).

Since the nineteenth century, as evidenced in papal writings on social justice, Catholic social thought "has emphasized the need and importance of structural change" (Curran 1978, 153). Through the preferential option for the poor, Christians are called to look at the world from the perspective of the marginalized and to work in solidarity for justice, to "begin to reassess policies and social institutions in terms of their impact on the poor" (Toth 2007, 874). It is as Roger Haight (2001, 233) describes: "theology is a public enterprise that must be responsible to the world situation, to other religious traditions, and to all people in the concrete intellectual culture and social situations in which they exist." Jesuit and health services researcher Michael Rozier (2016) posits that public health as a field focuses often on improving the health behaviors of persons, and resonates with a virtue-based approach which emphasizes the inner character or motivation of an individual (Slote 2013, 395), such as both helping persons make healthy eating choices and by organizations making healthy food choices normative. Lee (2018) proposes four basic ethics skills necessary for all public health personnel: 1. identify the ethical dimensions of public health action; 2. articulate ethical dimensions and dilemmas; 3. determine a path forward, especially when values and motivations conflict; and 4. implement and evaluate the solution to allow for course corrections and the addition of new information should the need arise. And yet public health ethics "while emphasizing the importance of community and considering the wellbeing of vulnerable individuals, makes no claim as to why it is significant that human beings share lives together or why we create communities" (McCarthy et al. 2020, 10). Put another way, if one personally cannot recognize the injustices of the public health problem, the vulnerable persons being further victimized, or that one's community is hurting, how could one expect outside persons to take the helm in addressing the problem?

We all can recall the various voices protesting the COVID-19 vaccine. Some declined vaccination due to an inaccurate understanding of cooperation with evil due to the source material of vaccines (Moschella 2021) but many of the loudest dissenters were those who objected to a mandate to be vaccinated. Some believed that the State was overstepping its role and authority. Still others had a general mistrust of health care professionals which led to vaccine hesitancy (Luo et al. 2022; Newman et al. 2023). Local health departments attempted to cajole, convince, and even coerce individual community members, especially health care workers of the importance of vaccinating. Each of the typical ethical theories from a duty to vaccinate to a utilitarian goal of herd immunity were employed and, to the dismay of many ethicists and public health professionals, people still rejected the vaccine. We failed to account for the rugged individualism vis-à-vis patient autonomy "for which our country is famous" (Shannon 1997, 25). Thomas Schindler (1989, 23) comments that perhaps the most problematic feature of individualism in the United States is that "individuals often have a deep sense of powerlessness that leads to passivity and recognition." He goes on to say that any group that attempts to change the political situation or take political action is also viewed as suspicious, as is often the barrier to effective public health action. Schindler also struggles to understand how a Church which has consistently opposed individualism expects that individuals passively accept specific moral directives (Schindler 1989, 75). However, as others have pointed out, much of the Catholic moral tradition has required adherence to magisterial teaching as evaluated by the individual conscience (Hamel 2017, 10–11; Kelly et al. 2013, 58–59) with an limited sightline to the Catholic social tradition.

6.4 A Catholic, Virtue Ethics Approach to Public Health

For public health to advance its mission, especially during a tumultuous time when the bureaucracy itself has turned against science and evidence-based practice (Jetelina 2025), a renewed focus on the common good and human flourishing necessitates itself. While many scholars situate virtue ethics as a non-consequentialist theory as the theory itself is character-based (Athanassoulis 2016, Hursthouse and Pettigrove 2016), public health must focus on both the intervention as good and just *and* the outcome as good and just in order to be considered successful. I often use the analogy that neither the ends justify the means nor the means justify the ends; both must be good. Jennifer Prah Ruger (2012, 36) states this more eloquently when she writes "human flourishing is a morally central aim shared by all persons by virtue of their humanity, with human dignity at its heart."

Might we then consider the three basic elements of a virtue ethic, as it applies to persons in community: Who ought we become? How ought we act? How ought we get there? Instead of decisions being about my individual behavior and my individual character, we must recognize that "any personal choice is a social act" (Astorga 2014, 35) that affects the community. Astorga's comments surface in a discussion about the personal and social aspects of ethics insofar as "there is no decision that is separated from its social context" (Astorga 2014, 34). This concept resonates with many public health issues such as the availability and use of tobacco products. While an adult over the age of 18 can legally purchase tobacco products, those purchases are often relegated to convenience stores or certain lanes at grocery stores due to the influence of groups such as Tobacco-Free Kids. Many private businesses restrict the use of tobacco on their property or within a certain number of feet from entrances. Municipalities have passed laws prohibiting smoking both inside and outside of restaurants and bars. Insurers place a higher life insurance premium on those who use tobacco products. Therefore, while a person might have the right to be unhealthy, that right is constrained by other stakeholders who recognize the primary, secondary, and even tertiary effects of tobacco use. If we can agree that health, well-being, and creating conditions for human flourishing are what it means to be in community and multiple groups have a stake in the community, then a Catholic-based virtue ethic seems accessible to public health professionals.

A local health department could reconcile secular core values and mission statements with the dynamics of the preferential option of the poor that "challenges doctors and other health providers to make an option for the poor by working on their behalf" (Farmer 1995, 13). "Christians have no monopoly on social ethical wisdom or insight" (Curran 1978, 158) including the principle of preferential option for the poor. A Catholic public health ethic responds to social inequalities through emphasis on justice, integrity, human dignity, and solidarity and acting with compassion to bring about the Kingdom of God here on earth. This occurs when we exercise solidarity and subsidiarity. Solidarity expresses the moral obligation of communities to secure the well-being of their members, in particular the weak and vulnerable (Pijnenburg et al. 2010, 315).

Pope Francis (2013, n. 187) similarly reminds us that "[e]ach individual Christian and every community is called to be an instrument of God for the liberation and promotion of the poor, and for enabling them to be fully a part of society." This also requires making decisions at the closest level possible to those impacted by a decision. Pope Pius XI (1931, n. 80) instructs that the State must "let subordinate groups handle matters and concerns of lesser importance, which would otherwise dissipate [the State's] efforts greatly." Lisa Sowle Cahill (2005, 121) reminds us that "[t]heological bioethics in the twenty-first century must make a new commitment to public participation. It must work hard to transform local and global access to the basic goods of health care, so that justice, equity, and the common good move much further from utopia toward reality."

6.5 Putting it into Practice

Public health ethics should include the study of and deliberation about moral norms that guide public health decision-making. Such decision-making supports public health's mandate to promote health and prevent disease, disability, and injury in the population. An enriched ethics capacity helps to clarify, prioritize, and justify possible courses of public health action. There is greater transparency in decision-making which enhances public trust and relationship building. Engaging in public health decision-making grounded in more than a rights-based or utilitarian ethic, but rather in a virtue-based ethic, fosters respectful deliberation and can alleviate ethical tensions. A default application of harm avoidance and other utilitarian approaches, as evidenced by major public health interventions such as smoking bans or smoke-free initiatives hinders the potential for human flourishing. While for many of us it is difficult to envision eating dinner or attending a movie while neighboring patrons enjoy smoking cigarettes, comprehensive tobacco-free laws are still new in many geographies (Tatalovich et al. 2017) and even in certain types of workplaces (Hyland et al. 2012) despite numerous public health efforts at reducing access to tobacco products (Voigt 2019). I recall working on a smoke-free campaign and there was a constant tension and disagreement between two tobacco-free community coalitions. One group was willing to wait on casinos and bars going smoke-free in order to achieve smoke-free restaurants as a first step whereas the other group believed that all workplaces should be smoke-free; that it was unjust to casino workers and bartenders to be exposed to what everyone agreed was an unsafe amount of second-hand smoke. The first group began from a "greatest good for great number approach" (utilitarian) whereas the second group espoused an approach closer to virtue ethics. The first group would criticize the lobbying efforts of the other as idealistic or unrealistic. At the same time, a group of vocal citizens lambasted local officials supportive of smoke-free legislation as an affront to personal liberty. The change in smoke-free regulations did end up being incremental across the metropolitan area, taking well over a decade, and interestingly there was a shift towards a virtue ethics approach when the casino workers would be exempted from a proposed smoking ban in a neighboring county. "We believe that every single worker has the right to breathe clean air" (Lippman 2018) became the message.

The common good serves as language and method for public health ethics. When a large portion of our community suffers from avoidable injustices, the responsibility to bring about the common good becomes more urgent. The Catholic moral tradition teaches that "each and every member of the community must flourish for the common good to be met" (O'Rourke and Boyle 2011, 19). The question we need to answer is *how* will we help persons to flourish and *how* will we work with community, in community, to bring about the good, not just the right. The success of public health relies on individuals recognizing the value of community interest as well as self-interest. Virtue ethics can provide the process for ethical analysis and evaluation of the common good, the outcome. As such, a public health ethic formed by the Catholic moral and social tradition focused on both forming persons to possess good character so as to achieve a shared sense of human flourishing seems a worthy experiment to try.

References

American Public Health Association. 2019. *Public health code of ethics* (p. 30) [Issue Brief]. American Public Health Association. https://www.apha.org/-/media/files/pdf/membergroups/ethics/code_of_ethics.ashx.

American Public Health Association. n.d.-a. *Social justice and health*. American Public Health Association. Retrieved Sept 1, 2024, from https://www.apha.org/what-is-public-health/generation-public-health/our-work/social-justice.

American Public Health Association. n.d.-b. *What is public health?* Retrieved July 6, 2018, from https://www.apha.org/what-is-public-health.

Ashley, B. M., J. DeBlois, and K. D. O'Rourke. 2006. *Health care ethics: A Catholic theological analysis*. 5th ed. Georgetown University Press.

Astorga, C. A. 2014. *Catholic moral theology & social ethics: A new method*. Orbis Books.

Athanassoulis, N. 2016. Virtue ethics. In *Internet Encyclopedia of Philosophy*. http://www.iep.utm.edu/virtue/.

Avila, C. 2020, June 23. "Vulnerable" is a public health euphemism we need to stop using. *The Incidental Economist*. https://theincidentaleconomist.com/wordpress/vulnerable-is-a-public-health-euphemism-we-need-to-stop-using/.

Bernheim, R. G., M. Stefanak, T. Brandenburg, A. Pannone, and A. Melnick. 2013. Public health accreditation and metrics for ethics: A case study on environmental health and community engagement. *Journal of Public Health Management and Practice: JPHMP* 19 (1): 4–8. https://doi.org/10.1097/PHH.0b013e31824acb25.

Boyce, D. (Director). 2024, May 14. Pueblo city council votes to ban syringe exchange programs [Broadcast]. In *Colorado Public Radio*. Colorado Public Radio. https://www.cpr.org/2024/05/14/pueblo-city-council-votes-to-ban-needle-exchange-programs/.

Byron, W. J. 1998. Ten building blocks of Catholic social teaching. *America* 179 (13): 9–12. https://doi.org/Article.

Cahill, L. S. 2005. *Theological bioethics: Participation, justice, and change*. Georgetown University Press. http://catdir.loc.gov/catdir/toc/ecip0511/2005010167.html.

Callahan, D., and B. Jennings. 2002. Ethics and public health: Forging a strong relationship. *American Journal of Public Health* 92 (2): 169–176.

Carman, A. L., and L. Timsina. 2015. Public health accreditation: Rubber stamp or roadmap for improvement. *American Journal of Public Health* 105 (Suppl 2): S353–S359. https://doi.org/10.2105/AJPH.2015.302568.

Center of Concern. 2008, Fall. *The principle of the common good*. http://www.coc.org/files/principle.pdf.

Centers for Disease Control and Prevention. 2020. *10 essential public health services*. Public Health Professionals Gateway. https://www.cdc.gov/public-health-gateway/php/about/.

Cochran, C. E. 1999. The common good and healthcare policy. *Health Progress (Saint Louis, Mo.)* 80 (3): 41–47.

Curran, C. E. 1978. Social ethics: Agenda for the future. In *Toward Vatican III: The work that needs to be done*, 146–162. Gill and Macmillan.

Daniels, N., B. P. Kennedy, and I. Kawachi. 2002. Justice, health, and health policy. In *Ethical dimensions of health policy*, ed. M. Danis and C. Clancy, 19–50. Oxford University Press.

Danis, M., and D. Patrick. 2002. Health policy, vulnerability, and vulnerable populations. In *Ethical dimensions of health policy*, ed. M. Danis, C. Clancy, and L. R. Churchill, 310–334. Oxford University Press.

Emanuel, E. J., G. Persad, R. Upshur, B. Thome, M. Parker, A. Glickman, C. Zhang, C. Boyle, M. Smith, and J. P. Phillips. 2020. Fair allocation of scarce medical resources in the time of COVID-19. *New England Journal of Medicine* 382 (21): 2049–2055. https://doi.org/10.1056/NEJMsb2005114.

Faden, R. R., S. Shebaya, and A. W. Siegel. 2019. Distinctive challenges of public health ethics. In *The Oxford handbook of public health ethics*, ed. A. C. Mastroianni, J. P. Kahn, and N. E. Kass. Oxford University Press. https://www.oxfordhandbooks.com/view/10.1093/oxfordhb/9780190245191.001.0001/oxfordhb-9780190245191-e-2.

Farmer, P. 1995. Medicine and social justice. *America* 173 (2): 13–17. aph.

Francis. 2013. *Evangelii gaudium*. https://www.vatican.va/content/francesco/en/apost_exhortations/documents/papa-francesco_esortazione-ap_20131124_evangelii-gaudium.html.

Gostin, L. O., and M. Powers. 2006. What does social justice require for the public's health? Public health ethics and policy imperatives. *Health Affairs* 25 (4): 1053–1060. https://doi.org/10.1377/hlthaff.25.4.1053.

Gutiérrez, G. 2009. The option for the poor arises from faith in Christ. *Theological Studies* 70 (2): 317–326. aph.

Haight, R. 2001. *Dynamics of theology*. Orbis Books.

Hamel, R. 2017. Reciprocity of consciences and Catholic health care. In *Conscience and Catholic health care: From clinical contexts to government mandates*, ed. D. E. DeCosse and T. A. Nairn, 1–16. Orbis Books.

Holland, S. 2015. *Public health ethics*. 2nd ed. Polity Press.

Hursthouse, R., and G. Pettigrove. 2016. Virtue ethics. In *The Stanford Encyclopedia of Philosophy*, ed. E. N. Zalta. https://plato.stanford.edu/archives/win2016/entries/ethics-virtue/.

Hyland, A., J. Barnoya, and J. E. Corral. 2012. Smoke-free air policies: Past, present and future. *Tobacco Control* 21 (2): 154–161. https://doi.org/10.1136/tobaccocontrol-2011-050389.

Jennings, B. 2016. Right relation and right recognition in public health ethics: Thinking through the republic of health. *Public Health Ethics* 9 (2): 168–177. https://doi.org/10.1093/phe/phv032.

Jennings, B. 2020. Ethics codes and reflective practice in public health. *Journal of Public Health* 42 (1): 188–193. https://doi.org/10.1093/pubmed/fdy140.

Jennings, B., J. Kahn, A. Mastroianni, L. S. Parker, H. Alvarez, R. Bayer, R. G. Bernheim, R. Bonnie, M. Garland, L. O. Gostin, E. Heitman, B. Jennings, J. Kahn, P. A. Marshall, A. Mastroianni, L. C. McKieran, P. Nieburg, L. S. Parker, K. Shrader-Frechette, J. Stull, J. Thomas, and S. Thomas. 2003. *Ethics and public health: Model curriculum*. Association of Schools of Public Health (ASPH).

Jetelina, K. 2025. Vaccine policy meeting: The essentials. https://yourlocalepidemiologist.substack.com/p/vaccine-policy-meeting-the-essentials.

Jotterand, F., R. Spellecy, M. Homan, and A. R. Derse. 2023. Promoting equity in health care through human flourishing, justice, and solidarity. *The Journal of Medicine and Philosophy* 48 (1): 98–109.

Kelly, D. F., G. Magill, and H. ten Have. 2013. *Contemporary Catholic health care ethics*. Georgetown University Press.

Kurz, R. S., Homan, M. E., Homan, S. M., Wilson, K. D., Sunshine-Hankins, M., and Saunders, J. 2007, November 5. *Asthma H.E.L.P.: Assessing the effects of home visits by asthma-educated social workers on children's health outcomes* [Poster Session]. "Politics, Policy and Public Health" APHA 135th Annual Meeting, Washington, D.C. https://apha.confex.com/apha/135am/techprogram/paper_147465.htm.

Lebacqz, K. 1986. *Six theories of justice: Perspectives from philosophical and theological ethics*. Augsburg Pub. House.

Lee, L. M. 2012. Public health ethics theory: Review and path to convergence. *The Journal of Law, Medicine & Ethics: A Journal of the American Society of Law, Medicine & Ethics* 40 (1): 85–98. https://doi.org/10.1111/j.1748-720X.2012.00648.x.

Lee, L. M. 2018. Ethical competencies for public health personnel. *Ethics, Medicine and Public Health* 4: 21–26

Lippman, R. (Director). 2018, October 18. *Anti-smoking advocates urge "no" vote on limited ban in St. Charles County* [Broadcast]. St Louis Public Radio. https://www.stlpr.org/government-politics-issues/2018-10-18/anti-smoking-advocates-urge-no-vote-on-limited-ban-in-st-charles-county.

Lumas, E. M. 2011. It takes a whole village to … do just about everything!: Embracing a preferential option for the poor. *New Theology Review* 24 (2): 72–84.

Luo, H., H. Qu, R. Basu, A. P. Rafferty, S. P. Patil, and D. M. Cummings. 2022. Willingness to get a COVID-19vaccine and reasons for hesitancy among medicare beneficiaries: Results from a national survey. *Journal of Public Health Management and Practice: JPHMP* 28 (1): 70–76. https://doi.org/10.1097/PHH.0000000000001394.

McCarthy, M., M. Homan, and M. Rozier. 2020. There's no harm in talking: Re-establishing the relationship between theological and secular bioethics. *The American Journal of Bioethics* 20 (12): 5–13. https://doi.org/10.1080/15265161.2020.1832611.

McCollum, C. G., G. Ventrelli, K. W. Gagnon, A. Loftis, A. Famurewa, C. Wimberly, and E. Eaton. 2024. Rapid response to the legalization of fentanyl test strips in Alabama: An academic–community partnership. *American Journal of Public Health* 114 (8): 785–788. https://doi.org/10.2105/AJPH.2024.307681.

McLeroy, K. R., D. Bibeau, A. Steckler, and K. Glanz. 1988. An ecological perspective on health promotion programs. *Health Education Quarterly* 15 (4): 351–377. https://doi.org/10.1177/109019818801500401.

Mkandawire-Valhmu, L., L. Lathen, M. J. Baisch, Q. Cotton, A. Dressel, J. Antilla, O. Olukotun, R. Washington, L. Jordan, and A. Hess. 2018. Enhancing healthier birth outcomes by creating supportive spaces for pregnant African American women living in Milwaukee. *Maternal and Child Health Journal* 22 (12): 1797–1804. https://doi.org/10.1007/s10995-018-2580-4.

Moschella, M. 2021. Dignitas personae, HEK 293, and the COVID vaccines. *The National Catholic Bioethics Quarterly* 21 (1): 107–121. https://doi.org/10.5840/ncbq202121110.

Newman, P. A., D. A. Dinh, T. Nyoni, K. Allan, S. Fantus, C. C. Williams, S. Tepjan, L. Reid, and A. Guta. 2023. Covid-19 vaccine hesitancy and under-vaccination among marginalized populations in the United States and Canada: A scoping review. *Journal of Racial and Ethnic Health Disparities* 12:413. https://doi.org/10.1007/s40615-023-01882-1.

Olson, N., and B. Pauly. 2021. Homeless encampments: Connecting public health and human rights. *Canadian Journal of Public Health = Revue Canadienne de Santé Publique* 112 (6): 988–991. https://doi.org/10.17269/s41997-021-00581-w.

O'Rourke, K. D., and P. Boyle. 2011. *Medical ethics: Sources of Catholic teachings*. Georgetown University Press.

Ortmann, L. W., D. H. Barrett, C. Saenz, R. G. Bernheim, A. Dawson, J. A. Valentine, and A. Reis. 2016. Public health ethics: Global cases, practice, and context. In *Public health ethics: Cases spanning the globe*, ed. D. H. Barrett, L. H. Ortmann, A. Dawson, C. Saenz, A. Reis, and G. Bolan, 3–35. Springer International Publishing. https://doi.org/10.1007/978-3-319-23847-0_1.

O'Toole, B. 1998. Four ways people approach ethics. A practical guide to reaching consensus on moral problems. *Health Progress (Saint Louis, Mo.)* 79 (6): 38–41, 43.

Pijnenburg, M. A. M., B. Gordijn, F. J. H. Vosman, and H. A. M. J. Ten Have. 2010. Catholic healthcare organizations and how they can contribute to solidarity: A social-ethical account of Catholic identity. *Christian Bioethics: Non-Ecumenical Studies in Medical Morality* 16 (3): 314–333. https://doi.org/10.1093/cb/cbq021.

Pius XI. 1931. *Quadragesimo anno*. https://www.vatican.va/content/pius-xi/en/encyclicals/documents/hf_p-xi_enc_19310515_quadragesimo-anno.html.

Powers, M., and R. R. Faden. 2006. Chapter 4: Social justice and public health. In *Social justice: The moral foundations of public health and health policy*, 80–99. Oxford University Press.

Robeznieks, A. 2021, December 29. *Variations in mortality data show health inequity isn't inevitable*. https://www.ama-assn.org/delivering-care/health-equity/variations-mortality-data-show-health-inequity-isn-t-inevitable.

Rozier, M. D. 2016. Structures of virtue as a framework for public health ethics. *Public Health Ethics* 9 (1): 37–45. https://doi.org/10.1093/phe/phv036.

Ruger, J. P. 2012. Global health justice and governance. *American Journal of Bioethics* 12 (12): 35–54.

Schindler, T. F. (with Internet Archive). 1989. *Ethics—the social dimension: Individualism and the Catholic tradition*. Wilmington, Del. : M. Glazier. http://archive.org/details/ethicsthesociald0000schi.

Shannon, T. A. 1997. *An introduction to bioethics*. Paulist Press.

Slote, M. 2013. Virtue ethics. In *The Blackwell guide to ethical theory*, ed. H. LaFollette and I. Persson, 2nd ed., 394–411. Wiley Blackwell.

Tatalovich, Z., D. G. Stinchcomb, J. A. Lyman, Y. Hunt, and J. E. Cucinelli. 2017. A geo-view into historical patterns of smoke-free policy coverage in the USA. *Tobacco Prevention and Cessation* 3 (December): 134. https://doi.org/10.18332/tpc/80135.

Tieman, J., and R. Hamel. 2008. Raising our voice for health reform: A national priority and moral imperative. *Health Care Ethics USA* 16 (4): 6–8.

Toth, W. J. 2007. Preferential option for the poor. In *Encyclopedia of Catholic social thought, social science, and social policy*, ed. M. L. Coulter, R. S. Myers, and J. A. Varacalli, vol. 2, 873–874. Scarecrow Press.

Voigt, K. 2019. Tobacco as a matter of public health: Ethical issues. In *The Oxford handbook of public health ethics*. Oxford University Press. https://doi.org/10.1093/oxfordhb/9780190245191.013.43.

Young, I. M. 1990. *Justice and the politics of difference*. Princeton University Press.

Chapter 7
Public Health Care Ethics: Applying Catholic Social Teaching Using the G-O-A-LS® Method

Claudia R. Sotomayor and William F. Sullivan

7.1 Introduction

The dynamic and interacting activities constituting the structure of human knowing, deciding, and loving can be affirmed if we are attentive to what we are doing in making any concrete decision. By engaging in these activities well, by continuously learning from past mistakes and oversights, and by cooperating with authentic others, we have the potential to reach correct judgments of fact and ethically good decisions. In this chapter, we describe a G-O-A-LS® method specifying the components on the level of human deciding and loving that need to be integrated for adequately addressing ethical questions.[1] These are based on long-acknowledged determinants of the goodness of moral acts in Catholic teaching (Thomas Aquinas 1948, I-II 18.2–4 and I-II 55.4). We will show how certain principles associated with Catholic social teaching (human dignity, the common good, solidarity, subsidiarity, and the preferential option for the poor and the vulnerable in society) that can guide decision making in health care are grounded on the integration of these components. In accepting that "according as a man is, such does the end seem to him" (Thomas Aquinas 1948, I-I 9.2), we affirm the crucial role of agapeic love in the moral development of agents, which can transform the horizon of ends guiding public health decisions from one based primarily on fear and a sense of scarcity to one inspired by hope and generosity.

---

[1] William F. Sullivan has used this method as a basis for teaching courses in bioethics at Georgetown University, Washington, D.C., and gratefully acknowledges and thanks many students for their inputs and affirmations.

C. R. Sotomayor (✉) · W. F. Sullivan
Georgetown University, Washington, DC, USA
e-mail: crs145@georgetown.edu; william.sullivan@georgetown.edu

J. T. Eberl (ed.), *Emerging Issues in Catholic Bioethics*, Philosophy and Medicine 153, https://doi.org/10.1007/978-3-032-08862-8_7

7.2 The Relevance of Catholic Social Teaching to Public Health Care

The Centers for Disease Control and Prevention defines public health care ethics as a systematic process involving the clarification, prioritization, and justification of potential public health actions based on ethical principles, stakeholder values, and relevant scientific information (CDC, n.d.). While various frameworks are utilized, discussions tend to be predominantly guided by utilitarian principles. In health care ethics, there has been greater attention paid to ethical questions regarding clinical health care of individual patients than to those regarding public health care, which is concerned with the health of communities and groups. An important question for ethicists is whether principles developed for clinical health care ethics are suitable or adequate for public health care ethics. Schröder-Bäck et al. (2014) proposed six mid-level principles for public health care—non-maleficence, beneficence, health maximization, efficiency, respect for autonomy, and justice—as practical tools for navigating ethical dilemmas. However, according to Holland (2002), addressing these issues requires a more complex engagement with both moral philosophies (e.g., utilitarianism and deontology) and political philosophies (e.g., liberalism and communitarianism).

Catholic social teaching can contribute additional fundamental principles for public health ethics, such as: (1) respect for human dignity; (2) the common good; (3) solidarity; (4) subsidiarity; and (5) the preferential option for the poor and the vulnerable in society. Dignity is the guiding principle, which is extremely important because it conveys a deep regard for every human's inherent value (Wright 2017). The common good, as defined in *Gaudium et spes*, is the total of social conditions that enable individuals and groups to achieve their full potential (Paul VI 1965, no. 26). It involves collaborative efforts among various agents to enhance everyone's well-being while balancing individual rights with the community's responsibilities to manage resources effectively. Two key principles are crucial to achieving the common good: subsidiarity and solidarity. Both principles serve to bridge the gap between secular public health approaches and Catholic social teaching by advocating for collaborative efforts and shared responsibility for society's well-being and flourishing. Subsidiarity, on the one hand, promotes decision-making at the most appropriate level, ensuring that local needs and contexts are addressed effectively while fostering coordinated efforts among various agents. Solidarity, on the other hand, emphasizes a collective commitment to justice and the common good, recognizing interconnectedness and mutual responsibility among all individuals.

7.3 From Principles to Method in Health Care Decision Making

Principles can offer an integrated framework for navigating ethical questions encountered in health care on a general level. Principles by themselves, however, are insufficient for thinking through particular intervention or policy decisions in health care. Such decisions usually concern situations that are unique, complex, and evolving. A *method* by which subjects and agents can incorporate these principles to reach concrete decisions is hence needed. Attending to the components of this method is crucial for discovering how to decide well. But what is the basis of this method? It cannot simply be any arbitrarily chosen method. It must be founded on the experiences of human agents themselves in making decisions. We suggest that there is a normative structure of activities of human knowing, deciding, and loving that one can affirm by being attentive to these activities in one's own experiences of making concrete decisions (Lonergan 1972).

These activities can be grouped together and organized sequentially on four levels. We could call these activities: experiencing, understanding, judging, and deciding. Organizing these activities into four levels emerges spontaneously from the order in which humans raise questions: (1) some awareness of sense stimuli (data of sense) or awareness of ourselves (data of consciousness) directs and sustains our attention. (2) We seek to interpret or understand 'what is this?' (or some such similar question) when we inquire into the presence or absence of a pattern or relation in these data. Not every insight we have in answer to such types of questions, however, is correct. Hence (3) we need to ask regarding our insights still further types of questions, such as, 'is it so?' That is, we must check the data or seek further data to affirm our insight or reach a different understanding. If we are confident of the sufficiency of evidence, then conditions are met for making a judgment of fact, e.g., this exists; this is real (Aquinas 1948, I.12.4). But such judgments could be revised in the future should new data emerge to shift the weight of positive evidence. (4) Our judgments of fact give rise to still further questions regarding the implications of our discoveries: 'is what I have judged *good*?'; what ought I *to do* in light of this knowledge?' (Aquinas 1948 1.87.4) Here the knower moves beyond judging fact to deciding and willing, i.e., to making practical judgments and exercising agency. Human decisions are always concrete, and they involve, crucially, the agent's apprehending something concretely good to protect or promote, or some harm or lack to avoid or mitigate, regarding realities discovered through prior judgments of fact.

Lonergan (1972) proposed that such apprehensions of good are given in moral feeling or an experience of contentment that some concrete decision is aligned with an agent's deep aspirations, core commitments, and priorities. In contrast, uncertain or misaligned decisions can be manifested by felt restlessness, uneasiness, or what some ethicists call 'moral distress' (Campbell et al. 2016; Hardingham 2004). Lonergan distinguishes moral feeling from other kinds of feelings that he calls 'satisfactions', which manifest desires for what is momentarily agreeable or disagreeable to the agent. Moral development of the agent requires expanding a capacity for moral feeling. Moral development also involves lifelong learning to distinguish moral feeling from satisfactions, to expand the horizon of an agent's commitments and to adjust priorities to approximate aspirations for an abundant and flourishing life.

In later works (Lonergan 2004, 2017a, 2017b), Lonergan clarifies that such shifts in horizon occur to the degree that an agent is open to the healing and transformative effects of agapeic love, which is experienced to some degree in authentic social relationships and, unrestrictedly, in one's relationship with God. Agapeic love quells the restlessness, emptiness, and fear that are part of the psychic realities of the human condition. These have the potential to impede or limit our knowing and deciding. In contrast, agapeic love uplifts us with a sense of belonging, fullness, and security, which can enhance our capacities for knowing and deciding and enable us to act lovingly towards others as we ourselves have been loved. The persistence of our questioning and the degree to which we are disposed to experience attentively, seek understanding creatively and intelligently, weigh evidence and judge facts reasonably, and decide responsibly and lovingly will impact the quality and adequacy of our knowledge and decisions. These dispositions to enhanced knowing may be described in traditional terms of virtue ethics as 'intellectual virtues', and those dispositions to enhanced decision making and to carrying out decisions as 'moral virtues' (Osborne 2022). The 'theological virtues' are our disposition to accept agapeic love, to make it effective in our desiring and acting in ways that are motivated by horizons of faith, hope, and love.

All decisions and actions of agents, at whatever stage of moral development, will shape the character of an individual or of a community, and the things, persons, and systems that will be affected by those decisions and actions. As finite agents, we continually learn, through trial and error, how to make better decisions. Ethical principles, on this understanding, are *generalizations* of past good decisions that are affirmed by individuals and communities. Method is fundamental to principles. To regard principles as abstract directives separated from the concrete activities of knowing, deciding, and loving of agents is to put the cart before the horse, so to speak.

There remains, for the purpose of this chapter, the need to elaborate in greater detail what must be considered in attaining good practical judgments and decisions. Lonergan did not elaborate on this. Neither did he apply the normative human structure of activities of knowing and deciding to questions regarding clinical or public health care. These are tasks to which we next turn.

7.4 The G-O-A-LS® Method in Health Care

What is it that we are doing whenever we make a decision to act or refrain from acting? We are *integrating* four determining factors: (1) our apprehending and intending of some concrete good(s) to be protected or promoted (or some harm(s) to be avoided or mitigated) in (2) circumstances known through our judgments of fact, and (3) our deliberating on one or more practical means to achieve these ends. As noted above, and we may verify this in our own experiences, our apprehensions of good are grasped and affirmed by means of moral feeling, which moves us to enjoy it or spurs us to realize it. The extent to which we engage in the activity of deciding depends fundamentally on (4) our opportunities and supports for exercising agency, and crucially, on who we are, based on our history and moral development. Agency and moral development are realized by individuals, but we are always dependent on, inspired, and helped to a certain extent by relationships with others and helped or limited by the structures and systems in which we live.

G-O-A-LS® is an acronym, a mnemonic for remembering the above four factors: Goods, Options, Agency, and Life Systems or circumstances, and the contexts of relationships ® in which this method unfolds. This is not a list of considerations external to the agent that has been devised as a framework to facilitate decision making. These are factors that reflect how our activity of deciding is structured any time we are making a decision. Thomas Aquinas, in the thirteenth century, had already described these factors that we seek to integrate in our moral acts, using the language of his times (Aquinas 1948, I-II 18.2–4 and I-II 55.4). Catholic moral teaching too, building on Thomistic ethics, refers to these same factors. The Catechism of the Catholic Church (2000, no. 1750) establishes that every moral act involves three key components: the subjective intention, the objective act, and the concrete circumstances. Each component must be considered to assess fully the ethical nature of a decision and action. Implicitly there is also a fourth component: the makers and actors of these decisions are moral agents who are embedded in relationships and who are expected to decide and act in ways that promote the good and avoid evil.

Let us specify the four factors as components integrated by the G-O-A-LS® method of healthcare deciding:

> G: In clinical health care, the goods intended are the health or integral development and flourishing of the patient. In public health care this is the common good or "total of social conditions that enable individuals and groups to achieve their full potential" (John XXIII 1961) for health or human flourishing.
>
> O: The options or means are health care interventions or policies that are selected to be most suited and effective for attaining these ends.
>
> A: The agent or agents who should be involved in making health care decisions are those whose health, development, and flourishing are at stake. Such agents should be adequately supported to exercise capabilities for making decisions and for developing the virtues that dispose them to decide well. In public health care, the agents are many, and they need to cooperate harmoniously and effectively to integrate efforts to promote the common good.
>
> LS: 'Life Systems' are the concrete health relationships, structures, and systems that need to be protected, restored, or promoted by health care interventions and policies.

Attending to this G-O-A-LS® method brings to light the reality that our decisions in health care are always concrete and based on considering goods, options, agents, and life systems that can vary. The G-O-A-LS® method, implicit in our decision making, is dynamic, structured, and relational. This method enables us to respond nimbly to unique, complex, and evolving healthcare contexts, while making decisions in diverse contexts in a consistent manner. Making explicit the G-O-A-LS®

method can help us to be more self-aware, systematic, comprehensive, and loving as decision-makers as we more fully realize what is required to make such decisions well.

Note that the same four factors related to making decisions in the G-O-A-LS® method are operative in both clinical and public health care. If we advert to the method of decision making that is common to both domains, we can better harmonize concrete decisions regarding clinical and public health care (See Sect. 5 below).

It could be argued that some of the main normative moral theories that are often employed to guide decisions regarding public health care focus on one or more of the factors in the G-O-A-LS® method but less often on how they are interrelated and integral. For example, deontologists tend to focus on maxims and duties governing the selection of options for acting, while abstracting from considering the dispositions of agents, the particularities of life systems, and the concrete good(s) to be protected, restored, or promoted in them. Utilitarians, in contrast, seek to maximize concrete goods that the majority of agents apprehend and desire, while tending to exclude members of minority groups from participating in decision making and overlooking options that regard specific needs in their life systems. Method, however, is more basic, concrete, and dynamic than ethical theories. Adverting to the G-O-A-LS® method can clarify the grounds on which various ethical theories stand, and it also enables one to better assess their adequacy.

Principlism is an approach in public health ethics that purports to "bypass intractable disagreements at the level of normative ethical theory and the resulting lack of agreement about how to proceed" by means of a framework for decision making based on general, mid-level principles on which there can be agreement among proponents of various ethical theories (Keeling and Bellefleur 2016). As we proposed above in Sect. 3, however, ethical principles are generalizations of good decisions arrived at through authentically applying the G-O-A-LS® method. We next turn our attention to how each component of the G-O-A-LS® method can ground certain principles of Catholic social teaching, such as the dignity of the human person, the common good, solidarity, subsidiarity, and preferential option for the poor and vulnerable in society, to guide decision making in public health care. By doing so, we aim to highlight the method's capacity to connect abstract and static ethical principles with how they can be applied in practice, thereby enriching public health strategies based on the principles of Catholic social teaching with a more robust ethical foundation.

7.5 Harmonizing the Good in Clinical and Public Health Care

Let us begin by focusing on the good ends to be pursued in public health care, as a component of the G-O-A-LS® method. In a society characterized by moral diversity, there will be varying interpretations of what is good to pursue. Aristotle argues that all humans seek *eudaimonia* or happiness (Kraut 2022). Thomas Aquinas (1948, I.5) interprets happiness as realizing full human potential or flourishing, which includes communion with God. On the other hand, hedonists (Moore 2004) identify the good with pleasure, while utilitarians (Driver 2022) seek to maximize the distribution and enjoyment of individual goods for the greatest number of people in society.

Edmund Pellegrino (2008) asserted, however, that the context of health care is unique because it has an "internal morality" conditioned by the patient's experience of illness or vulnerability to illness and the healthcare practitioner's profession to help. Pellegrino held that there can be agreement by all involved parties on the good ends to be pursued in health care, which is to restore or promote the patient to health and a sense of overall well-being, or to protect health by preventing illness and injury. In doing so, the healthcare practitioner, as a moral agent, is obligated to forgo incompatible self-interest in order to promote the patient's good. Pellegrino, however, specified that the patient's good has several dimensions. These include: the patient's *ultimate or spiritual good*, which is the patient's fundamental standard for making life decisions based on the patient's deepest value commitments, such as loyalties, sources of meaning, and purpose*; the biomedical good*, which results from the correct application of medical knowledge and skill; *the patient's own perception of good*, which is the patient's weighing of the benefits of health interventions against risks and burdens in relation to life goals; and *the human good*, which acknowledges the intrinsic worth of the patient and supports the patient's capabilities in decision making. In health care, determining what is right and good involves a nuanced integrating of these various dimensions.

It could be objected that Pellegrino's concept of the fourfold good of the patient as the ends of health care is most readily realized in the context of clinical health care where there is, ideally, a close health care practitioner-patient relationship. Applying this concept to the context of public health care could present significant challenges due to the involvement of multiple agents in decision making, possible disagreements regarding the goods at stake and their priority, and differences in understanding what constitutes the common good. Indeed, tensions could be anticipated between the ends of clinical health care of certain individuals and public health care. For example, the disproportionately high economic costs associated with addressing complex health and social support needs of certain persons with frailty due to old age or with disabilities might conflict with a strictly egalitarian understanding of the common good as equal distribution of health care resources to all. This way of allocating health care resources could also be incompatible with a utilitarian understanding of the common good that involves maximizing benefits for the greatest number of people in society.

We contend, however, that an adequate understanding of the common good recognizes that the same ethical demand for responsible decision making applies both to clinical health care and public health care. Since the same integrative G-O-A-LS® method is basic for decision making in both clinical and public health care, the goods to be sought in both ought to be, and can be, harmonized. Good decisions in clinical health care require considering the implications of healthcare decisions of individuals for the broader community since these are connected. Similarly good decisions in public health care also require considering the implications of public health decisions for the health and flourishing of all individuals affected by such decisions. Flourishing entails fulfilling human potential for integral development and attaining a sense of overall well-being that necessitates but goes beyond attaining the biomedical good (Sullivan et al. 2022).

A well-articulated principle generalizing this harmony of good ends in clinical and public health decision making is the understanding of the common good found in Catholic social teaching. This offers a compelling vision of the common good as the sum total of social conditions that allow people, either as groups or as individuals, to reach their fulfillment more fully and more easily (Paul VI 1965, no. 26). This understanding of the common good opposes other understandings of the common good, such as a strictly egalitarian or utilitarian understanding, on *methodological* grounds. If integrating the four components in G-O-A-LS® entails attending to the particular 'life systems' of groups and individuals, which vary, then an egalitarian understanding of the common good, which demands that society's benefits be distributed equally, and not according to needs, is inadequate. If the good ends sought in engaging in the G-O-A-LS® method is the good of all involved, then a utilitarian understanding of the common good that seeks to maximize benefits for the majority of agents is also inadequate.

In considering the common good in public health care, we should recognize that promoting health and well-being of all is intrinsically linked to cultivating basic life systems, which includes human relationships to Earth's physical and biological ecosystems. In this context, the pursuit of the common good in public health care extends beyond social and political dimensions to encompass protecting the Earth, which is our common home. This commitment necessitates active engagement in both environmental stewardship and social advocacy, reflecting a broader theological understanding of the common good as a response to the divine mandate to care for all creation. Pope Francis, in his encyclical *Laudato Si'* (Francis 2015), calls for a profound reassessment of our perspectives on humanity, life, society, and the natural world, advocating for an integrated understanding of these interconnected domains. He emphasizes that human actions have far-reaching effects on both individuals and the global community, highlighting the imperative to respect and nurture the intricate relationships between people, communities, and Earth's ecosystems. Attaining a balanced harmony among these interrelated domains is essential for achieving sustainable health and integral development and flourishing.

7.6 Selecting Options to Address the Needs of the Poor and Vulnerable in Health Care

A component that is incorporated in the integrated G-O-A-LS® method when applied to public health care is considering concrete options for interventions or policies. These would preserve, restore, or promote aspects of the life systems of members and contribute to the health and flourish of a group or entire population.

As these life systems vary, public health care involves a wide range of interventions. These interventions can vary greatly in complexity, from straightforward measures such as vaccination programs to complex, long-term initiatives like the 12-Step Program for alcohol dependence recovery. Additionally, such programs may be delivered both within the public health care system and in external settings, such as offices of family physicians. They may include incentives, such as performance-based bonuses tied to cancer screening rates to encourage uptake and compliance. Implementing these interventions effectively requires a thorough understanding of the particularities of the life systems of affected or at-risk individuals and communities. As we showed in Sect. 3 above, good decisions build upon correct judgments of fact. This involves not only evaluating empirical data regarding the life systems of groups and populations but also incorporating feedback from and promoting participation in policy decision making by individuals and groups who might be negatively impacted by such decisions, including marginalized individuals and groups in health systems. The G-O-A-LS® method reminds us that such groups must be included as agents in developing public health policies and implementing them for the overall well-being and flourishing of the whole community.

7.7 Promoting Agency Through Respecting Human Dignity and Fostering Solidarity

In broad terms, an agent is a person who is capable of knowing, deciding, and freely acting. ‘Agency’ also can extend to groups who, together, exercise this capacity to know, decide, and freely act. This concept applies across various contexts, including individuals, organizations, and institutions, all of whom can influence outcomes through their decisions and actions (Schlosser 2019). The source of this ability to act, according to the Catechism of the Catholic Church (2000, nos. 1730–1744), is rooted in the divine gift of free will. Such free will enables individuals, or groups, to make responsible decisions and to freely act, or freely refrain from acting, as appropriate. This capacity for choice is intrinsic to human nature. True freedom is realized when the agent’s decisions and actions align with the pursuit of the good. In Catholic social teaching, the exercise of human freedom and responsibility reflects the divine image. Promoting the capabilities of agents to make decisions freely and responsibly, i.e., supporting agents to undertake well the activities of knowing and deciding, as described above in Sect. 3, expresses respect for intrinsic human dignity.

Focusing on human dignity highlights the importance of placing individuals and their capabilities for decision making at the core of public health care. This involves not only addressing immediate health needs but also offering supports for decision making and programs such as education in ethics, in collaboration with faith-based and other community groups, which foster the moral development of agents.

Cahill (2018) underscores that attaining the common good in a diverse society hinges on our ability to unite and collaborate, striving collectively to enhance the well-being of every person within our communities and beyond. The principle of solidarity describes what occurs in communities that function well when agents support one another in pursuing their common good. This principle necessitates recognizing others as integral members of the same human family. In *Sollicitudo rei socialis*, Pope John Paul II (1987, no. 38) articulates that, "Solidarity is not a feeling of vague compassion or shallow distress at the misfortunes of so many people, both near and far. On the contrary, it is a firm and persevering determination to commit oneself to the common good, that is to say, to the good of all and of each individual because we are all really responsible for all." The principle of solidarity is a key component to integrate into decision making in public health care because, in this realm, there is often a tendency to overlook agents and the subjective aspect of the common good, focusing instead solely on objective and empirical evidence, such as epidemiological data. Decisions in public health care, however, are ultimately made by agents who are called to cooperate with one another, but should base their decisions on the common good.

Understanding that morality fundamentally involves relationships between those who act and those who are affected by those actions showcases the relational interdependence of humans (Gray et al. 2009). In *Gaudium et spes*, Pope Paul VI (1965, nos. 23–25) underscores the need to recognize this interdependence among human beings. The social nature of humans makes it clear, he states, that the advancement of both individuals and society is interdependent. For this reason, he insists that social institutions must prioritize the human person, who, by their very nature, relies on social interactions for their development. In *Solicitudo rei socialis*, Pope John Paul II (1987) further developed this concept by asserting that such interdependence is the basis of 'solidarity' and that the goods of creation should be for everyone's benefit. Most recently, in *Fratelli tutti*, Pope Francis (2020a, no. 112) highlights the importance in the Catholic tradition of seeking and pursuing the good of others, which extends to the entire human family and (as we noted in Sect. 5) all of creation. He asserts that this pursuit involves helping individuals and societies mature in moral values that promote development of the whole human person and an integral ecology.

The notion of interdependence is fundamental for ethical public health care. It highlights the link between individual health, which is the focus of clinical health care, and the conditions for such health (e.g., social and ecological determinants), which is the focus of public health care and related systems. What is specific to public health care is that the processes of knowing and deciding, and of implementing decisions, is shared among several agents, each with different contributions to make in these processes. Some agents, for example, make value judgments on the basis of interpreting epidemiological and other facts regarding what life systems to preserve, restore, or promote for the good of health and well-being of all in society, and what harms ought to be avoided or mitigated. Other agents decide on options regarding public health interventions and policies. Still others communicate those policies to the public, and others develop public health care plans to implement those policies, to monitor related practices, and to evaluate outcomes. All these agents, then, are interdependent and need to collaborate effectively with a singular purpose to promote the common good. Successive decisions and actions have cumulative and possibly different impacts on the health and well-being of various members of a community.

7.8 Considering Life Systems and Subsidiarity

Public health care decisions always regard some concrete circumstances (what we called 'life systems' in Sect. 4) that involve the health and well-being of a community. Life systems refer to the concrete interconnected health relationships, structures, and systems that public health care decisions and related policies, plans, and practices seek to protect, restore, or improve. These are concrete, dynamic, complex, and adaptive. Before deciding on possible public health care interventions, agents need to assess adequately these life systems and merits and difficulties of possible interventions and policies that adjust them. As outlined by Pope John XXIII in *Mater et magistra* (1961, no. 140), the principle of subsidiarity in Catholic social teaching recommends considering the most effective level at which decisions should be made, which is typically at the level where the relevant life systems would be best understood and interventions monitored in order to foster a well-ordered society that genuinely prioritizes the common good. In the modern context of epidemiological and mathematical modelling, techniques used to great advantage in public health care, the principle of subsidiarity reminds us that data generated at the grassroots level can best flag the complexity and chronicity of life systems on which public health decisions need to be based.

Subsidiarity is not opposed to the common good. On the contrary, the practice of subsidiarity in decision making promotes an optimal vantage point for assessing the complex interrelations between individuals and public health authorities, ensuring that personal needs and interests are adequately and dynamically supported by health and related life systems. The principle of subsidiarity also facilitates the necessary alignment of collaborating agents within and across relevant sectors of the state, including individuals, institutions, civil organizations, and faith communities (Pope John XXIII 1961, no. 56). Promoting this cooperation helps to generate solidarity by motivating and mobilizing concerted efforts within society to advance the common good, as described in Sect. 7.

It is important to highlight the necessity of integrating solidarity and subsidiarity. Wee (2022) helpfully pointed out that solidarity, as a principle in public health ethics, emphasizes the state's role in fostering a collective responsibility among citizens to prioritize conditions for promoting the welfare of all, extending beyond concerns of individuals or subgroups. This collective orientation is foundational to public health efforts, as the attainment of societal well-being is unlikely without coordinated action. In contrast, the principle of subsidiarity advocates that governance decisions should be made at the lowest feasible level, empowering individuals, families, institutions such as schools, workplaces, and associations representing marginalized groups, to assess and manage acceptable risk levels independently. When solidarity is prioritized without integrating subsidiarity, it risks creating an approach to public health of total risk control, thereby neglecting the inherent uncertainties of life, the impact of public health policies on other aspects of Life Systems than the biological, and the responsibility of individuals and subgroups to navigate acceptable levels of risk in view of the totality of their Life Systems. What we highlight in Wee's analysis is that integrating solidarity and subsidiarity occurs, not on the conceptual level, but on the level of concrete decision making. This is *through the method of integrating the components of G-O-A-LS®* both for arriving at policy decisions for the entire community and in their implementation by individuals and subgroups.

7.9 Agapeic Love and Preferential Option for the Poor and the Vulnerable in Society

The integration of solidarity and subsidiarity should also take into consideration the specific health and related needs of subgroups that are disadvantaged in our health care systems. In Sect. 3, we described the healing and transformative effects on human knowing and deciding of agapeic love or charity, which is experienced to some degree in authentic social relationships and, unrestrictedly, in one's relationship with God. Despite the diverse range of theories and proponents in public health care ethics (Anand et al. 2006; Dawson 2011; Holland 2002), none positively incorporates charity or love in any way foundational to public health care. In contrast, charity or agapeic love is foundational and integral to Catholic social teaching as the cornerstone of Catholic engagement in health care (Wright 2017). For example, Pope Benedict XVI (2009, no. 9) emphasized that, "only in charity, illuminated by the light of reason and faith, is it possible to pursue development goals that possess a more humane and humanizing value."

Agapeic love, however, should not simply be regarded as another principle of Catholic social teaching. Rather, it is the fundamental basis for all the principles, *perfecting the intellectual and moral virtues that agents require to make correct judgments of fact and ethically good decisions*. It also provides the impetus for *moral development* of agents, enabling agents in turn to diffuse this love to others through their various social relationships. On the level of deciding, i.e., in engaging in the G-O-A-LS® method, it can expand the horizons of our apprehensions of concrete goods (G) and the prioritizing of these goods to approximate what contributes to conditions for integral development and the flourishing of all creation.

For example, in a General Audience, Pope Francis (August 19, 2020b) exhorted the faithful to embody the principles of charity and justice, articulating that the virtues of faith, hope, and love inherently necessitate extending aid to those in greatest need. This exhortation underscores the principle of the preferential option for the poor and vulnerable in society. This stance is pivotal within Catholic social teaching. This principle highlights the ethical obligation to prioritize the needs and welfare of marginalized and disadvantaged individuals and groups, thereby reflecting a profound commitment to justice and compassion within the Christian tradition.

The objective of Catholic social teaching with regard to the preferential option for the poor and vulnerable is to create conditions that enable people who are poor and marginalized to flourish. Perkins (2021) described human flourishing as fundamentally linked to recognizing and valuing an individual's uniqueness across biological, psychological, cultural, and spiritual dimensions. It involves achieving human dignity, diversity, freedom, happiness, and holistic well-being, which together represent the ultimate good for a person. This aspiration cannot be realized if fundamental barriers are not addressed, such as hunger, environmental degradation, and insufficient safety nets for people who live with illnesses, frailty, or disabilities, or are unemployed or homeless.

Ethical vulnerability is another type of vulnerability that is often not adverted to in health care. By ethical vulnerability we mean the lack of appropriate supports to make ethically good decisions. Such supports to agents making public health policy decisions are necessary for discerning wisely how the components of G-O-A-LS® ought to come together concretely such that solidarity, subsidiarity, and preferential option for the poor and the vulnerable in society are integrated. In providing such decision-making support, agapeic love can positively influence the moral development of agents, which, as we proposed in Sect. 3, occurs by means of an *affective* transformation. This transformation shifts making public health decisions from being motivated primarily by fear and a sense of scarcity to being inspired by hope and generosity. The effects of this transformation can include transforming understanding the common good in public health care from, for example, maximizing the best outcomes for the majority to a heightened concern for the well-being of all, especially the most vulnerable in society. Below, we provide an example of the difference such transformations could make to the ethical framing of public health interventions and policies during a pandemic and the implications thereby for promoting conditions for the health and well-being with people with intellectual and developmental disabilities as valued members of society. The health needs of such persons tend to be inadequately or inappropriately addressed, which puts them at risk of 'death by indifference' (Mencap 2007, 2012).

During the COVID-19 pandemic, a review by the Care Quality Commission (2021) found evidence that health care practitioners and organizations routinely applied 'do not attempt cardiopulmonary resuscitation' (DNACPR) orders to persons with disabilities, including people with intellectual and developmental disabilities in the United Kingdom. In Oregon, U.S.A., there were reports of such practices also during the same period (Shapiro 2020). In some instances, the existence of these orders led to people with intellectual and developmental disabilities being excluded from emergency intensive care interventions and experiencing excessive delays in accessing regular health care. In Oregan, Senator Sara Gelser, a mother of a young man with intellectual and developmental disabilities, introduced a bill, which became law, to prohibit healthcare practitioners and health care organizations from developing policies requiring all patients with disabilities to sign do-not-resuscitate orders. Gelser stated in an interview with Shapiro (2020) that, in Oregon, "before we even needed to triage, medical systems were deciding on reserving resources for nondisabled people that the system valued more in case they ran out of resources later."

What Gelser suggested was that public health care decisions in Oregon were being made out of an undue fear of scarcity of health care resources. The agents involved in making these decisions did not include people with intellectual and developmental disabilities and their loved ones. Individuals and health care systems decided to ration health care resources based on an understanding of the common good that aimed at maximizing availability of those resources for the greatest number of people in the state, namely, those who do not have intellectual and developmental disabilities.

Understanding the G-O-A-LS® method helps to clarify that the basis of such policies is a focus primarily on certain good outcomes deemed to be the common good (G). But this understanding of the common good resulted in policy options (O) that exclude the health and well-being of people with intellectual and developmental disabilities because, in the words of Gelser, others in society are "valued more" as agents of public health care decisions (A). Haque and Stein (2020) surmise that such value judgments are based on negative attitudes against people with disabilities that stem from ableist bias and ignorance regarding the lived experiences, or life systems (LS), of people with disabilities.

Gelser, who advocated for a law to prohibit exclusionary public health care policies regarding access to resources, had a different understanding of the common good. As a mother who loves her son with intellectual and developmental disabilities, she apprehended the intrinsic dignity and value of people with such disabilities as co-agents and co-beneficiaries of the common good. Although she understood that do-not-resuscitate orders are sometimes ethically appropriate for anyone with a life-limiting illness (i.e., as we have argued in this chapter, when all the components of G-O-A-LS® are adequately considered and integrated in such decisions), she recognized that the exclusionary public health care decisions were also motivated by an excessive fear of scarcity of resources. This tipped the scales in weighing options for rationing. We propose that loving relationships does quell such fears regarding an uncertain future by motivating solidarity (a faith and hope that adaptations to the use of health care resources can be accomplished by cooperations across health care systems and society that would not necessitate exclusionary policies), and generosity that extends to include caring for the health needs of the neglected and disadvantaged in health care systems and society.

7.10 Final Remarks

We have proposed ways in which bioethicists can mobilize the perspectives and resources of Catholic social teaching to critically engage public health care decisions. These involve attending to the basic and normative activities of human knowing, deciding, and loving that are the fundamental grounds for the principles of Catholic social teaching. These principles should not be considered abstractly apart from some method that applies such principles to reach concrete decisions regarding public health care questions that can be adapted dynamically to address unique, complex, and evolving issues. We showed how a correct methodological grounding highlights that wise decisions in public health care require both correct judgments of fact and an adequate understanding of the common good. The adequacy of such an understanding of the common good in turn depends on an adequate account of the interrelated activities of human knowing, deciding, and loving. We discussed, specifically in relation to deciding, the G-O-A-LS® method, which is a normative method based on what agents typically do when they make decisions well. That is, we *integrate* the four interrelated components referred to in the acronym G-O-A-LS® in a context of loving relationships: intending and pursuing some Good (in public health, this involves harmonizing of the individual and common good); selecting appropriate and effective Options that align with this good; developing and exercising our intellectual, moral, and theological virtues as Agents, with supports as needed, and attending to the particularities of complex-adaptive Life Systems. We contend that ethical principles, including those of Catholic social teaching that can guide public health decision making (such as human dignity, the common good, solidarity, subsidiarity, and preferential option for the poor and the vulnerable in society), are generalizations of wise decisions reached through this integrated method. Finally, we described implications for public health care when activities of knowing and deciding are transposed to the key of agapeic loving relationships.

References

Anand, S., F. Peter, and A. Sen. 2006. *Public health, ethics, and equity*. Oxford University Press.

Benedict XVI. (2009). *Caritas in veritate*. http://w2.vatican.va/content/benedict-xvi/en/encyclicals/documents/hf_ben-xvi_enc_20090629_caritas-in-veritate.html.

Cahill, L. S. 2018. Laudato si': Reframing Catholic social ethics. *Heythrop Journal* 59:887–900. https://doi.org/10.1111/heyj.13010.

Campbell, S. M., C. M. Ulrich, and C. Grady. 2016. A broader understanding of moral distress. *American Journal of Bioethics* 16 (12): 2–9. https://doi.org/10.1080/15265161.2016.1239782.

Care Quality Commission. (2021). *Protect, respect, connect: Decisions about living and dying well during COVID-19*. Care Quality Commission. https://www.cqc.org.uk/sites/default/files/20210318_dnacpr_printer-version.pdf.

Catechism of the Catholic Church. 2000. *Catechism of the Catholic Church*. 2nd ed. United States Catholic Conference.

Centers for Disease Control and Prevention. (n.d.). About CDC's public health ethics activities. https://www.cdc.gov/os/integrity/phethics/index.htm.
Dawson, A. 2011. *Public health ethics: Key concepts and issues in policy and practice*. Cambridge University Press.
Driver, J. (2022). The history of utilitarianism. In E. N. Zalta & U. Nodelman (Eds.), *The Stanford Encyclopedia of Philosophy*. https://plato.stanford.edu/archives/win2022/entries/utilitarianism-history/.
Francis. (2015). *Laudato si'*. https://www.vatican.va/content/francesco/en/encyclicals/documents/papa-francesco_20150524_enciclica-laudato-si.html.
Francis. (2020a). *Fratelli tutti*. https://www.vatican.va/content/francesco/en/encyclicals/documents/papa-francesco_20201003_enciclica-fratelli-tutti.html.
Francis. (2020b). *General audience*. https://www.vatican.va/content/francesco/en/audiences/2020/documents/papa-francesco_20200819_udienza-generale.html.
Gray, K., L. Young, and A. Waytz. 2009. Moral typecasting: Divergent perceptions of moral agents and moral patients. *Journal of Personality and Social Psychology* 96 (3): 505–520.
Haque, O. S., and M. A. Stein. 2020. COVID-19 clinical bias, persons with disabilities, and human rights. *Health and Human Rights* 22 (2): 285–290.
Hardingham, L. B. 2004. Integrity and moral residue: Nurses as participants in a moral community. *Nursing Philosophy* 5 (2): 127–134.
Holland, S. 2002. *Public health ethics*. John Wiley & Sons.
John Paul II. (1987). *Sollicitudo rei socialis*. https://www.vatican.va/content/john-paul-ii/en/encyclicals/documents/hf_jp-ii_enc_30121987_sollicitudo-rei-socialis.html.
John XXIII. (1961). *Mater et magistra*. https://www.vatican.va/content/john-xxiii/en/encyclicals/documents/hf_j-xxiii_enc_15051961_mater.html.
Keeling, M., and Bellefleur, O. (2016). "Principlism" and frameworks in public health ethics. National Collaborating Centre for Healthy Public Policy. http://www.ncchpp.ca/docs/2016_Ethics_Principlism_En.pdf.
Kraut, R. (2022). Aristotle's ethics. In E. N. Zalta & U. Nodelman (Eds.), *The Stanford Encyclopedia of Philosophy*. https://plato.stanford.edu/entries/aristotle-ethics/.
Lonergan, B. J. F. 1972. *Method in theology*. Longman and Todd: Darton.
Lonergan, B. J. F. 2004. Philosophy and the religious phenomenon. In *Philosophical and theological papers, 1965–1980: Volume 17*, ed. R. Croken and S. J. R. Doran, 391–409. University of Toronto Press.
Lonergan, B. J. F. 2017a. A post-Hegelian philosophy of religion. In *A third collection: Volume 16*, ed. S. J. R. Doran and J. Dadosky, 194–213. University of Toronto Press.
Lonergan, B. J. F. 2017b. Natural right and historical mindedness. In *A third collection: Volume 16*, ed. S. J. R. Doran and J. Dadosky, 163–176. University of Toronto Press.
Mencap. (2007). *Death by indifference*. https://www.mencap.org.uk/sites/default/files/2016-07/DBIreport.pdf.
Mencap. (2012). *Death by indifference: 74 deaths and counting*. https://www.mencap.org.uk/sites/default/files/2016-08/Death%20by%20Indifference%20-%2074%20deaths%20and%20counting.pdf.
Moore, A. (2004). Hedonism. In E. N. Zalta & U. Nodelman (Eds.), *The Stanford Encyclopedia of Philosophy*. https://plato.stanford.edu/archives/win2019/entries/hedonism/.
Osborne, T. M., Jr. 2022. Intellectual and moral virtue. In *Thomas Aquinas on virtue*, 44–81. Cambridge University Press.
Paul VI. (1965). *Gaudium et spes*. https://www.vatican.va/archive/hist_councils/ii_vatican_council/documents/vat-ii_const_19651207_gaudium-et-spes_en.html.
Pellegrino, E. D. 2008. *The philosophy of medicine reborn: A Pellegrino reader*. University of Notre Dame Press.
Perkins, I. 2021. Human flourishing in Christ: A journey of hope in the Christian community. *The Linacre Quarterly* 88 (2): 175–183. https://doi.org/10.1177/0024363920920409.

Schlosser, M. (2019). Agency. In E. N. Zalta & U. Nodelman (Eds.), *The Stanford Encyclopedia of Philosophy*. https://plato.stanford.edu/entries/agency.

Schröder-Bäck, P., P. Duncan, W. Sherlaw, C. Brall, and K. Czabanowska. 2014. Teaching seven principles for public health ethics: Towards a curriculum for a short course on ethics in public health programmes. *BMC Medical Ethics* 15 (1): 73. https://doi.org/10.1186/1472-6939-15-73.

Shapiro, J. (2020). Oregon hospitals told not to withhold care because of a person's disability. *NPR*. https://www.npr.org/2020/12/21/948697808/oregon-hospitals-told-not-to-withhold-care-because-of-a-persons-disability.

Sullivan, W. F., J. Heng, J. T. Eberl, et al.. 2022. Ethics and pandemics: A statement of the IACB on integrating ethical approaches to clinical care and public health. *National Catholic Bioethics Quarterly* 22:337–352.

Thomas Aquinas. (1948). Summa theologica (Fathers of the English Dominican Province, Trans.). Christian Classics.

Wee, Michael. 2022. Solidarity and subsidiarity as principles for public health ethics. *National Catholic Bioethics Quarterly* 22:221–229.

Wright, K. S. 2017. The principles of Catholic social teaching: A guide for decision making from daily clinical encounters to national policy-making. *The Linacre Quarterly* 84 (1): 10–22. https://doi.org/10.1080/00243639.2016.1274629.

Chapter 8
Global Health Ethics

Henk ten Have

8.1 Introduction

The word 'bioethics' was introduced in the intellectual discourse in the early 1970s. The term was first used by Van Rensselaer Potter (1911–2001), a scientific researcher in oncology at the University of Wisconsin in Madison in the US (Potter 1970).[1] Around the same time, it was used by Andre Hellegers as the initial name of the first institute in this new area at Georgetown University at Washington, DC. The term 'bioethics' became quickly adopted and widely used. In 1973, for example, Dan Callahan published "Bioethics as a Discipline" (Callahan 1973). In fact it was an ideal term to designate a new movement, away from the traditional medical ethics, and referring to an innovative discipline that was open for experts from a broad range of other disciplines. For Potter, the new discipline of bioethics is modelled on his experiences in oncology, which essentially is an interdisciplinary field. In explaining cancer, it is necessary to go beyond the level of individual persons and beyond the medical perspective, since cancer is often associated with social conditions, as well as lifestyle and environmental influences. In the fight against cancer, there will be some limited progress at the individual level (in terms of alleviation of suffering and improved treatment) but much more can be accomplished at the level

[1] Hans-Marin Sass (2007) has argued that the term 'Bio-Ethik' has been first used by German Pastor Fritz Jahr in 1927, indicating that ethics should apply to all forms of life, and not only humans. However, the term had no impact outside of limited German circles at that time.

H. ten Have (✉)
Anahuac University, Mexico City, Mexico

Duquesne University, Pittsburgh, PA, USA
e-mail: tenhaveh@duq.edu

J. T. Eberl (ed.), *Emerging Issues in Catholic Bioethics*, Philosophy and Medicine 153, https://doi.org/10.1007/978-3-032-08862-8_8

of populations (in terms of prevention, for example, through restrictions on smoking). Similarly, in health care in general many challenges in the 1970s also transcended the individual and medical levels (Potter 1971, 150). Although he did not systematically discuss them, Potter listed the priority problems as: population (and thus malnutrition and hunger), war, pollution (or broader, environmental degradation and biodiversity loss), poverty, politics, and the negative side effects of the idea of progress. He regarded these problems, which are in fact also dominating the current agenda of global problems, as jeopardizing the survival of humankind, and their urgency induced in him a growing concern regarding the future. What was necessary, therefore, according to Potter, was a new science of survival, a new discipline that he called 'bioethics'. This discipline should replace or supersede traditional medical ethics, which is primarily focused on individual and medical perspectives, and mostly practiced and controlled by medical professionals alone.

8.2 From Bioethics to Global Bioethics

The new discipline should bring together knowledge from different domains: biological knowledge or the science of living systems (hence "bio"), and knowledge of human value systems (hence "ethics"). The goal of this discipline would be wisdom. Already in his first publication on bioethics, Potter defined wisdom as "knowledge of how to use knowledge" for human survival and for improvement of the quality of life (Potter 1970, 127). Wisdom is action-oriented; it is a guide for action.

The rapidity with which the word 'bioethics' was disseminated in both academic and public discourse surprised Potter. However, he also saw that it was used to demarcate the activities of ethics experts from the traditional discourse of medical ethics without incorporating a really new approach as he had advocated. He complained that, although using the word 'bioethics' suggested innovation, the ethical practice remained business as usual. Rather than a new approach, bioethics developed as an "outgrowth of medical ethics" (Potter 1988, 1). First, it was solely concerned with the perspective of patients: how can their lives be enhanced, maintained, and prolonged through the application of medical technologies? Second, it was exclusively interested in the short-term consequences of medical and technological interventions as well as the prolongation of our current individual existence. Third, it was unrelated to social, cultural, political and environmental determinants of human life.

To emphasize that a broader and more inclusive approach in ethics is needed, Potter started to use a new term: 'global bioethics' (Potter 1988). He applied the idea of his former university colleague Aldo Leopold, an American pioneer in wildlife conservation, that there are three stages in the development of ethics. In the first stage, ethics concerns the relations between individuals, in the second stage it focuses on the relations between individual and society, and in the third stage, which does not yet exist, ethics would deal with the relations of human beings with their environment, i.e., land, animals and plants. Potter was convinced that the rise of global bioethics heralded the emergence of Leopold's third stage of ethics.

8.3 The Global Dimension of Bioethics

Global bioethics in the vision of Potter unites two meanings of the word 'global' (Potter 1988). First, it is a system of ethics that is worldwide or planetary in scope. Second, it is broad and comprehensive. The first meaning emphasizes that contemporary bioethics is confronted with a different type of challenge as before. Bioethical discourse used to focus on problems that emerged in the clinical setting of individual patient care, as well as those that were associated with medical research and health policy within a specific country or group of countries. Due to the globalization of health care and medical research, numerous new moral problems were recognized that impacted health and disease: poverty, hunger and malnutrition, migration, environmental degradation and climate change, war and violence, deficient governance, corruption and ethno-nationalism, organ trade, medical tourism, and pandemics. Such problems affect the whole of humankind, regardless where people live. What is at stake is the health and survival of humanity, not simply individual health and well-being. Challenges not only cross borders but concern and threaten the planet as a whole. Even if such problems exist only in few countries, or emerge first in a specific region of the world, the way they are addressed will have consequences for other countries. Usually the transactions and interconnections between developed and developing countries can either exacerbate or diminish the impact of such problems on society and culture. Often national legislation, regulation, or policies will not be sufficient but global cooperation and action will be required. Even if the moral values in specific countries and regions differ, a common ground has to be found as a world community in order to cope with these global challenges (Ten Have and Gordijn 2014). The global nature of contemporary ethical challenges is well-illustrated during the recent COVID-19 pandemic. Even if the SARS-CoV-2 virus originated in a specific locality, it cannot be contained locally but spreads around the globe. Like other global phenomena, the pandemic is not a local, national, and regional but global threat, requiring global responses. Other challenges such as climate change, poverty, and loss of biodiversity do not emerge in a specific place but are disproportionately affecting some countries and regions which can only cope with them through global cooperation and action. These problems are fundamentally different from the ones that have confronted bioethics thus far, not only because of their global nature but also because they are intimately connected to processes and practices of globalization which have characterized international relations.

This last observation refers to the second meaning of 'global'. The planetary dimension of today's moral challenges requires a new, broader, and inclusive approach to ethics. To examine and cope with these issues bioethics should develop an encompassing and comprehensive perspective, combining traditional professional (medical, clinical, and nursing) ethics with ecological concerns and the larger problems of society. The need for a transformation of bioethical discourse is forcefully underlined in *Laudato Si'* (Francis 2015). Imagining the planet as our common home, it argued that ethical concerns should not only focus on individualistic and anthropocentric issues but should include social and environmental concerns in order to preserve the planet. In a address to the Pontifical Academy for Life, Pope Francis (2018) advocated for a global vision of bioethics. He pointed out that we need an ethics today that recognizes the vulnerability and frailness of each individual, that is based on the dignity of every person in every phase and condition of existence, and that provides responses of solidarity and justice. With this vision, Pope Francis exactly reflects the transformation of contemporary bioethics.

This implies more than simply declaring that today's problems are global and affect everyone. It implies that in ethical reflections we connect individual, social, and environmental issues. First, it requires interdisciplinary cooperation. Global problems such as poverty, climate change, and inequities in health care can only be addressed by obtaining and applying different types of knowledge. It is unavoidable to bridge the gap between science and humanities. Secondly, it requires that diverse perspectives must be used to explain and understand complex phenomena. Global problems can no longer be approached only from an exclusively Western or Eastern perspective. Health care will not be improved by simply importing and applying medication and technology; we need to understand the existing value systems in which they are employed, as well as the global mechanisms of economic policies and practices that promote their use across the world. Various methods and theories will therefore be used in global bioethics. It also needs input from cross-cultural, empirical studies as well as philosophical analysis. This reflects the idea that the global and local levels in ethics are connected. Global bioethics principles need to be implemented at local levels, which assumes that some principles will be more important than others when the local context is considered. For example, in developing countries, principles of justice, benefit sharing, and social responsibility are primary perspectives to examine moral issues, while in other countries with broad and inclusive health care systems, the principle of respect for autonomy will receive priority, although vulnerability will also be a major principle. Some principles, such as protection of the environment and protection of future generations, will however be significant in all countries given the global threat of climate change.

These two dimensions of bioethics as a global enterprise determine the distinctive nature of global bioethics. It is not simply international bioethics. Ruth Macklin (2022), for example, argues that global bioethics explores "the ethical aspects of relations between and among nations or regions of the world" (Macklin 2022, 11). She correctly points out that specific ethical problems arise in international relations and gives several examples, such as the use of "double standards" in international health research. However, what makes global bioethics different from international bioethics is the context of globalization which not only intensifies these relations but transforms them with pervasive implications for the lives of people, wherever they are located. More is needed therefore than simply exploration of the relations between nations and regions, because moral challenges arise due to the varying impact of processes of globalization on citizens across the world. Of course, solidarity, cooperation, and diplomacy are crucial but not merely between and among nations. The World Health Organization (WHO), given a central place in Macklin's article, clearly has a role in promoting global bioethics, but many other international organizations can make essential contributions.[2] At the same time, the impact of such organizations is limited, and global bioethical perspectives are often advanced at regional, national, and local levels by non-governmental organizations, civil initiatives, ethics committees, and individual experts.

Restricting global bioethics to international bioethics is inadequate for two reasons. One is the nature of contemporary ethical challenges. As the COVID-19 pandemic illustrates, these challenges have a specific character as global phenomena. To address these problems demands critical analysis of these processes and their inherent moral values and perspectives, and thus a broader bioethical approach. It is true that the dominant discourse of bioethics nowadays is likewise concerned with international relations, assuming that its methodology and ethics frameworks apply to all nations and regions. It is no longer, as Macklin rightly observes, concerned with ethical issues arising in a particular country or in the developing world but has a global outreach. Nonetheless, recognition that nowadays bioethics is facing another set of problems due to globalization does not necessarily imply the recognition that a different methodology with a wider set of theoretical perspectives is required. The need of a broader approach is manifested in current publications that question the dominant discourse of bioethics during the pandemic with its emphasis on individual autonomy, and neglect of solidarity, vulnerability, and inequity (Ho and Dascalu 2020; Fins 2020), while it is also argued that anthropocentric approaches are no longer sufficient to resolve public health problems (Bishop and Fitzgerald 2020). Global bioethics differs from mainstream bioethics since, as a response to a specific set of problems, it has a different moral orientation. Previously, I have defined global bioethics as "The study of global ethical problems related to health, healthcare, health science and research, and health technologies and policies, and the activities, practices and policies to influence and resolve these global problems" (Ten Have 2016, 243). But this definition only specifies the first dimension; it describes the field of work rather than the broad moral orientation. It does not articulate that global bioethics is inspired by the ideals of cosmopolitanism: the unity of humanity, solidarity, equality, openness to differences, and focus on what human beings have in common. These ideals consider each human being as a citizen of their own community or state (*polis*) as well as at the same time as citizen of the world (*cosmos*). In the first, they are born, sharing a common origin, language and customs with co-citizens. In the second, they participate because they belong to humanity; all human beings share the same dignity and equality. Being a citizen of the world liberates the individual from captivity within categories such as culture, tradition, and community, but also gender and race. Cosmopolitanism

[2] For example, the Food and Agriculture Organization (FAO), the United Nations Educational, Scientific and Cultural Organization (UNESCO), and the World Intellectual Property Organization (WIPO).

acknowledges that human beings are connected to other beings and the environing world. Potter employs the image of a "bridge" to refer to connectedness as the familiar experience of globalization. This anthropological experience (the 'togetherness' frequently highlighted during the pandemic) should be the starting-point for ethical reflections that take human relationality more seriously. If humans do not merely interact with each other but belong together and are mutually dependent, taking responsibility and shaping their lives in sharing the world, relationships with others thus define who a person is. This relational context is often expressed in non-Western cultures, for example in the African philosophy of Ubuntu: "I am because we are" (Chuwa 2014). Against this background, individual autonomy should then be redefined as a relational concept. Community, mutual support, social responsibility, cooperation, and solidarity should have a significant role in inclusive and comprehensive bioethical discourse. Furthermore, being situated in a web of connections is a precarious experience. Because their bodies position them in the world, human beings are exposed to the world and other persons, necessarily implying vulnerability.

8.4 A Global Ethical Framework

It has been suggested that global ethics is a two-level phenomenon (Kymlicka 2007). At the abstract level there is the international human rights discourse defining a minimum set of standards agreeable to all. At the contextualized level, there is a multiplicity of different ethical traditions. These 'local' traditions define what is ethically required beyond and above human rights. The same distinction can be used for global bioethics. On the one hand, there is a set of standards on which traditions and cultures agree; this is elaborated into specific bioethics principles in connection to international human rights language. On the other hand, there are many efforts to articulate more specific bioethics standards in the context of particular religious and cultural traditions. Members of these traditions also bring their views in the global debate through constructive dialogues and sometimes negotiations, so that the dialectic of global and local also helps to construct and produce global bioethics. Thus, the universal principles of global bioethics are the result of continuous and multilateral articulation, deliberation, and production. This is exemplified in the *Universal Declaration on Bioethics and Human Rights* adopted by UNESCO member states in 2005 (UNESCO 2005; Ten Have and Jean 2009). The request to develop a common framework of ethical principles in bioethics was explicitly made by developing countries. They were afraid that, with the rapid evolution and globalization of medical science and research, they would insufficiently benefit from the advances and suffer too many harms and risks. A major concern was that international medical research and healthcare endeavors would proceed along double standards so that people in developing countries would receive substandard care and be involved in clinical trials without the ethical protections that exists in developed countries. The adoption of the Declaration shows that agreement could be reached on fifteen global bioethics principles. However, these universal principles need to be interpreted and applied in specific local settings of different cultures and traditions. The abstract and contextualized levels are therefore interacting along bottom-up and top-down lines of communication. That this may create difficulties is clear, for example, in cooperation between Institutional Review Boards between different countries, or in applying informed consent in cultures where community consent is the prevalent practice (Meslin et al. 2014; Al 2021). Global platforms and local contexts mutually help each other to construct and produce global bioethics. Thus, global bioethics is the result of continuous and multilateral articulation, deliberation, and production. While global and local levels can be distinguished as two levels of ethical discourse, they are not separated but entangled. The local and global, according to French philosopher Edgar Morin, are implicated in each other (Morin 1999). When the local is isolated from the whole of which it is a part, it becomes abstract. Likewise, as soon as the global is detached from its localized parts, it is converted into an abstract notion. Since global and local are dialogically and recursively related, there is no opposition between the two. Global policies do not imply that local approaches cannot be heterogeneous, while at the same time local policies are contributing to global approaches. The same is true for people who are citizens of a particular nation whilst citizens of the world; being cosmopolitan does not entail that we are rootless, abstract individuals.

As argued above, in Potter's vision contemporary bioethics should be oriented towards the future; a bridge between the present and the future because, for the survival of humankind, it is vital to focus on long-term interests and goals. This is also a major concern in *Laudato Si'*; since everything is connected, and the living and non-living world are entangled, a holistic perspective is needed, relating scientific, environmental, economic, social, cultural, and everyday life ecologies, in other words an 'integral ecology' (Francis 2015). In this perspective, the ultimate goal of bioethics is long-term global human survival. For this purpose, new knowledge and old wisdom should be combined. Otherwise, the pressing global problems of today cannot be properly addressed. Another point is that the goal of human survival can only be accomplished when there is a balanced social and ecological system for human life to flourish. Bioethics therefore should do more than present an ethical framework focused on the individual good; it should also refer to the common good as an anchor for social justice and environmental sustainability, so that it can help to shape the future of our planet (Francis 2015).

The ethical principles declared in the *Universal Declaration on Bioethics and Human Rights* provide precisely such an expanded framework (UNESCO 2005). The principles determine the different obligations and responsibilities of the moral subject (*moral agent*) in relation to different categories of moral objects (*moral patients*). The principles are arranged according to a gradual widening of the range of moral objects: the individual human being (human dignity; benefit and harm; autonomy), other human beings (consent; privacy; equality), human communities (respect for cultural diversity), humankind as a whole (solidarity; social responsibility; sharing of benefits), and all living beings and their environment (protecting future generations; protection of the environment, the biosphere, and biodiversity).

Although no hierarchy among principles is implied, this ordering means that the focus of ethical concern is not restricted to an individual point of view; ethical concern should equally take into account common social and ecological perspectives. Global bioethics therefore cannot be merely medical, social, or environmental; it should include all perspectives at the same time. At a global level, bioethics should have a wider focus and be concerned with humanity as a whole.

8.5 A Different Philosophical Anthropology

The framework of ethical principles articulated in the UNESCO Declaration obviously depart from dominant neoliberal values such as self-interest, individualism, competition, rational choice, and free trade, or at least frame them in a wider conception of what is ethically relevant (Ten Have and Pegoraro 2022). A broader notion of bioethics first of all emphasizes the notion of human person. Human beings are not regarded as abstract and de-contextualized individual entities. They are related to each other and socially connected. This interpretation does not reject the significance of individual autonomy, but uses a broader concept of autonomy as relational and as constituted through interactions in social life. The context of personal existence (especially human vulnerability) should be taken into account when bioethical analyses are produced. Second, society is not an aggregate of individuals. It provides the conditions that make human flourishing, particularly health, possible. A positive image of society articulates the importance of cooperation (rather than competition), social responsibility (rather than individual responsibility), and solidarity (rather than private interests). Third, the focus on the common good emphasizes that human beings are living together and sharing resources that are essential to all of them in order to flourish and survive. Common interest refers not to the summation of private interests. The neoliberal assumption of *homo economicus*, i.e., that human beings are driven by rational self-interest, is rejected; they are citizens concerned with the common good. Fourth, if human beings can only flourish within relationships, it is important to continuously change and improve the conditions in which humans exist. This requires collective action and engagement, not only at the global level (with increased cooperation and global networking) but also at local levels (with local activism). Finally, ethics is primarily a social enterprise. Human interactions are not simply communications and transactions between autonomous individuals. They are intersubjective and reciprocal. They involve the sharing and exchanging of perspectives. Ethics as social ethics designates not only a different field but also a different frame for thinking and acting. The main purpose is to widen the scope of ethical considerations from individual to social, environmental, cultural, and economic concerns.

The different normative frame of global bioethics is based on a specific philosophical anthropology, implying a distinct view of humans and human existence, emphasizing relationality, the common good, and solidarity.

8.5.1 Relationality

Global bioethics articulates that human persons are essentially characterized by relationality. As integrated wholes of body and soul they are embedded within communities and they exist in a web of relationships with other beings and the environing world. This is why the notion of 'sphere' is more appropriate than 'globe.' Relationality is a more fundamental characteristic than relatedness and connectedness. A person is continuously engaging in relations but this is often conceived from the viewpoint of the individual. The notion of relationality expresses that individuals not merely connect and interact with each other but belong together and are mutually dependent, taking responsibility and shaping their lives together. The first experience of humans is that the world is shared with others. From this perspective, individual autonomy is redefined as 'relational autonomy.' A human person is constituted through encounters and dialogues with other beings. Authentic human being is being-together; in the words of Gabriel Marcel: being present and available to others (Marcel 1962). Relationality and being situated in the world implies vulnerability since it exposes humans to other persons and the environing world. Relationality is not an option and we cannot make ourselves immune to the world. The world, as Pope Francis (2015) has said, is a web of relationships.

It is evident that relationships and relatedness have become problematic during the COVID-19 pandemic. Other people are presented as a threat, and relations may have lethal outcomes since humans are the principal vectors of the virus. The main objective of public health measures is to prevent connections and interactions. Distancing, masking, prohibition of visits, working remotely, and sheltering at home obstruct being too close together with other persons. COVID-19 therefore seems to affect the anthropological condition of human beings. They risk to have their presence and availability reduced, and thus to lose what is specific for humanity. All people face the same dilemma between being secluded or being open to the world since relationships are disrupted, but their fundamental relationality is not annulled. For many people public health measures created significant problems, physical ones because they have difficulties in providing for their basic needs, and mental ones because they are lonely and depressed. This rendered the continuation of isolation policies increasingly problematic.[3] It also explains why the term 'social' distancing is considered inappropriate, 'bubbles' appeared in which closeness and intimacy with at least some others was allowed, and many other ways of interaction and communication emerged (Kaminsky 2020).

[3] Aaron Kheriaty (2020) has called this the "other pandemic," i.e., a mental health epidemic resulting from lockdowns.

8.5.2 Common Good

The opposition between individual and common interests that often dominates in pandemic discourses ignores the fundamental relationality of human beings. Individuals are not isolated, abstract entities but social beings. This point of view is not accepted in the ideology of individualism, prevailing especially in the West, according to which human beings are independent and self-reliant, masters of their own life, choosing their own values, and thus as unique individuals separated and demarcated from other beings. The normative implication of this view is that respect for individual autonomy means non-interference: individual decisions and actions should be respected as long as they do not harm other human beings. In this perspective, public health measures should first appeal to individual responsibility; any interference with personal liberty is problematic, and lockdowns and curfews are unacceptable. In the perspective of global bioethics, however, the opposition between individual and common interests is false because the first type of interests must be reinterpreted, while the last type should be taken seriously. One argument is that personal autonomy is a relational notion. Not only has it originated and been nurtured within a context of dependency but it is also always exercised in interaction with other people, dependent on social and cultural conditions (Mackenzie and Stoljar 2000). Another argument is that preferences, values, and beliefs are not merely individual but conditioned by the social context. Societies transmit values across generations because norms are internalized. The human capacity to internalize norms means that human preferences are socially 'programmable' and human behavior is guided by the moral values of social life. Because human agents are socially entangled and networked, their conduct cannot be explained by self-regarding rationality directed at maximizing self-interests but by social rationality, i.e., taking into account the well-being of other people and the needs of larger society (Gintis 2017). A further argument, especially expressed in global bioethics documents is that autonomy is intrinsically connected to responsibility. Individual actions and decisions have social consequences, so individual autonomy and social responsibility cannot be opposed. Personal autonomy is not abstract and decontextualized but has impacts on concrete other people (WeCope 2020).

The COVID-19 pandemic clearly illustrates that individual behavior affects the well-being of the community. Widespread use of face masks will protect not only the individual but also other people against possible infection. Testing will identify whether someone is infected, but it is a warning signal that others may be at risk. The aim of vaccination is not just to protect individuals but society as a whole. In a public health emergency, appeals to self-interest cannot be separated from concerns with the interests of others. Individual decisions whether or not to adhere to public health measures have an inherently social dimension. Appeals to individual responsibility will therefore not be sufficient without articulating social responsibility, and without creating the social, political, and economic conditions for the exercise of responsible autonomy. That individualistic policies fail without this dimension of social responsibility is evident in debates concerning quarantine, isolation, lockdown, and distancing where it is argued that human dignity and human rights are violated. In these debates, dignity and rights are frequently regarded as notions that apply strictly to individuals. Human dignity is considered as a theoretical and abstract construct, an intrinsic quality that applies equally to every human being. It does not depend on human characteristics or conditions such as age, gender, or disease. It cannot be diminished or taken away by any authority or political system, or disregarded in emergency conditions. Dignity does not depend on whether it is recognized or respected since it continues to exists even in the most dismal or cruel circumstances. However, this is only part of the story of human dignity. It is also a practical experiential phenomenon, a lived experience. It refers to how humans behave and are treated; human dignity is thus a relational quality. In this perspective, dignity can be disrespected, lost, or destroyed. In certain situations and practices, humans experience threats to their dignity and are confronted with undignified conditions (Bieri 2017). Human rights are the focus of similar discussions. Sometimes they are interpreted as rights of individuals, especially emphasizing non-interference to protect individuals against the state, and thus regarding civil and political rights as more important than social and economic rights. In this view, moral individualism is at the core of human rights language as the discourse of individual empowerment. However, all human rights are interdependent. Civil and political rights cannot be exercised if basic conditions for human existence are not provided, as expressed by social and economic rights. Individual persons can only be empowered within a relational context with others. The right to health illustrates that the individual dimension of human rights is connected to a dimension of solidarity and collective good: if appropriate conditions such as access to health care and quality health services do not exists, individuals cannot enjoy their right to health. Just health care policies therefore require that governments take action to realize this right (Eberl et al. 2011). Like human dignity, human rights are based on the recognition that human beings share fundamental needs and vulnerabilities (Ten Have, 2016, 113ff).

The above arguments apply to another opposition which has intensified during the pandemic: nationalism versus globalism. In emergency conditions, national interests dominate the approach to COVID-19. Countries are first concerned for their own citizens, and try to seize as much protective equipment, masks, and testing materials as possible, often in competition with each other, and without consideration for the needs of other, less powerful and economically weaker nations. Vaccines are pre-ordered and purchased in enormous quantities by high-income countries, leaving other countries at the end of the queue. The WHO (2021) and international actors have argued multiple times that national policies will not be sufficient to control the pandemic as long as a global approach is missing. The arguments are familiar: nationalism will hurt everybody and is self-defeating because all people are connected. The virus does not recognize borders and nations but affects the global population and requires global solutions. Even if the virus can be eliminated in one country, trade and travel will remain affected, economies will not recover, and stability and prosperity will not return as long as the virus is rampaging across the world. It is therefore in the interest of each nation to engage in global efforts to address COVID-19 (Eurasia Group 2020). Nationalist approaches are not only narrow-minded but also self-defeating in the longer run when the virus continues to disseminate and mutate in some parts of the world.

8.5.3 Solidarity

In the context of public health, solidarity has since long been endorsed as a key ethical value. Because health systems are interdependent, and disease threats are global, collaboration between health care institutions is necessary at national, regional, and global levels, requiring open communication, sharing of information, and coordination of policy responses. In the COVID-19 pandemic, international bodies have repeatedly emphasized solidarity as a core concept. The ethical committees of UNESCO call it "an ethical duty to build solidarity and cooperation" (UNESCO 2020, 4) while the ethics advisors of the European Union refer to solidarity as "a social vaccine" against indifference and exclusion (European Group on Ethics in Science and New Technologies 2020, 4). Remarkably, the WHO's Working Group on Ethics and COVID-19 lists solidarity as the first ethical principle to apply (Dawson et al. 2020). The Vatican COVID-19 Commission states that the principle of solidarity must be the basis of any specific and concrete intervention in response to the pandemic, which implies that vaccines must be available and accessible to all (Pontifical Academy for Life 2020).

Solidarity is often explained with references to the same grounds as the notion of relationality: it is based on the mutual recognition that human beings share the same needs, that their destiny is interconnected, that vulnerabilities are crucial human features but not equally experienced, and that the well-being of all citizens of the world should be the primary concern of global policies. Yet, what is typical for solidarity are not just these theoretical explanations, but its practical implications. Rather than a feeling of connectedness, and intentions to act, solidarity shows itself in supporting a specific cause and in common action, recognizing that capacities to cope with global threats are unequal. It requires that understanding interdependency and willingness to assist others translates in public action, demonstrating that one's own interests are subordinated to those of others. Such action can be motivated by mutual self-interest, especially in a pandemic where it is in everybody's interests to reduce and eliminate infections; but the core of solidarity is moral concern for others, selfless commitment to the other rather than the expectation of unilateral benefit. It is not manifested because other people are a threat to our health, but because global health is connected and interdependent. Solidarity differs from charity, aid, and generosity: it signifies mutuality, a symmetrical relation between equals, and implies therefore inclusion and cooperation (Jennings and Dawson 2015).

Although there have been many examples of solidarity at interpersonal and institutional levels, the absence of solidarity at the global level during the pandemic is striking. This is not surprising since the conditions for solidarity have been eroded in the past few decades. Global policies and international cooperation have primarily focused on economic interests. For example, in the European Union, protection of human health has not received priority since the organization and delivery of health services and medical care is the primary responsibility of individual member states. Global institutions such as the WHO have been systematically weakened by budget cuts and attempts to delegitimize its work (Brown and Susskind 2020). In most countries, public health infrastructure has been reduced, and health is first of all regarded as an individual rather than collective responsibility. The main driving force for cooperation is the neoliberal ideology of the free market, emphasizing competition, free trade, and commercialization of all aspects of human life. In this ideology, government interference must be reduced as much as possible, and deregulation, privatization, and reduction of taxes and public expenditures encouraged. In this philosophy of rational egoism, societies are mere collections of individuals, and solidarity is rejected or regarded as a superfluous value. The same processes have undermined solidarity within societies. The dominance of individualism and the view of the human person as *homo economicus* have diminished the experience of human beings as embedded within communities, cultures, and environments, and the consciousness that their destiny is connected to distant others as citizens of the world. Since solidarity cannot be imposed unilaterally or top-down, it will not emerge in these conditions (West-Oram and Buyx 2017).

This analysis clarifies that the failure of global solidarity is the result of policies which advance specific values at the expense of others. The COVID-19 pandemic demonstrated the inadequacies of these policies. Public health infrastructures in most countries proved incapable of coping with the virus. Appeals to individual responsibility alone did not manage to control viral transmission. Massive government interventions were necessary to support the health care system, the economy, and all sectors of social life. The free market was not able to produce sufficient quantities of protective equipment, medication, and vaccines without substantial public support. At the same time, mainstream bioethics, relying on the language of autonomy, interests, utility, efficiency, and negative rights presented a myopic view of relevant ethical concerns. Starting from the point of view of the autonomous individual, it could not recognize the connectedness of human beings, and the global dimensions of the pandemic, and thus the need for global responses. After COVID-19, bioethics can no longer assume that autonomy is the dominant ethical principle; it must recognize that taking human relationality seriously implies enhancing and embracing social and structural conditions that make solidarity possible.

8.6 Future Challenges

Like many other international ethical standards in bioethics, such as the *Declaration of Helsinki* (World Medical Association 1964) and the *European Convention on Human Rights and Biomedicine* (1997), the *Universal Declaration on Bioethics and Human Rights* (UNESCO 2005) is embedded in the human rights tradition, while it is unique as the first global instrument that endeavors to cover the entire field of bioethics (Andorno 2007, 2009). Consequently, the future perspectives of global bioethics, thus framed, are closely connected with the prospects of the human rights tradition itself (Gordijn and Ten Have 2014). The appeal to human rights makes a lot of practical sense in order to seek avenues for a more effective governance of global health since contemporary challenges are often crossing borders and demand international solutions. Additionally, bioethics and international human rights are held to have similar historical roots: World War II, the Nazi concentration camps, and their follow-up events triggered the establishment of both (Annas 2004, 2010; Baker 2001). The challenge today is that human rights discourse is weakened. One reason is that theoretical lack of agreement persists as regards the justification of human rights, the anthropocentrism involved in the exclusiveness of human beings as the sole bearers of human rights, and the focus on rights without consideration of corresponding obligations. Another reason is that currently processes of globalization are increasingly criticized and fragmented. In response to the COVID-19 pandemic and challenges related to massive migration, more and more governments have implemented national measures, for example, reinforcing border controls. But perhaps the main reason is that many governments, and particularly autocratic regimes, no longer respect human rights. In some cases they argue that human rights should be regarded as Western imperialism, even though all governments or members of the United Nations have adopted international human rights law as a guiding framework for policy-making.[4] In a significant number of countries, human rights activists are persecuted and imprisoned.

The globalization of bioethics furthermore demonstrates the ambivalent character of modern bioethics: the tension between academic scholarship and theoretical commitment on the one hand, and activism and practical engagement on the other. Bioethicists cannot avoid being involved in policy-making at either local, national, or international levels where bioethicists are involved in a wide range of activities that go beyond the exclusive domain of academic enquiry.

A final challenge is that in some countries bioethics has achieved a strong institutional base and a high level of sophistication, but in many other countries it has clearly not yet reached a full-fledged state of development. Efforts are needed to strengthen bioethics in most countries worldwide, so that global ethical standards might be more effectively communicated and implemented. In the area of medicine, health care, and the life sciences this can be done by promotion of bioethics education and capacity building. Existing initiatives such as the Ethics Education Program of UNESCO must be further expanded. Bioethics committees can play an important role in developing bioethics infrastructures. As interdisciplinary bodies of expertise, bioethics committees can initiate and expand ethics education at various levels (for example, information and discussion sessions in secondary schools, and

[4] The universal nature of human rights was underlined by the "Philosophers Committee" chaired by Jacques Maritain to provide ideological support for the Universal Declaration on Human Rights. The Committee itself had a cosmopolitan nature and concludes that sources of human rights are to be found in almost all religious and cultural traditions (Eberl et al. 2011).

encouragement of universities to use the UNESCO resources and materials in ethics teaching) and in a range of disciplines. Health and well-being are in the interest of everybody, so the assumption is that every citizen in interested in learning about the advancements and possibilities of modern medicine. It is not merely the policy-makers who need to be advised, but the population as a whole that is in need of engagement and involvement. This is especially relevant in democratic societies, since elected leaders may or may not reflect health-related interests, and in which adherence to public health measures is sought by measures other than police force.

8.7 Conclusions

Global bioethical problems require global answers. To provide answers a global ethics perspective is needed that extends beyond the usual focus on clinical and individual moral challenges in health and disease. Potter has proposed a concept of global bioethics that encompasses individual, social, and environmental perspectives focused on addressing global problems. The *Universal Declaration on Bioethics and Human Rights* has elaborated this concept into an articulated ethical framework with a broad range of ethical principles. The adoption of the Declaration in 2005 was the result of deliberation and negotiation based on a dialectic process of interaction between global and particular moral views. The outcome is provisional since the dialectical nature of global bioethics will not deliver a closed set of ethical principles. The third contribution to a global perspective on ethical issues is articulated in *Laudato Si'*, emphasizing the harmony between the Creator, humanity, and creation, as well as the common human responsibility for individual, social and environmental well-being (Francis 2015). More important than the ethical principles will be their application in practical settings, requiring a change of mindset, leading to their continuous debate, modification, and transformation. The same dialectical process that has given rise to the Declaration will be required in its implementation. This is a task of the governments that have unanimously adopted this document, but much more so of actors and stakeholders in countries, regions, and contexts that are faced with the same global bioethical challenges as all of us.

References

Al, Pepijn. 2021. The value of communities and their consent: A communitarian justification of community consent in medical research. *Bioethics* 35 (3): 255–261.

Andorno, Roberto. 2007. Global bioethics at UNESCO: In defence of the universal declaration on bioethics and human rights. *Journal of Medical Ethics* 33 (3): 150–154.

Andorno, Roberto. 2009. Human dignity and human rights as a common ground for a global bioethics. *Journal of Medicine and Philosophy* 34 (3): 223–240.

Annas, George J. 2004. American bioethics and human rights: The end of all our exploring. *Journal of Law Medicine and Ethics* 32 (4): 658–663.

Annas, George J. 2010. Human rights and American bioethics: Resistance is futile. *Cambridge Quarterly of Healthcare Ethics* 19 (1): 133–141.

Baker, Robert. 2001. Bioethics and human rights: A historical perspective. *Cambridge Quarterly of Healthcare Ethics* 10:241–252.

Bieri, Peter. 2017. *Human dignity: A way of living*. Cambridge: Polity Press.

Bishop, Jeffrey, and Martin Fitzgerald. 2020. Norming COVID-19: The urgency of a non-humanist holism. *Heythrop Journal* 63 (3): 333–348.

Brown, Gordon, and Daniel Susskind. 2020. International cooperation during the Covid-19 pandemic. *Oxford Review of Economic Policy* 36 (S1): S64–S76.

Callahan, Daniel. 1973. Bioethics as a discipline. *The Hastings Center Studies* 1 (1): 66–73.

Chuwa, Leonard. 2014. *African indigenous ethics in global bioethics: Interpreting Ubuntu*. Dordrecht: Springer.

Dawson, Angus, Ezekiel Emanuel, Michael Parker, Maxwell Smith, and Teck Chuan Voo. 2020. Key ethical concepts and their application to Covid-19 research. *Public Health Ethics* 13 (2): 127–132.

Eberl, Jason T., Eleanor D. Kinney, and Matthew J. Williams. 2011. Foundation for a natural right to health care. *Journal of Medicine and Philosophy* 36 (6): 537–557.

Eurasia Group. 2020. *Ending the Covid-19 pandemic: The need for a global approach*. November 25; https://www.who.int/docs/default-source/coronaviruse/act-accelerator/2020-summary-analysis-of-ten-donor-countries-11_26_2020-v2.pdf.

European Convention of Human Rights and Biomedicine. 1997; https://www.echr.coe.int/documents/d/echr/Convention_ENG

European Group on Ethics in Science and New Technologies. 2020. *Statement on European solidarity and the protection of fundamental rights in the Covid-19 pandemic*, April 2; https://ec.europa.eu/info/sites/default/files/research_and_innovation/ege/ec_rtd_ege-statement-covid-19.pdf.

Fins, Joseph. 2020. Covid-19 makes clear that bioethics must confront health disparities. *The Hastings Center*, July 9; https://www.thehastingscenter.org/covid-19-makes-clear-that-bioethics-must-confront-health-disparities/.

Francis. 2015. *Laudato Si*. Vatican City: Libreria Editrice Vaticana. https://www.vatican.va/content/francesco/en/encyclicals/documents/papa-francesco_20150524_enciclica-laudato-si.html.

Francis. 2018. *Address to participants in the plenary assembly of the Pontifical Academy for Life*, June 25; https://www.vatican.va/content/francesco/en/speeches/2018/june/documents/papa-francesco_20180625_accademia-provita.html.

Gintis, Herbert. 2017. *Individuality and entanglement: The moral and material bases of social life*. Princeton and Oxford: Princeton University Press.

Gordijn, Bert, and Henk ten Have. 2014. Future perspectives. In *Handbook of global bioethics*, ed. Henk A. M. J. ten Have and Bert Gordijn, 829–844. Dordrecht: Springer Publishers.

Ho, Anita, and Iulia Dascalu. 2020. Global disparity and solidarity in a pandemic. *Hastings Center Report* 50 (3): 65–67.

Jennings, Bruce, and Angus Dawson. 2015. Solidarity in the moral imagination of bioethics. *Hastings Center Report* 45 (5): 31–38.

Kaminsky, Carmen. 2020. Normality "ex post": Social conditions of moral responsibility. In *Medicine and ethics in times of corona*, ed. Martin Woesler and Hans-Martin Sass, 63–74. Zürich: LIT Verlag.

Kheriaty, Aaron. 2020. The other pandemic: The lockdown mental health crisis. *Public Discourse*, October 24; https://www.thepublicdiscourse.com/2020/10/71969/.

Kymlicka, Will. 2007. The globalization of ethics. In *The globalization of ethics. Religious and secular perspectives*, ed. William Sullivan and Will Kymlicka, 1–16. New York: Cambridge University Press.

Mackenzie, Catriona, and Natalie Stoljar. 2000. *Relational autonomy: Feminist perspectives on autonomy, agency, and the social self*. New York: Oxford University Press.

Macklin, Ruth. 2022. A new definition of global bioethics: COVID-19, a case study. *Global Bioethics* 33 (10): 4–13.

Marcel, Gabriel. 1962. *Homo Viator. Introduction to a metaphysic of hope*. New York: Harper & Row.

Meslin, Eric, David Ayuku, and Edwin Were. 2014. "Because it was hard ...": Some lessons developing a joint IRB between Moi university (Kenya) and Indiana University (USA). *American Journal of Bioethics* 14 (5): 17–19.

Morin, Edgar. 1999. *Homeland earth. A manifesto for the new millennium*. Cresskill (N.J.): Hampton Press.

Pontifical Academy for Life. 2020. *Vaccine for all. 20 points for a fairer and healthier world*. December 29; https://press.vatican.va/content/salastampa/it/bollettino/pubblico/2020/12/29/0697/01628.html#notaing.

Potter, Van Rensselaer. 1970. Bioethics, the science of survival. *Perspectives in Biology and Medicine* 14:127–153.

Potter, Van Rensselaer. 1971. *Bioethics: Bridge to the future*. Englewood Cliffs, NJ: Prentice-Hall.

Potter, Van Rensselaer. 1988. *Global bioethics—Building on the Leopold legacy*. East Lansing: Michigan State University Press.

Sass, Hans-Martin. 2007. Fritz Jahr's 1927 concept of bioethics. *Kennedy Institute of Ethics Journal* 17 (4): 279–295.

Ten Have, Henk. 2016. *Global bioethics. An introduction*. London and New York: Routledge.

Ten Have, Henk, and Bert Gordijn. 2014. Global bioethics. In *Handbook of global bioethics*, ed. Henk A. M. J. ten Have and Bert Gordijn, 3–18. Dordrecht: Springer Publishers.

Ten Have, Henk A. M. J., and Michele S. Jean, eds. 2009. *The UNESCO universal declaration on bioethics and human rights: Background, principles, and application*. Paris: UNESCO Publishing.

Ten Have, Henk, and Renzo Pegoraro. 2022. *Bioethics, healthcare and the soul*. London: Routledge.

UNESCO. 2005. *Universal Declaration on Bioethics and Human Rights*. https://www.unesco.org/en/ethics-sciencetechnology/bioethics-and-human-rights.

UNESCO. 2020. *Statement on Covid-19: Ethical considerations from a global perspective*. Paris: UNESCO. https://unesdoc.unesco.org/ark:/48223/pf0000373115.

WeCope (World Emergency Covid-19 Pandemic Ethics Committee). 2020. Statement on individual autonomy and social responsibility within a public health emergency. In *Medicine and ethics in times of corona*, ed. Martin Woesler and Hans-Martin Sass, 419–425. Zürich: LIT Verlag.

West-Oram, Peter, and Alena Buyx. 2017. A. Global health solidarity. *Public Health Ethics* 10 (2): 212–224.

World Health Organization. 2021. WHO *Director-General's statement on Tanzania and Covid-19*; https://www.who.int/news/item/20-02-2021-who-director-general-s-statement-on-tanzania-and-covid-19.

World Medical Association. 1964. *Declaration of Helsink.*; https://www.wma.net/what-we-do/medicalethics/declaration-of-helsinki/doh-jun1964/

Chapter 9
Taking Responsibility for Consequences in Faith-Based Global Mission Work: Insights from Modern Catholic Missionaries

Erica L. Smith and Rachelle Barina

> *"[O]ur spirituality is the wellspring of moral living. Ethics without spirituality is rootless; spirituality without ethics is disembodied."*
>
> *—Richard Gula (2002)*

9.1 Introduction

Stories of failures or successes in global mission work are not difficult to find. On one hand, there are successes—just Google "medical surplus recovery organization" and click on any link that comes up to see faces of real people who have benefited from international collaboration. On the other hand, there are multiple accounts of failures: expired pharmaceuticals sent after a tsunami; broken medical equipment dumped in a rural African community; winter coats and 'one rollerblade' donated to Haiti after an earthquake in 2010 (Kempton 2015; Boulet-Desbareau 2013); and beyond supplies, college students traveling on a short-term mission trip posting photos of themselves partying.

Such a binary, simplistic view of global mission work as either "good" or "bad" is unproductive; in most cases, a complete view of the actual impact of global mission work is more elusive. We often do not have enough (or even any) evidence-based research to guide and illuminate ongoing work. In fact, while millions of dollars are spent in the United States each year to move donated medical equipment to low-resource settings, no research exists to collect data about a specific donation of medical equipment from a high-resource country to a low-resource country from

E. L. Smith · R. Barina (✉)
Hospital Sisters Mission Outreach and Hospital Sisters Health System, Springfield, IL, USA
e-mail: Erica.Smith@mission-outreach.org; Rachelle.Barina@hshs.org

J. T. Eberl (ed.), *Emerging Issues in Catholic Bioethics*, Philosophy and Medicine 153, https://doi.org/10.1007/978-3-032-08862-8_9

origin to disposal of equipment (Perry and Malkin 2011). Instead, anecdotes often fill the headlines, and the quantitative and qualitative data that do exist are limited. In short, it is extremely difficult to get a clear and comprehensive grasp on the full impact within global mission work.

The Catholic moral tradition affirms the dignity of every human person and calls each of us to embrace our shared responsibility to advance the common good (Pontifical Council for Justice and Peace 2004). Nevertheless, engaging in ways that truly uplift human persons and advance communal conditions proves exceptionally challenging. This chapter invites reflection on these challenges and the dynamics and spirituality that often undergird actions intended to help people in low-resource settings.

In this chapter, we explore why it is so difficult to gain a complete understanding of what impact results from global mission work and why it is even more difficult to actually achieve lasting, positive impact. We then explore how returning to modern stories of Catholic missionary figures may challenge us to reexamine our own motivations and reconsider our actions within global mission work in light of a deeper spirituality. Drawing on the themes of humility, sacrifice, solidarity, and relationship, we suggest that engaging such stories may provide a deeper source of spiritual insight for faith-based stakeholders who seek to reduce detrimental consequences and advance positive impact.

Before proceeding, we offer two clarifications. First, we use the phrase "global mission work" to refer to cross-national engagements intended to advance health of people in low-resource settings. Currently, two common manifestations of global mission work that we are seeing within contemporary churches and health systems are short-term mission trips and the donation of health-related supplies and equipment. Research on the ethical complexities of medical mission trips has seen meaningful and substantive growth in the last decade.[1] As such, while we use the phrase "global mission work" generally, we have particular interest in engagements wherein supplies or equipment are donated either as a standalone project or as part of short-term mission trips. Further, we address faith-based organizations and people acting out of their own faith commitments. Second, we write from the perspective of being residents of the United States. Because of where we see many challenges originating, we write specifically for an audience in the United States and specifically regarding the work of individuals, churches, and NGOs, rather than major government programs. While much could be said to and within the wider audience inclusive of partners, recipients, and governments, that conversation is beyond our scope.

[1] For example, see the June 2011 volume of *HEC Forum* (vol. 23, no. 2), which was entirely dedicated to ethical considerations of short-term medical volunteers.

9.2 Why Is Impact So Elusive?

9.2.1 The Problem Is Real

Scenario 1: Four clinicians in their health system travel to Peru annually to volunteer in a pediatric clinic. The community is seeing a rise in diabetes and obesity, alongside continued evidence of nutritional deficiencies. While in country, the personnel experienced a pregnant woman who came to the clinic with severe illness. It took 2 hours for transportation to be arranged to take her to a hospital with lab, obstetrics, and surgical services. She was diagnosed with sepsis upon arrival and later died. Troubled by the experience, the clinicians asked the local parish that runs the clinic if they would like an ambulance, and the parish resoundingly agreed. Upon return to the United States, the clinicians begin to work to source an ambulance that could be donated and transported to be owned by the clinic.

Scenario 2: A local church has a sister parish in a rural community in Kenya. A decade ago, the US parish funded a school. In addition to providing financial donations, each year the parish also has done a "back to school" supply drive for the Kenyan school—which now has grown to more than 200 elementary students. A small council of parishioners voted this year not to do the supply drive, and instead raise funds for a local non-profit instead. Now, the sister parishes are in discussion about building a clinic, and the US parish has secured a private Foundation grant to fully fund construction of the clinic. The US parish priest serves on the board of a local US hospital and has requested the donation of any extra equipment that would help in establishing the clinic. There have already been several drop-offs at the parish of extra material that was pulled from hospital storage, ranging from a filing cabinet to an EKG unit.

On face value, these scenarios seem like good efforts; however, if we dig a little deeper, questions of due diligence, motivation, and impact surface. In scenario 1, it is good that the US stakeholders have an established ongoing relationship from which this idea comes; however, is there a plan to operate the ambulance when the clinic is not open 24 hours a day? Does the clinic have a budget for staff, gas, and repairs? Are there security concerns with such a valuable vehicle? Is there a cheaper way to accomplish the same outcome of enabling transportation from a remote community to the nearest hospital? It seems possible, and not all that improbable, that thousands of dollars will be spent on providing an ambulance, and the ambulance may end up being used infrequently or only for a short time. Even if the ambulance is used, does it actually lead to improvements in maternal health outcomes or accessing emergency medical care? Similar questions arise regarding scenario 2.

These are neither exaggerated concerns nor are they small in scope. A study from the Catholic Health Association found that 600,000 tons of medical donations were made by Catholic health care organizations on an annual basis. Sixty percent of the 432 hospitals that participated in the study reported they donated broken equipment, and 90 percent reported donating expired supplies. Only 8 percent of the hospitals said they work with organizations capable of ensuring the donations are functional and appropriate (Compton 2012). This has been corroborated by research on the recipient side, as only 10–30 percent of donated equipment ends up being put into effective use in developing countries (WHO 2011). And the equipment that does end up being operationalized may be used only for a short time, if repair impediments or support barriers occur. Over the last 20 years, many poignant accounts have illustrated the stories and faces behind these statistics. Ken Stern's *With Charity for All* (2013) and Robert Lupton's *Toxic Charity* (2010) offer an array of vivid examples of the harmful or ineffective impact of interventions intended to promote health and advance overall social, educational, and economic interests.

While we must pause at the gravity of these sobering facts, we do not want to put forth an overly negative view. The whole reason we write this chapter is the unwavering belief that good can be, and often is, accomplished through global mission work; for example, while overall the donation of pharmaceuticals after a tsunami poses significant cost, waste, and environmental detriments, if even some of those drugs reach people and prevent or treat illness, that would be a good outcome. The issue US stakeholders must face is that when we pursue one area of intended positive impact in global health work, such as making drugs accessible to those who need it, there is inevitably a ripple-effect of wider impacts that occur. We cannot lose sight of the *staggering* statistic that the *majority* of equipment that is transported to low-resource countries ends up not being used. Not only is the impact of sending equipment sometimes unrealized, but also sending unusable material creates a problem of disposal in the recipient communities—a negative impact that often goes overlooked.

Ultimately, concern about the impact of global mission work is a matter of integral human development—the concept well established in Catholic social doctrine that development is multifaceted. To advance the good of human persons and communities, we must pursue the whole good of each human person, in all dimensions—such as spiritual, social, physical, cultural, economic, environmental, and political.[2] Insofar as Catholic health care makes hundreds of thousands of tons of medication donations, and only rarely works with organizations that confidently ensure those donations can be appropriately handled (Compton 2012), it bears an important responsibility for leading the way towards a better understanding of the intended and unintended impact of global mission work.

---

[2] Paul VI (1967) first used the phrase "integral human development" in *Populorum progressio*. Interestingly, this conceptualization of integral human development was a response to the emphasis on economic development in the prior eight decades of Catholic social teaching. Louis-Joseph Lebret, the French Dominican who heavily influenced *Populorum progressio* was worried about a growing dependency and exploitation between Latin America and what he saw as the United States' neo-imperialism. Almost six decades later, Lebret's worry finds an interesting echo with charitable endeavors that are inadequately accountable for ensuring positive impact. Further, in *Laudato Si'*, Pope Francis (2015) offered a dynamic development upon the concept of integral human development insofar as he illustrated that creation, climate change, and environmental conditions are inextricably connected to the development and flourishing of individuals and communities.

9.2.2 The Challenges of Nebulous, Hazy Motivations and Unanticipated Consequences

The notion of "unintended consequence" in Catholic bioethics is prominent. With the towering presence of the principle of double effect, ethics in clinical medicine widely depends on the conclusion that human agents may not be morally culpable for foreseen but unintended consequences (Sulmasy and Pellegrino 1999). Perhaps, however, aspects of this widely engaged principle have made the language of "unintended consequences" something so common that it may be glossed over; we may assume unintended consequences are tolerable because they come with a sense of inevitability. When it comes to global mission work, however, we need to renew focus and heighten concern around the notion of undesirable or unintended consequences. As such, we turn to a sociological perspective in order to consider unintended consequences from a fresh angle.

The theory of unanticipated consequences of purposive social action was prominently articulated by Robert Merton (1936) and has spurred subsequent thought and application in many fields.[3] Merton's thought has been applied to the reality that today's community development efforts have a variety of consequences, some of which are not intended or anticipated and some of which are bad and harmful. Merton (1936, 895) defines unanticipated consequences as "limited to those elements in the resulting situation which are exclusively the outcome of the action, i.e., those elements which would not have occurred had the action not taken place." The consequences may not always be negative, but they are always the result of a choice between alternatives. Merton's theory and those who build on his scholarship lead to four points of particular relevance for faith-based global mission work in the twenty-first century.

First, people make choices and choices have consequences, including consequences that are unanticipated or unintended. The notable point here as we reflect on global mission work is that the consequences in question are not stemming from simple behaviors or random events. Merton emphasizes the action under consideration involves the motivation of a human actor, or an explicit purpose. Merton (1936, 895) defines purpose as "concerned with 'conduct' as distinct from 'behavior,' that is, with action which involves motives and consequently a choice between various alternatives." Those engaged in global mission work hold a motivation or explicit purpose, such that there is a choice between multiple available alternatives. Perhaps sometimes we hold multiple motivations. Perhaps sometimes we have never clearly defined our motivation.

[3] Of note, Hanna and Kleinman (2013, 15) argue that the application of social theories to global health work "can help students and practitioners understand and interpret the nature, effects, and limitations of medical and public health interventions." They point to four social theories that can inform global health efforts, including Merton's theory of unanticipated consequences. In this chapter, we focus on Merton because it helps to illustrate some fundamental challenges that are often overlooked.

Critically, Merton reminds us that insofar as we choose to engage in global mission work, the actions taken could, with a different choice, have a different outcome and impact. In our first scenario, perhaps our intended purpose is to reduce maternal mortality and other causes of mortality. Certainly, one case of need presented and motivated a tremendous and laudable desire to respond on the part of the four volunteers, and that motivation likely comes out of both a sincere desire to aid others in need as well as the psychological difficulty of witnessing a preventable death. That being said, there are many actions that could be taken to reduce preventable deaths (which, notably, was a problem of unquantified and unknown magnitude for the generous volunteers willing to give their own money). We could instead buy a car. We could consider virtual infrastructure for this remote community that happens to have remarkably strong cell service. We could decide the resources are better spent on education and planning with pregnant women about the warning signs of obstetrical emergencies and what to do in the event of an emergency given the actual resources available. While providing an ambulance may appear like the most direct solution, there are multiple potential actions, and these actions produce different outcomes—including different benefits as well as different problems.

Second, we must recognize that the consequences and outcomes of our actions can be subtle or hidden. Although writing nearly a century ago, Merton laid the groundwork to recognize that consequences can affect the actor, people, or community, particularly through social and cultural structures. Hanna and Kleinman (2013, 20–21) provide numerous examples of how Merton's theory can be applied to global projects intended to help people in low-resource areas with disastrous consequences. From a case of a physician who misunderstands a patient's cultural tradition for non-compliance, to the risk of overburdening a current hospital by preemptively announcing new services, significant harm can be done by ignoring the outcomes of actions or failing to consider all perspectives, namely that of those most affected by the actions: the local people who are being acted upon. While exceeding what is immediately evident to the vantage of a person in the US, cultural and social impacts are essential to ensuring work truly advances integral human development.

US stakeholders engaging in global mission work have responsibility to pursue integral human development, which means taking responsibility for not only the obvious impact of work but also the more subtle consequences. Reflecting again on the first scenario, the US volunteers talk with local members of the community about the need for an ambulance. Naturally, the action of donating an ambulance likely seems better than not donating an ambulance. But what might be some of the potential impacts to consider, both positive and negative? These volunteers come back repeatedly with a gentle, compassionate, and humble presence. Yet, they are increasingly perceived as a wealthy keeper of foreign resource. Further, great care is needed to involve leaders of the clinic in understanding the degree to which avoidable maternal death is an issue and the nature of transportation barriers. This scenario presents an opportunity to think critically about how an infusion of resources might be best spent, such that local leadership owns and chooses what needs are most critical and collaborate with the US volunteers in partnering to address those needs. Otherwise, it is a lost opportunity to strengthen local leadership and unite the community, and likely reinforces an attitude of dependence and a "we want anything you give us" mentality. In a low-resource setting, the starting point for collaboration around donating resources is not "Do you want an ambulance?"

Third, as human agents, we may fail to anticipate consequences for a variety of reasons, and naming these may prove helpful in reflecting on our own actions. Merton (1936, 898–904) notes four sources from which unintended consequences stem: lack of adequate knowledge; error originating in inadequacy of understanding the present, inferring what will happen in the future, or attending to only some relevant aspects of the situation; imperious immediacy of interest, leading to focus on foreseen immediate consequences rather than more remote or ambiguous consequences; and introduction of a new element into the concrete situation, which alters the relevance of planning and predictions.

Without any doubt, there is no ability for human persons to anticipate every consequence and mitigate all bad consequences. Yet Merton's insights a century ago resonate with the more recent study from the World Health Organization (2011) on medical donations, which explains why inappropriate medical donations persist: Donors tend to lack understanding of recipients' specific and particular challenges and needs, and communication is inadequate. These factors can be addressed. But today as in centuries past, we continue to be distracted by that which is immediate, neglecting the difficult work of following impact over time and through the whole lifecycle of resulting consequences.

Fourth, a continual diligence in attending to purpose and impact is needed. This point resonates within the Catholic moral theological tradition and is a critical follow-up on the prior three points. Merton (1936, 896) notes that actions, when repeated to the point of becoming a habit, can start with explicit purpose that is lost in repetition: "[T]he aim of action is more often than not nebulous and hazy." Habitual action "may originally have been induced by conscious purpose," but then over time "is characteristically performed without such awareness."

In the realm of global mission work and the challenge of refining focus on and understanding of impact, too often what is being pursued is nebulous, hazy, or unexamined. Often, we hold the vague goal of helping others, which does not set up clearly and precisely what kind of impact is desired: If I travel to Peru each year to help the clinic, I may make a general conclusion that it will help the community to get an ambulance since they do not have one. Similarly, if I have been working with our sister parish for years, I may intuitively conclude that it will help the Kenyan parish to receive anything that the US parish can send them because they have so little. It will help them have equipment to open their clinic and offer care for which there is a gap. Faith-based communities in the US often tend to absorb the inspiration of their traditions and develop a habitual sense of charity, helping, and service: that all giving is good and we ought to give whenever possible.

However, these kind of habitual motivations around helping and giving fail to position us to reduce unwanted consequences and optimize intended impact. These examples are not facetious or uncommon. Tens of millions of dollars are spent around the world based on work that is driven by inadequately specified objectives and hoped for impact, although the work and need may seem extremely proximate, obvious, or familiar. In the midst of inadequately specifying our purpose in partnering on global mission work, we lose track of the first point from Merton: that helping involves a choice between alternatives, and those alternatives result in different outcomes.

Global mission work requires us to assume responsibility for unintended consequences. In global mission work, we voluntarily choose to intervene. While we may do so to advance human dignity and the common good, our voluntarily actions have consequences which may create a net harm. We choose our actions, and thus different actions will lead to different consequences. Admittedly, some consequences can be impossible to anticipate due to unavoidable changes in the world; however, avoidable limitations in our understanding and from our habits of thought may lead us to act on autopilot rather than with specifically delineated motivations and goals. These factors contribute in a very fundamental way to the elusive nature of pinning down the impact we hope to have and articulating the impact we actually have.

9.2.3 The Challenge of Grasping the Full Scope of Impact

As we have reflected on the theory of unintended consequences of purposive action, we have challenged global mission stakeholders to consider or reconsider our choices, make explicit our motivations, and recognize the consequences of our actions. A further step is to engage in more rigorous evaluation of impact. Work to develop qualitative and quantitative data on impact is critical. The impact of global mission work reaches across economic, social, cultural, financial, health, and other spheres. The work of rigorously pursuing assessment of wide-reaching consequences and impact can be daunting. It can also be technically challenging to assess in a meaningful way, whether quantitative or qualitative.

Further, organizations carrying out global mission work have not typically been highly interested in collecting objective, empirical data regarding the outcomes of projects, partnerships, and activities. This is especially the case for smaller NGOs or entities like churches whose primary objective is not global work. There are reasons for this lack of willingness or interest: the cost of conducting studies, particularly those that must use vetted research instruments, as Biehl and Petryna (2014) show; the dependence on philanthropic funding that incentivizes NGOs to share only the "good" that is being done and downplay any negative outcomes, amidst a focus on the needs of philanthropists to the neglect of the recipients needs.[4] As a result, as Hanna and Kleinman (2013) reiterate, many NGOs attempt to measure the "effectiveness" of their programs and initiatives but fail to acknowledge the full impact of their interactions on the people and communities being served; put simply, they fail to consider and measure the degree to which they are truly advancing integral

[4] A helpful summary from a philanthropic perspective can be found in Moody (2022).

human development. Even using evidence-based research as the standard for knowledge about global health work prompts questions, because the methods of obtaining data can exclude vital cultural and social information that may not fit the research design and goals (Adams 2013).

When we acknowledge the multifaceted scope of impact stemming from global mission work, the challenge of grasping impact becomes even more elusive. Defining what is a positive or negative outcome must encompass both empirical evidence and "all the things that simply can't be counted, or that, even if they could be counted, in being counted would be turned into misrepresentations of their true relationship to health." (Adams 2013, 82) While it is elusive, it is critical, and global mission work must continue to strive to refine impact and reduce unwanted consequences in the realms of economy, environment, workforce, culture, and health.

For short-term mission trips and medical donations, impact to the local economy warrants further comment. It may be satisfying for individuals traveling from the US to cram suitcases full of clothing, toys, and school supplies, and then to make a connection with smiling children who receive the gifts with excitement. However, regional vendors sell these items, and local workers and vendors earn income based on sale of these items. Similarly, painting walls, doing building maintenance, planting gardens, or constructing a playground may seem to be terrific week-long volunteer projects for teens on spring break, but these are jobs that skilled local workers do to support their families and benefit their communities. Donations of medical equipment that is broken, missing parts, or lacking the necessary consumable supplies to make it work become costly nuisances for doctors and administrators in rural hospitals to manage. All these actions inspired by charitable intentions result in lost opportunity or even harm—all of which is preventable if the US-based partner sought mutual understanding and consideration of goals and impact of actions. And drawing on Merton's first and simple point, these effects are neither necessary nor inevitable; different actions would lead to different impact.

In considering impact across many spheres of community life, one area of critical differentiation involves goals involving long-term development and goals pursuing immediate aid, which has short and long-term consequence. Too often, US stakeholders take action within communities ready for long-term development, but the stakeholders intervene with tactics of immediate aid, particularly in the aftermath of a disaster. When people are in crisis and lack basic resources for survival—food, clothing, shelter, safety—immediate aid is appropriate and required. For example, a natural disaster where homes are destroyed, displacement because of war or violence, or famine brought on by crop failure are all situations in which immediate aid is needed and must be provided until the person or family can return to stability.

The problem comes when aid continues to be given past the point of crisis, and is extended when the actual need is for *development*. Infrastructure and socioeconomic systems administrated and staffed by local people are examples of development; they are major investments into regions and communities that build or restore systems that promote individual and collective sustainability. Appropriate and effective development happens when people act for themselves and can take years, if not decades, to achieve goals. Short-term, foreign-driven projects that exclude local leadership, people, and customs are never appropriate to provide development.

The irony, as Richard Lupton (2011) argues, is that most of our charitable efforts through short-term help cause harm because we do not understand our aid is exacerbating problems: "[T]he overwhelming majority of our mission trips are to places where the needs are for development rather than emergency assistance." For example, the "free labor" and "donations" of volunteer groups and NGOs displace local businesses and entrepreneurs; local entities compete to receive charity instead of collaborating, and local entities devise project after project to keep foreign visitors happy (and monetary donations coming in) instead of engaging or employing local people.

Mission trips of skilled professionals meeting a specific need in response to, and partnership with, locally established groups, can be effective to provide aid in time of crisis. Similarly, medical donations may sometimes be appropriate as aid to a crisis, especially when supply chains are disrupted or there is a temporary shortage of vital resources. Nevertheless, when such donations extend too long without productive collaboration, or unwanted resources are sent, aid may prevent development and runs the risk of engraining lasting dependencies.

9.2.4 Turning Toward Spirituality, the Wellspring of Our Moral Lives and Actions

Ultimately in this chapter, we do not take on tactical suggestions of how to better pursue and quantify impact. Our assumption, stemming from our own professional experiences, is that one cause (not the only cause) of these challenges is the internal state of faith-based stakeholders—meaning fundamental motivations, beliefs, and self-perceptions.

This turn toward spirituality may seem like an unexpected turn; however, let us remember that actions and habits are embodied expressions of who we are and what we value. Catholic health care organizations are frequently inviting their employees to make connections between the organization's identity, the values being put into action, and what it means to and for all people involved. As we note in the epigraph of this chapter, spirituality is the wellspring of our moral lives, and without it, our ethics is rootless. Richard Gula (2000) writes:

> Ethics is about more than interpreting principles and assessing consequences. Ethics is also interested in personal character—our stable identity, our deepest desires and ideals, our loyalties and commitments, our beliefs and attitudes, our motivation and perspective, our values and virtues. When we move from the decision making/act-centered interests of principles and consequences to the person-centered aspects of character, we are reaching into the inner dynamics of human action. These deep dimensions of the self are influenced by our spirituality.

In the Catholic imagination, a distinct source of spirituality involves the narratives of holy men and women, some of whom have been formally canonized as saints and many others who have not. Reaching toward the narratives that shape Catholic spirituality, we turn to stories of several missionaries from recent centuries. Retrieving these stories turns us toward the spiritual tradition and Christian imagination that shapes who we are and how we conceive of our own actions. It prompts consideration of our way of being engaged in mission work across national borders, and in exploring our way of being engaged, we may find a new source of insight and inspiration to ask questions, pursue impact, care about consequences, and develop tactical skills for planning and evaluation.

9.3 The Missionary Figure in the Catholic Imagination

For Christians, the concept of the missionary originates alongside the birth of the church. At a time when travel was slow and dangerous, Jesus's early followers traveled to foreign lands to talk about his life, death, and resurrection. They would preach in a community and then move on to the next. Over the centuries, Christians continued to travel to foreign lands and the vigor of missionary activities was part of the growth of Christianity (Sullivan 1994).

Today, the long tradition of missionary activity has led to relevant narratives from within more contemporary dynamics. After all, when Europeans crossed the ocean beginning in the fifteenth century with imperialist dreams, the missionary gained an association as the priest who goes to the "new land," sometimes converting or ministering amidst severe conditions and linguistic barriers. Some of history's Catholic missionaries have fallen under criticism for their imposition on local culture and their proximity to the use of violence. Contrasting the European missionaries of centuries past centuries who carry the embarrassment of their home nations' colonialist atrocities, a cadre of more recent missionaries have arisen within the Catholic imagination and provide spiritual insight for the present day. We turn to two prominent stories from more modern times.

9.3.1 The Story of Sister Mercy Hirschboeck

"During the Korean War, a woman correspondent happened upon a clinic in Pusan where she found a nun on her knees, cleaning the gangrenous sore on a man's leg. 'Sister,' she said, 'I wouldn't do that for a million dollars.' 'Neither would I,' the Sister replied. That story, told of Sr. Mercy Hirschboeck, Maryknoll's first Sister-Doctor, made headlines around the world. But most of her life was spent out of the limelight."[5]

[5] Give Us This Day, September 2024: https://giveusthisday.org

This "not for a million dollars" story has been often repeated about Sister Mary Mercy Hirschboeck, a Maryknoll missionary who was the first physician in her community. She was born Elizabeth Hirschboeck in 1903 and would gain national attention in the 1950s with her startling reply of "neither would I." This story would enter the American Catholic imagination as a portrait of the missionary spirit—and has reached almost mythical status. In 2017 *America* magazine printed the exact same story, but attributes the heroic line to Mother Teresa of Calcutta—not Sister Mercy.[6]

Sister Mercy's story entails a compelling journey of determination, setback, and affirmation. As a student at Marquette University, Elizabeth was seriously injured in a major automobile accident. She believed her life was spared only because God destined her to be a consecrated religious. After she recovered, she wanted to leave medical school to enter the Maryknoll missionary order; however, the founder of the Maryknoll Sisters and provincial of the community initially rejected Elizabeth's entry and told her she first must complete her studies to become a physician. Then, if she still wants to, she may return and become a Sister.

Upon graduation from Marquette in 1928, the determined Elizabeth returned to Maryknoll where she was welcomed into the order and given the name Sister Mary Mercy. That same year she was sent to Korea to serve as a physician; she gained the respect of the Korean people and upon being called back to the US in 1940 her departure was mourned. A Maryknoll father wrote, "The women especially will feel her loss … for she was to them not only a doctor but a saint."

Sister Mercy worked for a while in the US and then at another mission in Bolivia. Her mind remained on Korea, however, and she was asked to return there; but initially the assignment was halted because of the outbreak of the Korean War. Refusing to be told no, Sister Mercy personally petitioned the infamous General Douglas MacArthur to allow her and two other Sisters to return to Korea to care for the people of Pusan—including refugees from North Korea—during the long and brutal war.

It is during her service as a physician during the Korean War that she said the famous "neither would I" line that gained her national attention. She returned permanently to the US in 1955 and continued to lead a remarkable, although quiet, life. Not only does she have a book written about her, she also was the administrator of the first integrated hospital in Kansas City. In 1964, she was at the scene on the Ossining, New York, campus to provide emergency medical care to one of the bandits who attempted a dramatic heist of the Maryknoll post office and was injured in a shoot-out with police (Keane 2020). After spending her last years working in an impoverished New York City community, Sister Mercy died unexpectedly on September 20, 1986, on the feast of the Korean Martyrs and just 4 days before her own feast of Our Lady of Mercy—a most appropriate conclusion to her life story.

[6] https://www.americamagazine.org/faith/2017/02/02/what-mother-teresa-wouldnt-do-million-dollars

An examination of Sister Mercy's biography as articulated widely over the last 50 years reveals a kind of hagiographical treatment of her life story—even a kind of archetypal "Hero's Journey" as first conceptualized by Joseph Campbell (1949). What characterizes Sister Mercy's heroic story is the overcoming of obstacles; the power of her faith in sustaining her determination; her willingness to travel to foreign lands in selfless service of others in need; the reception and love of those she served, signifying success of her mission; the regret of her departure from foreign lands, and the wide mourning of her death. Sister Mercy's story is not a standalone narrative.

9.3.2 The Story of Father Damien of Molokai

The story of St. Damien of Molokai blossomed in the late twentieth century, before he was canonized in 2009. James Kroeger (2015, 2018) offers a comprehensive account of his life. Joseph de Veuster was born in Belgium in 1840 near Louvain. While his father hoped he would inherit the family's farm, his family's Catholic devotion inspired him at age 19 to become the second boy in his family to enter the priesthood. His brother Pamphile had entered religious life prior to Joseph, and two of his sisters became Ursuline nuns. Joseph took the name Damien, inspired by Saints Cosmas and Damien, who were medical doctors and twins.

When Pamphile became ill, Damien was not yet ordained. Nevertheless, Damien wrote to the superior general and asked to become a substitute for Pamphile's mission to Hawaii. In 1863, Damien set sail with five vowed religious men and ten sisters. Knowing that he would never return to Europe and acutely aware of the sacrifice he was embracing, Damien sent a letter to his parents as he departed:

> We are now, my dear parents, on the eve of leaving not only our fathers and mothers, sisters and brothers, not only our second family, our brethren of Louvain and Paris, but also the fair continent of Europe … The sacrifice is great indeed for one who tenderly loves his parents, his family, his brethren, and the land of his birth. But the voice that has called upon us to make a generous sacrifice of all is the voice of God Himself … So, my dear parents, do not trouble yourselves in the least about us. We are in the hands of God. (Kroeger 2018, 11)

After a 5-month journey, he arrived in Hawaii on the Feast of St. Joseph, and was ordained in May 1864. Without any period as an assistant to a more seasoned priest, Father Damien was sent to two parishes in Puna, where he was pastor for the subsequent 9 years. He survived illness, hurricanes, and an earthquake. He also built eight chapels, learning to work with the local materials of the Hawaiian area.

At this time in Hawaii, the community was beginning to suffer from an epidemic brought by Europeans who traveled to the islands: leprosy. Today, leprosy is known as Hansen's disease, a curable bacterial skin infection that causes deformities, blindness, and progressive disabilities if not treated. Transmission occurs through droplets from the nose and mouth amidst prolonged contact.[7] However, at the time, the Hawaiians lacked any natural resistance, and diseases like leprosy, flu, pertussis, and smallpox were spreading rapidly. While the population of Hawaii was around 275,000 in 1778 when Captain James Cook arrived in the islands, it fell to 31,000 in 1896. Ninety percent of the population died because of European diseases.

The government established a Board of Health in 1850. Shortly after and amidst fear of the severe disease, King Kamehameha V enacted laws that allowed for the removal and isolation of people with leprosy. Under these 1865 laws, more than 8000 children and adults were taken from their families and shipped to the island of Molokai, where a 10 square mile area on a peninsula surrounded by steep cliffs had been designated for those relocated. Over the years, a community formed on this peninsula and infrastructure was established. These laws lasted over 100 years, being lifted in 1969.

Father Damien knew the reality of leprosy. From his parishes in Puna in 1869, he wrote to his parents that "the native population is continually decreasing." In 1872, he describes his knowledge in even more detail: "Leprosy is beginning to be very prevalent here. There are many men covered with it. It does not cause death at once, but it is very rarely cured. The disease is very dangerous, because it is highly contagious. The population of our islands consists of some sixty-two thousand at present. It was larger formerly." And after 6 months of living on Molokai, Father Damien wrote to his brother Pamphile that only 800 of the 2000 who have been sent remain alive: "A priest who should be placed here must consider himself shut up with the lepers for the rest of his life." (Kroeger 2018, 15).

In May of 1873, Father Damien volunteered to travel with Bishop Maigret to Molokai alongside 50 people who were being exiled to the isolated area. Father Damien slept under a tree for the first 5 months on the island as there was not yet a house for a priest. For Father Damien, the "harsh realities of Molokai quickly became apparent—there was chaos in the settlement, with no organization of the community, minimal health and social services, no communal order, and nothing to promote a dignified human life, even if the conditions were rough. The anguished cry of the lepers that greeted Damien was the axiom *Aole kanawai ma keia wahi* ("in this place there is no law"). (Kroeger 2018, 6) Residents tended to be in competition and drunkenness was pervasive. Many people died alone, and the dead were thrown into a ravine where while boars consumed remains.

Father Damien became a champion of order and structure, amidst the multifaceted needs of the community. He offered the sacraments to those who were sick. He cleaned and bandaged wounds. He began cleaning homes. He sought to improve the cemetery, erecting a fence to prevent scavenging animals from entering and disturbing those buried. Using carpentry skills he had gained in Pena, he started making coffins and organized residents to provide dignified burials for the approximately 200 people who died annually during his initial years on Molokai. He secured benefactors to fund two homes, one for boys and one for girls. He communicated with

[7] For a summary, see https://www.who.int/news-room/fact-sheets/detail/leprosy

the government regarding a clean water supply. With his guidance, the residents imploded the coral on the shoreline to ensure a sufficient docking area. In an area that was suffering from chaos, he nurtured a sense of community and encouraged care for each other. A band and choir were organized, a small chapel was built, and games were encouraged (Kroeger 2018).

Father Damien has been described as having a strong personality: being "self-willed, enterprising, indefatigable, and creative yet practical. He could prove to be very assertive; some would say even stubborn or bullheaded." (Kroeger 2018, 12) Yet it is important to see that this perseverance and endless pursuit was in the service of those with whom he lived and done with deep fidelity to his faith and vocation—not for self-aggrandizement or attention. In fact, in 1876 when his family published one of his letters, he wrote, "I was a little annoyed at seeing my last letter printed in the *Annales*. Once for all, let me tell you I do not like that done. I want to be unknown to the world, and now I find, in consequence of the few letters I have written, that I am being talked about on all sides, even in America." (Kroeger 2018, 12) Father Damien, amidst his demanding personality and willingness to advocate for his co-residents even to the point of tension or reprimand, was deeply humble and motivated by endless compassion for the suffering of others.[8]

Father Damien lived with the daily knowledge that leprosy intimately surrounded him, and his way of living was to serve the residents of the community with humility and happiness. He wrote to his family in 1874, "As I told you last year, I am living here in a village consisting entirely of lepers. Though this disease is contagious, I am still enjoying perfect health, and am very happy in the midst of these unfortunate people." Father Damien took up the habit of smoking to avoid carrying the pervasive odor caused by leprosy.

Admittedly, Father Damien's sense of unity carried the cultural dynamics of his time. Particularly in his earliest years in Pena as a young priest, Father Damien sometimes looked down upon his co-residents. He initially described the people of Hawaii as uncivilized, and criticized them for cultural practices of eating, sitting, and dress. Nevertheless, Father Damien developed a profound sense of unity and solidarity with his co-residents on Molokai, though some of the language even later in his time living on Molokai would today would constitute condescension.

This sense of solidarity was not contrived: When visiting the mission headquarters in Honolulu, Damien was relegated and quarantined. He was not permitted to wear the same priestly vestments or drink from the chalice at Mass. Long before becoming ill, he described the feeling of being stigmatized with the cross of leprosy (Kroeger 2018, 32). In fact, within months of his arrival, he began to preach with the phrase "We lepers..." Later in 1883, he describes this phrase as he writes of his

[8] A visit from Father Damien's bishop offers another poignant story. The bishop arrived by ship, but the ship's captain would not permit him to disembark: "In order to see the bishop, Damien sailed out to the boat. The captain refused Damien's request to board. The priest pleaded in vain with the captain, saying that he wanted to confess his sins. 'Bishop,' the priest called to the boat, 'will you hear my confession from here?' The bishop consented, and Damien, in an exercise of humility that touched all who witnessed it, confessed his sins aloud to the bishop." (Kroeger 2018, 24-25)

sense of deep solidarity with the other residents: "I still feel some repugnance when I have to hear the confessions of those near their end, whose wounds are full of maggots. Often, also, I [scarcely] know how to administer Extreme Unction, when both hands and feet are nothing [but] raw wounds … I make myself a leper with the lepers, to gain all to Jesus Christ. That is why in preaching, I say, *We lepers*, not, *My brethren*, as in Europe." (Kroeger 2018, 16) This profound sense of solidarity and love for his co-residents developed across his life.

Living into this sense of love and solidarity, Father Damien did not live as a captive to his fear of contagion. Instead, he touched and embraced his co-residents. He dined with them and anointed the dying with oil. Eventually in 1884, the intimacy with which he lived caught up with him. As Father Damien soaked his feet in hot water, he discovered that he lacked sensation in his toes. Father Damien had contracted leprosy. In a poignant moment, he became bedridden on the Feast of St. Joseph in 1889, and died at the age of 49 several weeks later. He had lived on Molokai for 16 years.

9.4 Why and How We Engage in Global Mission Work

Far from the early Christian missionaries who went from community to community, these narratives develop immersive, mutual relationships and focus on building blocks that are locally driven and sustainable. In reality, these narratives border on hagiography and mirror a kind of hero archetype. The ordinary person faces near failure, overcoming challenges to act in extraordinary ways for the good of others. While there may be some hagiographical aspects of these narratives, they offer some central themes that invite deeper reflection by those engaged in global mission work.

9.4.1 From a Helping Reaction to Humble, Prepared Action

After her accident, Sister Mercy wanted to enter the Maryknoll order and go forward as a missionary. Instead, she spent time completing medical school in hopes that the provincial will then welcome her entry into religious life. Sister Mercy wanted the immediacy of dropping everything to go and serve people directly. With the guidance of the provincial, however, her impulse is moderated and guided. Sister Mercy had to work first to become a physician and gain medical competencies, such that she would be professionally trained as she considered her religious vocation with an order devoted to missionary work. The provincial embodied the virtue of wisdom in channeling and guiding Sister Mercy's desire to drop everything and go serve. Even after Father Damien's haste to venture across the sea, his capacity to advance the needs of people on Molokai began with learnings as a parish priest, including how to be a handyman with local resources and navigate Hawaiian society.

As we look to today, people in the US often have a similar impulse to take action in response to an event or opportunity. A parish is building a clinic; what can we send? A woman died of sepsis; how can we get them an ambulance? A natural disaster hit; let's have a drive to collect supplies. The impulse to respond and help is beautiful and admirable. But as embodied by the provincial who wisely guided Sister Mercy, sustainable and stable action usually takes significant preparation and planning.

This impulse to help may even become habitual. A habitual reaction to help is likely to lack explicit recognition of what we hope to accomplish and a deeper understanding of our own motivations. As Merton (1936) notes, when we act from habit, we may lack clear awareness of our own action. The inclination to engage in a helping action is good; it is part of the embodied spirituality that illustrates the depth and truth of faith.

Nevertheless, we need to learn the art of preparing and planning, such that we ensure we are carefully navigating the beautiful impulse that we may have to respond to need or suffering. As people of faith who may be biased to action, we may be prone to act prematurely, before we have prepared ourselves internally and explored our own motivations. We are all prone to lack humility, to enjoy a self-serving satisfaction that is too often ignorant of what transpires by our actions. We may also be at risk of acting without sufficient partnership, failing to listen to the needs and experiences of those living in low-resource settings. In pausing, we can ensure we are reflective about why we are taking action and how we want to be as we take action. In pausing, we can thoughtfully ensure understanding, preparation, communication, and consideration of overall goals through responsible and respectful collaboration that successfully advances human dignity and the common good.

9.4.2 Giving in Solidarity, Giving of More than Our Excess

Global mission work is often seen as that which we are able to do from our abundance or because of the fact that we have excess. When we reflect on the lives of Sister Mercy and Father Damien, we see a very different model of giving—where we give of ourselves or our essential resources. While Sister Mercy and Father Damien both gain energy from the adventurous experience of embracing missionary activities, they are sacrificing what is most essential to them—time, health, and resource. This is a far cry from those who want to go on a mission trip as a fun alternative spring break, or efforts to collect stuff we do not need (which sometimes may even actually be unsalvageable) and give it to those in low-resource settings.

To put it simply, the stories of these missionaries remind us that to give in meaningful ways, we must vulnerably give of our whole selves with a wholehearted affirmation of that each person has been created by God with inalienable dignity and value. This approach to giving of one's whole self embodies the principle of solidarity. Solidarity is: "not a feeling of vague compassion or shallow distress at the misfortunes of so many people, both near and far. On the contrary, it is a firm and persevering determination to commit oneself to the common good; that is to say to the good of all and of each individual, because we are all really responsible for all." (Sollicituduo Rei Socialis, 38). Giving of one's whole self means letting go of the focus on what this work does for me, and focusing instead on what is happening for the other and between us as partners.[9]

When we give of our whole selves with a deep sense of solidarity, we will also likely find that giving becomes much more complex. In our real-world experiences, we cannot overstate the frequency with which we see US stakeholders striving to give unusable stuff to stakeholders in low-resource communities. If we think that we can do effective global mission work by giving our leftovers with no further investment, we likely have an unrealistic view and a need to grow in our sense of solidarity. Obviously, there is something fundamentally different between donating supplies and moving to an isolated community of people with leprosy; however, as Sister Mercy and Father Damien show us, truly serving in solidarity is almost certainly going to involve some sacrifice that changes our lives or impacts our resources. Insofar as many individuals and health systems are engaging in global mission work today, no one is expecting such stakeholders to make the kinds of sacrifices that Father Damien made; however, effective partnership is not likely to come *only* from our excess.

9.4.3 Building Lasting and Equitable Relationships

If there is one overarching characteristic from the missionary lives of Father Damien and Sister Mercy, it is the fact that they live in relationship with the people they seek to serve. To build relationships requires an investment of time and being in proximity with others. Father Damian and Sister Mercy embraced a lasting presence with people, such that they are with a community in times of hope and amidst the reality of suffering. Facing the unpleasant realities of bodily smells or repulsive wounds, they nurture real and lasting connections with local people. As Sister Mercy illustrates, money could not motivate this action. This is born of a deep faith that inspires the desire to be in deep relationship, to the point that Father Damien describes himself as part of the community even before he has contracted the disease ("We lepers…"). Amidst the priority placed on being with people, in relationship, with a sense of deep and lasting solidarity, these missionaries show a kind of reverence for people.

Today, people engage in global mission work for a variety of reasons. The notion of spending years as a missionary is not a consideration for most people, and this reality does not eliminate a responsible space for global mission work. Yet, there are

[9] Fascinatingly, Pope John Paul II (1987, n. 39) subtly suggests solidarity as an approach in contrast with the tendency of the modern world towards neocolonialism and modern forms of imperialism: "Interdependence must be transformed into solidarity, based upon the principle that goods of creation are meant for all."

intrinsic issues that arise from episodic or short-term engagements. These trips—often less than 2 weeks and some even described as "voluntourism"—may lack meaningful immersion into the community or mutuality in relationship. Their focus may be centered too heavily on the life-changing experience of the traveler and too little on local experience and the impact of the work. Social media posts about their "service" add complexity, as Hamilton (2021) notes, suggesting that the North American participants have developed deep bonds with the vulnerable in the low-resource setting.

The episodic, or short-term, nature of global mission work heightens the likelihood of negative unintended consequences and creates a propensity for instability, ineffectiveness, and harm simply because there is not time invested to build equitable relationships between the host and visitors. Without great effort to achieve long-term communication and sustained partnership, the mission trip may lead to a focus on immediate activities designed to fit within the time constraint of spring break week, which may shift focus away from honest consideration of the actual needs or wants of the host organization or community.[10]

Perhaps reflection on the lives of Sister Mercy and Father Damien may prompt participants to consider whether the US-based organization/stakeholders demonstrate evidence of sustained, equitable relationships: Does the organization employ people who live permanently in the community, who are part of local culture, or who speak local language(s)? The lasting, sustained relationships of Sister Mercy and Father Damien illustrate how solidarity, service, and collaboration grow from equitable relationships. They even seem to echo within the "culture of encounter" for which Pope Francis repeatedly calls. A culture of encounter is characterized by friendship and dialogue, "a culture in which we find brothers and sisters, in which we can also speak with those who think differently, as well as those who hold other beliefs …" (Francis 2013). Central to a culture of encounter is nurturing the art of listening to allow space for thoughtful communication and real, lasting relationships.

Efforts to donate medical supplies and equipment require the cultivation of partnerships with ongoing communication between colleagues in both countries. Too often, medical donations are a "one-off" deposit of items, many of which were not specifically identified as needed by the recipient.[11] Again, the *majority* of donations end up not being used by those receiving the donations in low-resource settings. Another positive step would be to reduce, if not eliminate, donations that are

[10] The book *Toxic Charity* (Lupton 2011) cites several real-life, specific examples of such consequences of short-term work.

[11] It may be poignant to reflect on the sending of unvetted, unhelpful, and unrequested medical donations in light of Pope Francis' remarks: "This is one of the ugliest things: A Christian who is 'on the balcony,'" observing the person in need with detachment and distance (Brockhaus 2022). Instead of throwing to the poor from a balcony, Pope Francis (2022) asks, "do you touch the hand of the person you give the money to?… If you give alms without touching the reality, without looking into the eyes of the person in need, those alms are for you, not for that person. Think about this. Do I touch poverty, even the poverty that I am helping? Do I look into the eyes of the people who suffer, of the people that I help?"

"one-off" and pursue only projects that accurately respond to the actual need—aid or development. Misplaced aid where development is needed is one root problem of irresponsible surplus donation; but organizations involved in global mission work cannot know that if they just ship something and do not have the established relationships to know and hear the truth about what happened.

While a visit or donation is a short-term activity, the relationship must be one that is *committed*. The work must be driven by the needs and wants of the host community and oriented toward long-term outcomes as mutually defined; and the relationship is defined by trust, communication, and immersion in language, history, customs, and community. With these factors as central, there is improved opportunity to collectively consider unintended consequences, both positive and negative. It may be tempting for those who have an abundance of resources to assume that "anything is better than nothing" (particularly during a crisis), but this impulse to give can cause lasting burden and harm in both the short and long term (Compton 2012). Key differentiators of effective, respectful mission work that avoids harm is ensuring a thorough and mutual understanding of the current state, immediate and future needs, and resources/strengths within the low-resource setting and how proposed interventions will affect local people and systems now and in the future.

Without doubt, there are laudable examples where a long-term connection and relationship animates partnership—including when short-term mission trips are involved. A medical team that specializes in a type of surgery, for example, may be organized by a US-based organization that has a continuous, permanent partnership with a local organization in a low-resource community. While that particular team of clinicians may only be there for a few weeks, the overall project is stable and builds up local experts who work with their US counterparts to identify community needs and grow their skillsets. This way of working together to advance health in low-resource settings is centered on regular communication, two-way input, shared problem-solving, and deep respect—qualities that echo the lives of the missionaries we have explored.

9.5 Conclusion: Toward a Deeper Sense of Responsibility

While much can be written from an academic perspective to analyze ethical considerations of global mission work, our experience has been that the origin of the challenges in advancing positive impact in global mission work involves, at least in part, our most basic motivations and expectations; our sense of what it means to give to those in low-resource settings; our diligence—or lack thereof— to ensure we only make donations that can realistically be used; and our perceptions of the meaning of our own actions and the value of our actions for those living in low-resource communities. Ethical analysis alone will not sway our behaviors on these factors. We need a deeper spirituality involving what it really means to serve across borders as an expression of faith. We need models who inspire us to see more fully what it means to act in solidarity with people.

In drawing on the Catholic imagination and its remembrance of holy men and women from times past, we better position ourselves to embody our ideals. The narratives of holy men and women from decades or centuries past may help us unpack what it really means to embrace a sense of solidary and prompt another way of reexamining our actions from a deeper perspective. US stakeholders may not live with those in low-resource settings, but we can approach every engagement with the desire to deepen a sense of solidarity and act with reverence for the wisdom, experience, and daily life of people. Doing so is likely to prompt considerations, questions, and concerns about the quality and long-term impact of our actions.

Returning to the two scenarios from Peru and Kenya we started with, how would a renewed view of mission work look at each of these situations? We would still acknowledge that both scenarios have much that is positive about them, and there is good being done to promote greater access to healthcare and education. There are intended actions with positive outcomes that can be identified, and each has a degree of sustained relationship.

At the same time, US stakeholders in both scenarios may begin to understand and perceive the depth of complexities entangled in improving health: What do members of the community feel is their biggest need? Do the communities have the resources (human and material) to sustain the donations being offered? What happens if they don't? What dependencies have been (or may be) created by donation that would be better served by investment in local resources? When is aid being provided where development is instead required? How can the partnership evolve to reduce the need for donations and promote local economic and social solutions to fill gaps in services? All these questions—and more—are necessary to ensure the mission work is done not *for* others, but *with* others, in solidarity that is based in relationship and respect.

The consequences of engaging in global mission work are immense and often the answers to our questions are not what we had hoped to hear. To ignore those answers creates risk for people and communities that are already under-resourced and even vulnerable; this is why sometimes harm is done by faith-based actors seeking to help international communities. Often, opportunities are lost to truly achieve shared goals of stakeholders in the US and low-resource settings because we are more interested in "making a difference" than in understanding the need. Almost always, waste is created through and because of global mission efforts simply because we do not have the time to listen or consider the perspective of those we claim to serve.

To break the bad habits of charitable mission work, people of faith have a responsibility to examine their motivations, perceptions, and actions. This includes appreciating the complexities of achieving lasting impact and becoming aware of the problematic aspects of mission work throughout the history of the church, which continues today in some cases of global mission work.

The narratives of more recent missionaries like Father Damien and Sister Mercy offer renewed insight for those motivated to reduce and prevent the unintended negative consequences of global mission work. As people of faith, we are called to embody humility, act with solidarity, and invest in the primacy of relationship.

In our opening paragraph, we recalled some failures—expired drugs in foreign languages sent after a tsunami; broken medical equipment dumped in an African community; piles of stuff sent to Haitians after the earthquake. Having experienced isolation and stigma but not yet having contracted leprosy himself, Father Damien deeply embodied a sense of solidarity, seeing his life together with the people in Molokai. So deeply aware of the suffering of the people, Father Damien would often identify as part of the community, preaching in the first person plural: "*We lepers…*".

Perhaps it might be poignant to apply this rhetoric to ourselves today. If we as US stakeholders embody a humble and listening spirit with a resolute sense of solidarity, perhaps we might sound like this: "*We, as US and Haitian partners invested in advancing health in Haiti*, are devastated by an earthquake. We need water and cots for our Haitian communities. We need emergency health workers. We need to clear transportation routes. We don't want coats because we live in the Caribbean. We don't need rollerblades, and we don't have capacity to rummage through unwanted, discarded stuff."

References

Adams, Vincanne. 2013. Evidence-based global public health: Subjects, profits, erasures. In *When people come first: Critical studies in global health*, ed. João Biehl and Adriana Petryna, 54–90. Princeton: Princeton University Press.

Biehl, João, and Adriana Petryna. 2014. Overview. In *When people come first: Critical studies in global health*, ed. João Biehl and Adriana Petryna, 23–29. Princeton: Princeton University Press.

Boulet-Desbareau, Pierre. 2013. Unsolicited in-kind donations & other inappropriate humanitarian goods, strategic plan, May 2013, commissioned by OSHA.

Brockhaus, Hannah. 2022. Pope Francis wants us to do this when we give to the poor. *Catholic News Agency*, 26 March: https://www.catholicnewsagency.com/news/250804/pope-francis-wants-us-to-do-this-when-we-give-to-the-poor.

Campbell, Joseph. 1949. *The hero with a thousand faces*. New York: Pantheon.

Compton, Bruce. 2012. Thinking globally—Let's make donations more effective: Don't waste these important efforts. *Health Progress*, March-April.

Francis. 2013. Address on the vigil of Pentecost with the ecclesial movements, 18 May: https://www.vatican.va/content/francesco/en/speeches/2013/may/documents/papa-francesco_20130518_veglia-pentecoste.html.

Francis. 2015. *Laudato Si*. Vatican City: Libreria Editrice Vaticana. https://www.vatican.va/content/francesco/en/encyclicals/documents/papa-francesco_20150524_enciclica-laudato-si.html.

Francis. 2022. Angelus, 10 July: https://www.vatican.va/content/francesco/en/angelus/2022/documents/20220710-angelus.html.

Gula, Richard. 2000. Spirituality and ethics in healthcare. *Health Progress*, July-August.

Hamilton, L., L. M. B. Tokpa, H. McCain, and S. Donovan. 2021. #WhiteSaviorComplex: Confidentiality, human dignity, social media, and social work study abroad. *Journal of Social Work Education* 59 (2): 466–476.

Hanna, Bridget, and Arthur Kleiman. 2013. Unpacking global health: Theory and critique. In *Reimagining global health: An introduction*, ed. Paul Farmer, Jim Yong Kim, Arthur Kleinman, and Matthew Basilico, 15–32. Oakland: University of California Press.

John Paul II. 1987. *Solicitudo rei socialis*. Vatican City: Libreria Editrice Vaticana. https://www.vatican.va/content/john-paul-ii/en/encyclicals/documents/hf_jp-ii_enc_30121987_sollicitudo-rei-socialis.html.

Keane, James. (2020). *The Great Maryknoll Post Office Robbery of 1964 America Magazine*. New York.

Kempton, Kathryn G. 2015. Donating a single rollerblade is not going to help disaster victims. https://www.npr.org/sections/goatsandsoda/2015/01/12/376716063/donating-a-single-rollerblade-is-not-going-to-help-disaster-victims.

Kroeger, James H. 2015. Damien of Molokai: Witness of missionary joy and compassion. *Philippiniana Sacra* 50 (150): 263–294.

Kroeger, James H. 2018. Damien of Molokai and missionary compassion: 'He stretched out his hand and touched the leper' (Mark 1:41). *Landas* 32 (1): 1–41.

Lupton, Robert D. 2011. *Toxic charity: How churches and charities hurt those they help (and how to reverse it)*. New York: HarperOne.

Merton, Robert K. 1936. The unanticipated consequences of purposive social action. *American Sociological Review* 1 (6): 894–904.

Moody, Michael. 2022. Philanthropic harm: How 'doing good' can go bad. https://johnsoncenter.org/blog/types-of-philanthropic-harm-a-working-list/.

Paul, V. I. 1967. *Populorum progressio*. Vatican City: Libreria Editrice Vaticana. https://www.vatican.va/content/paul-vi/en/encyclicals/documents/hf_p-vi_enc_26031967_populorum.html.

Perry, L., and R. Malkin. 2011. Effectiveness of medical equipment donations to improve health systems: How much medical equipment is broken in the developing world? *Medical and Biological Engineering and Computing* 49:719–722.

Pontifical Council for Justice and Peace. 2004. *Compendium of the social doctrine of the Church*. Vatican City: Libreria Editrice Vaticana. https://www.vatican.va/roman_curia/pontifical_councils/justpeace/documents/rc_pc_justpeace_doc_20060526_compendio-dott-soc_en.html.

Stern, Ken. 2013. *With charity for all: Why charities are failing and a better way to give*. New York: Doubleday.

Sullivan, Richard E. 1994. *Christian missionary activity in the early middle ages*. Aldershot, UK: Variorum.

Sulmasy, Daniel P., and Edmund D. Pellegrino. 1999. The rule of double effect: Clearing up the double talk. *Archives of Internal Medicine* 159 (6): 545–550.

World Health Organization [WHO]. 2011. Medical device donations: considerations for solicitation and provision. https://iris.who.int/bitstream/handle/10665/44568/9789241501408-eng.pdf?sequence=1.

Part III
Pandemic Policies

Chapter 10
Triage and the Patient-Physician Relationship

Bernard G. Prusak

10.1 Introduction

The primary thesis of this chapter is that different accounts of the ethics of triage—that is, of how it should and should not be done—vary according to their answer to the question: Who does the physician work for? In other words (and more precisely), *whose agent* is she, or should she be? That is the underlying, critical question.

10.1.1 A First Thought Experiment

Imagine that you are a physician in your country's military: a medical officer, or a combat medic. There has just been a terrible battle that your country has won, but at great cost to life and limb. Either your unit rushes to the battlefield, or helicopters and the like evacuate wave upon wave of wounded soldiers to you. You are overwhelmed by the scale of injury. Your unit has neither the requisite personnel, nor the medical supplies to meet all the need. Thus, you must practice triage: according to the etymology of the word, you must sort through, separate out, or sift the wounded by some system or plan, based on an algorithm or set of criteria.

Historians tell us that "[t]he practice of triage arose from the exigencies of war," with the earliest documented systems dating to the eighteenth century (Iserson and Moskop 2007a, 276). Systems multiplied over the nineteenth and twentieth centuries, but so did differences over whose needs should be prioritized. The hopelessly wounded, whom it is not reasonable to expect to live even with extensive medical attention, are as a rule excluded from life-sustaining treatment. (Ideally, other forms

B. G. Prusak (✉)
John Carroll University, University Heights, OH, USA
e-mail: bprusak@jcu.edu

J. T. Eberl (ed.), *Emerging Issues in Catholic Bioethics*, Philosophy and Medicine 153, https://doi.org/10.1007/978-3-032-08862-8_10

of care would be provided.) But who should be first in line, so to speak? Should it be the severely wounded whose lives can be saved though they are unlikely to return to battle anytime soon, if ever? Or should it be soldiers with less pressing needs, but who can be treated quickly and returned to active duty?

One way or the other, in your role as a medical officer, the answer is not yours to give. For you stand under orders, in a military hierarchy, and it may happen that your commanding officer will determine that your established system, say of tending first to the immediate needs of the severely wounded, must be suspended in the interests of the overriding goal of winning the war. This will be a "state of exception" (Agamben 2005) to how you and your colleagues normally conduct triage, and you might find it emotionally difficult and even morally distressing, but you know what to do: follow lawful orders, as you agreed to do in enlisting in the military. For you are an agent of your country's armed forces. You work for your country, and the ethics of military medicine are different in this respect from the ethics of civilian practice: "Upon entering military service, physicians assume obligations to pursue military goals and abide by military procedures, with only limited freedom to resist these on medical or other grounds" (Moskop 1998, 79).

10.1.2 A Second Thought Experiment

Imagine now that you are a physician in a large hospital system. There is a pandemic on the horizon, possibly of an even greater magnitude than the COVID-19 pandemic that recently ravaged the globe. Your system is not yet overwhelmed, but the cases are coming, and you are worried that they will hit like a crashing wave. In anticipation of this eventuality, the ethics department has asked you and other physicians and health care professionals to help review the system's pandemic preparedness plan, including its protocols for rationing critical care resources, in particular its triage procedures under crisis standards of care. (Think of rationing as concerned with the allocation of scarce goods among a number of persons. Triage is concerned with which persons should be prioritized—or de-prioritized—for receiving those goods and why.)

In your role, you cannot simply dictate what the triage procedures should be, but you do have the freedom to give voice to your values. Further, if you end up strongly disagreeing with the procedures, you can protest or even leave the system and seek work elsewhere. In that respect, your situation is markedly different from that of a military medical officer, whose freedom is much more restricted. In any event, it is likely that you and all your colleagues will agree with the traditional triage goal of "striving to save the most lives while ensuring fairness and avoiding discrimination" (DeMartino et al. 2024, 1298). The nub will be determining what constitutes fairness and what makes for unjust discrimination.

No doubt some of your colleagues are utilitarians. They will argue that the hospital system's goal should be to maximize welfare or well-being, first by "giving highest priority to those with the highest chance of surviving and needing the lowest duration of treatment" (Savulescu et al. 2020, 626). Second, your utilitarian colleagues will argue, more controversially, that the system should seek to maximize both length of life, which is to say expected "life years" beyond discharge from the hospital, and quality of life, according to the metric of quality-adjusted life years (Savulescu et al. 2020; cf. MacAskill 2016, 34–39). As "operationalizing" these factors would likely be quite difficult in the chaos of an emergency department or ICU in a pandemic, the system might make it a rule to provide, for example, mechanical ventilation only "to patients predicted to survive for at least 5 years with normal quality of life" (Savulescu et al. 2020).

There is much to consider and dispute in that proposal, but focus for the moment on its implicit answer to the question of who the physician is working for, or more precisely whose agent she is. The utilitarian's answer here is: the wider public, or society at large, *not* the individual patient on a gurney. This is in strong contrast to the traditional claim that "physicians' ethical obligations are first and foremost to the individual patients entrusted to their care" (Eberl and Donovan 2020, n.p.). It has even been claimed that the value of fidelity to individual patients is "foreign to triage" (Iserson and Moskop 2007b, 283). If so, this would indeed be another "state of exception" from normal practice—though arguably the development of managed care and the like over the last several decades has already begun to cast the physician as "*primarily* an advocate for society rather than for his patient" (Pellegrino 1997, 324).[1]

Before we concede the utilitarian's answer, however, consider that, in a pandemic, patients typically present themselves sequentially rather than simultaneously, as they do after a battle (Truog 2021, 57). The image of the impartial triage officer, surveying the wounded and separating out those who will be treated from those who will not, belongs to the context of war. By contrast, it is typically out of place in the context of a pandemic in which patients present themselves one after another, and physicians initiate treatment without knowing who will come next and in what condition. It seems incontrovertible that physicians rightly initiate treatment in these circumstances, so long anyway as the patients in question are not beyond saving. After all, to withhold treatment from patients who could benefit from it would be to "abandon[] the responsibility for the patient in the absence of an immediate need to do so" (Bierer and Truog 2023, n.p.). But then we have a potential complication on our hands: What to do when a patient who ranks as higher priority for triage comes to the hospital, only for there to be no resources available for him? Should resources be withdrawn and reallocated from a lower priority patient, even when they are benefiting him and he has a fighting chance of making it out of the hospital and going home? Is it permissible in these circumstances to withdraw the resources without consent or even against the wishes of the patient or his surrogate decision-maker? Further, should the patient's physician go along with and even advocate reallocation in these circumstances, or should she resist on grounds of fidelity to him? In brief: Who does she work for? Whose agent is she, or should she be?

[1] Pellegrino is merely describing, and not at all endorsing, this development.

10.1.3 The Subject of This Chapter

With those questions, we come to the subject of this chapter: the controversy in Catholic bioethics over the ethics of triage for patients in need of life-sustaining treatment under crisis standards of care. For it was just these questions about reallocation that divided Catholic ethicists during the COVID-19 pandemic. To be sure, there were other controversies over rationing and triage during the pandemic. For example, much ink was spilled over the use of expected life years beyond hospital discharge as a factor in allocating critical care resources.[2] Quite early in the pandemic, however, Catholic ethicists generally rejected taking into account expected life years, as well as quality-adjusted life years, on the grounds that these notions reduce the value of a life to the goods it offers in the present and promises in the future and thereby discount some lives in relation to others (Prusak 2020; Sulmasy 2020a, 2020b).[3] In other words, they are forms of utilitarian reasoning, which traditional Catholic moral thought differs and dissents from in several regards (see Cahill 1981, 627–629). Taking into account expected life years beyond hospital discharge is also objectionably discriminatory both toward people with disabilities who accordingly have more limited life expectancy and toward people with life-limiting comorbidities rooted in systemic injustices such as racism (Elbaum 2020). So Catholic ethicists did not argue among themselves about the use of expected life

[2] See, for example, Bledsoe et al. (2020), Gaurke et al. (2021), and the accompanying commentaries and following exchange of letters; Scire et al. (2022), Jecker (2022), and Ennis et al. (2023).

[3] But see Daly (2020), who proposes that, "[i]n situations in which patients are expected to realize the same qualitative benefit from an intervention, the medical facility can consider the expected duration of the benefit" (p. 53). His first example is "if patient A is expected to survive for a few weeks on a ventilator, and patient B is expected to survive indefinitely, then the ventilator should be given to patient B" (pp. 53–54), which is surely uncontroversial. After all, the ventilator's benefit to patient A is quite limited: he does not even have a prospect of living without it. Daly goes on, however, to consider an example "in which two patients are expected to benefit from an intervention and [further] are expected to be discharged from the hospital post-intervention" (p. 54). Here too he proposes that "the scarce resource should be offered to the patient who is expected to enjoy the benefit for the longer duration" (p. 54), which he submits at least normally will recommend a younger patient, of say 40, over an older patient, of say 80. But this argument equivocates on the notion of the "expected duration of the benefit" of a medical intervention. Imagine that you take an antibiotic that saves your life from an infection. Does the antibiotic keep on benefiting you during your whole life? Inasmuch as it saved your life, there is a sense in which it does, but there is another, clearer sense in which the benefit was wholly realized in its work of saving your life. That, really, is when it benefited you; from this point of view, it is strange and even puzzling to suggest that the antibiotic keeps benefiting you over your whole life. The same point can be made with respect to the use of a ventilator to save a person's life from a respiratory disease. We might be willing to allow that there is a sense in which the ventilator goes on benefiting a person over his whole life, but that is a bit strange to say. Really, the benefit was helping the person through his respiratory disease. From that point of view, the benefit of the use of a respirator is the same to a 40-year-old who thereby survives COVID-19 as it is to an 80-year-old who thereby does the same. The upshot is that Daly must support his recommendation that "age indirectly guide[] the distribution of a scarce resource" (p. 54) on different grounds, whether in terms of expected life years, life-cycle considerations, or so-called fair innings.

years, though they did argue with others, to the extent that the arguments found common ground.[4]

By contrast, it is possible to distinguish two Catholic schools of thought on the questions having to do with reallocation of resources. On the one hand, there is what might be called the Saint Louis school, represented by Jason T. Eberl and Jeffrey P. Bishop. On the other hand, there is what might be called the Georgetown school, represented by Daniel P. Sulmasy and G. Kevin Donovan. The Saint Louis school holds that extraordinary crisis situations, like the COVID-19 pandemic, generate a "state of exception" in which ordinary ethical norms may not apply or may have to be applied atypically: to the point, "ventilators or other scarce life-sustaining resources may be withdrawn from a patient in order to be reallocated to another who is estimated to have a superior chance of survival or better satisfies triage criteria" (Eberl 2020, 210).[5] According to the Saint Louis school, this may be done even when the ventilator or the like is benefiting the patient and there is some likelihood that he will be discharged from the hospital. Further, the ventilator or the like may be withdrawn "unilaterally," which means "without, and perhaps against, the explicit consent of the patient or an appropriate surrogate" (Eberl 2020). The Georgetown school roundly rejects the permissibility of withdrawal and reallocation on those grounds and holds that "[v]entilator support ought not to be discontinued without consent unless the treating team has determined that ventilator treatment has become biomedically futile (that is, that the patient will not be discharged alive and will die even with maximal support)" (Sulmasy and Maldonado 2021a, 2168). This school does allow that "[t]ime-limited trials of intensive care are reasonable and ethically defensible" (Sulmasy and Maldonado 2021b, 2170), and it also concedes that a pandemic-like situation may justify change to normal practice, but this is merely "that we may need to change … the ability of a surrogate to overrule a strong medical determination" that treatment of a patient has become non-beneficial (Eberl and Donovan 2020, n.p.), which arguably surrogates should not be able to do even apart from a pandemic.

From here, the chapter proceeds in three steps. First, I examine the arguments that the two schools have leveled at one another. As will become evident, my sympathies lie in the main with the Georgetown school, though more because I do not find the Saint Louis school's challenges compelling than because I think the Georgetown school's own position is perfect in every way. Second, I propose that those arguments have not adequately engaged a deeper disagreement, namely, in the schools' answers to the critical question: Who does the physician work for? That is, *whose agent* is she, or should she be? Here again my sympathies lie in the main with the Georgetown school, but there is a case to be made, which is this chapter's subsidiary thesis, that the Georgetown school fails to reckon with a second critical question: Under crisis standards of care, in the context of a pandemic or the like,

---

[4] See again Gaurke et al. (2021) and the accompanying commentaries and following exchange of letters.

[5] The example of choice is nearly always ventilators.

when patients present one after another at a hospital, *whose patients* are they? Are they physicians' patients, or the hospital's patients, or both? If they are in some sense the hospital's patients, what are the hospital's obligations? Whose agent is it, or should it be? To close the chapter, I sketch this case and propose it for further thought.

10.2 Point, Counterpoint

10.2.1 Prima Facie Claim?

The Georgetown school's first salvo in this controversy long predates the COVID-19 pandemic. In a 1994 paper, Sulmasy and Jeremy Sugarman claim that, "[p]rovided that there is a genuine need and that therapy is biomedically effective, the historical fact that therapy has been initiated entails a claim to continue therapy" (Sulmasy and Sugarman 1994, 218). This holds, they claim further, even in a setting of scarce resources when a different patient has a better prognosis of survival with the resource in question, for example 90% versus 50% (Sulmasy and Sugarman 1994, 221). Sulmasy and Sugarman advance this claim in an argument against what they call the "equivalence thesis," namely (in brief), that "there is really no moral difference between withholding and withdrawing life-sustaining therapy" (Sulmasy and Sugarman 1994, 218). They hold that there is in fact a moral difference (withdrawing may require "more justification" [Sulmasy and Sugarman 1994, 221]), but we can abstract for present purposes from the argument over the equivalence thesis and focus, instead, on the grounding they propose for the claim that the historical fact that therapy has been initiated entails a claim to continue it. This is Robert Nozick's principle of "original acquisition of holdings," which states that "[a] person who acquires a holding in accordance with the principle of justice in acquisition is entitled to that holding" (Nozick 1974, 151). A way to acquire a holding unjustly is to steal it from someone who had justly acquired it. So long as there is no such injustice in the background, and further so long as a patient has a medical need for a therapy and that therapy is not biomedically ineffective, Sulmasy and Sugarman (1994, 220) submit that the principle of original acquisition of holdings "seems to explain" the intuition they have and they suggest others will share that "there is a claim to original acquisition that arises whenever one has already been started on a therapy, and this claim must be considered whenever one contemplates withdrawing therapy."

But the argument is remarkably weak. First, Sulmasy and Sugarman (1994, 219) reject the "broad application" of Nozick's principle. That is, they are skeptical that it is a basic principle of justice, though for Nozick it is. Thus, second, the principle supplies little more justification for the intuition that they invoke it to justify than does the intuition itself. For an intuition is not well supported by a principle that itself finds support only in a limited set of intuitions. In that case, invoking the principle comes to little more than reiterating the intuitions. Sulmasy more or less concedes the weakness of the argument in a 2021 paper that cites it: "Patients *seem* to acquire a claim to continued care once that care has been started" (Sulmasy and Maldonado 2021a, 2167, emphasis added). "From the point of view of the patient, having been triaged to [life-sustaining treatment] *seems* to constitute at least a *prima facie* expectation that such support will continue" (Sulmasy and Maldonado 2021a, emphasis added). "The commitment entailed by initiating treatment *seems* to matter morally" (Sulmasy and Maldonado 2021a, 2167, emphasis added). The tentativeness of the phrasing is striking.

In a 2020 paper, Eberl challenges Sulmasy and Sugarman's argument on several grounds. First, he writes, "even if a patient has a *prima facie* claim-right—akin to a property right—to continued ventilation once started, situations involving dire need on the part of others call this apparent property-like right into question" (Eberl 2020, 210–211).[6] Here Eberl invokes Thomas Aquinas's argument to the effect that (in Eberl's formulation) "one's claim-right to a resource of which others are in dire need may be abrogated in relevant circumstances" (Eberl 2020, 211). This argument, however, does not quite apply to the circumstances that Sulmasy and Sugarman have in view. In these circumstances, the patient on the ventilator that another patient direly needs direly needs the ventilator himself! If the first patient has a *prima facie* claim-right to the ventilator (as Eberl appears to concede), it is hard to see how or why the neediness of the second patient overrides the neediness of the first and abrogates his apparent claim-right.

Eberl also objects to disallowing unilateral withdrawal on the grounds that it "results practically in a *first-come/first-served* policy" for the distribution of scarce critical care resources, which is arbitrary and "could unjustly disadvantage individuals who lack ready access to health care services" (Eberl 2020, 211). Finally, though along similar lines, he claims that we must consider whether disallowing unilateral withdrawal conflicts with "other salient ethical principles, such as distributive justice and preferential consideration for those who experience disparities in health care access" (Eberl 2020). These two objections might be summarized by observing that, if a patient who has been started on a therapy does have a claim that "must be considered whenever one contemplates withdrawing therapy" (Sulmasy and Sugarman 1994, 220), or in other words "at least a *prima facie* expectation that such support will continue" (Sulmasy and Maldonado 2021a, 2167), that claim or expectation, to judge from those articulations of it, hardly seems to be absolute, and so it might justifiably be set aside in the context of a pandemic or other public health crisis in which triage has become necessary.

---

[6] To have a claim-right, as Stephen Darwall (2006, 18) nicely explains, "includes a second-personal authority to resist, complain, remonstrate, and perhaps use coercive measures of other kinds … if the right is violated." In other words, a claim-right gives you "a standing to make a special demand against people who might step on your feet—you have the authority to resist, claim compensation, and so on."

10.2.2 Double Effect?

Bishop and Eberl (2021a, 2166) propose two justifications for resolving what they call this "ethical dilemma," with values at stake on both sides, in a way that permits unilateral withdrawal. First, they appeal to double effect reasoning. As they formulate the so-called rule of double effect:

> [I]t may be permissible for an agent to undertake a directly intended action aimed at a moral good even if the agent foresees that this action will result in a negative outcome, provided that (1) the agent does not intend the negative outcome; (2) the negative outcome is not a necessary means to bring about the intended good; and (3) the positive value of the intended good is [favorably] proportionate to the negative value of the foreseen outcome (Bishop and Eberl 2021a).

Eberl speaks to each of these conditions. According to him, "the intended good end is to make a vital resource available to save a person's life," and (1) "the foreseen death of the patient who is extubated is neither intended for its own sake, nor (2) is it a means of bringing about the good of saving the patient to whom the ventilator is reallocated" (Eberl and Donovan 2020, n.p.). After all, if the patient does not die, that is all for the better, and his death is not a crucial part of the successful execution of the plan. Further, if the patient does die (as is admittedly foreseen), his death as such does not have the effect of saving the other patient. That is, his death does not, strictly, play any part in achieving that end. Finally, (3) "the proportionality criterion is satisfied insofar as what is gained/lost are two equally valuable lives of persons with intrinsic dignity"; the resource has been withdrawn and reallocated only because it has been judged that the second patient "is more likely to benefit physiologically" from its use (Eberl and Donovan 2020).

Several comments are in order before we review the Georgetown school's response. First, surely "mak[ing] a valuable resource available to save a person's life" is not the only end intended in this scenario. Extubating the patient is also intended as an end. That is a distinct object of action. Described no further, extubating a patient is not morally objectionable or problematic, but that description is clearly incomplete. The action here that the patient's team intends is extubating a patient who is benefiting biomedically from a ventilator and thus has a fighting chance of recovering and being discharged from the hospital. That action is at least morally problematic and may well be morally objectionable. Second, if the team extubating the patient fully expects him to die, then it is hard to see how this "negative effect" is not intended, even if the patient miraculously lives. It is true that the patient's death is not "intended *for its own sake*" (emphasis added), as a goal that might have been pursued independently under entirely different circumstances, but if the death is expected with near certainty, it is disingenuous and unconvincing to deny that it belongs to the plan of what is being done.[7] The team intends to do some-

[7] It might be objected that this claim entirely undercuts the logic of double-effect reasoning, but I disagree. Instead, it undercuts accounts that have "stretched the distinction between the intended and the foreseen beyond the point of credibility" (Sulmasy 2007, 117–118). In this case, extubating

thing that will cause the patient to die, or at the very least that highly risks his death. And that is again at least morally problematic and may be morally objectionable. Third, it is true that the patient's death as such is not a means to saving someone else's life, but extubating the patient is part of the means, and extubating him is an action that is expected to cause him to die—which is once more at least morally problematic and may be morally objectionable. Moreover, it is not plausible in this case to claim that withdrawing life-sustaining treatment merely allows the patient to die of his underlying condition.[8] That is true when a life-sustaining treatment is biomedically ineffective, but in the case in question it is stipulated that the patient is benefiting from the treatment, and it is undeniable that removing it is part of what sends him to his death.[9]

Sulmasy and Maldonado (2021a, 2168) likewise observe that "reallocation involves *two* intentional actions (removing a ventilator and connecting it to someone else), not just one action as required for correctly applying the rule of double effect." After all, double effect reasoning has to do with one action that has two effects, the second of which (the effect that follows "secondarily" from what the action is primarily about) is considered morally problematic and thus calls the action into question (that is, makes it a fit subject for casuistry). By contrast, "Taking one patient off a ventilator … does not, of itself, do anything to save another patient's life" (Sulmasy and Maldonado 2021b, 2170). Donovan observes that the first action, namely, of removing the ventilator, is properly described more fully as removing a ventilator "from a patient whose recovery is still in play" (Eberl and Donovan 2020, n.p.). Inasmuch as it is recognized "that the withdrawal from the first patient must result in [his] death," it must be admitted that bringing about his death is part of what the team removing the ventilator intends to do, though in different circumstances the team would not pursue his death as an independent goal (n.p.). As Elizabeth Anscombe acidly remarks, "Circumstances, and the immediate facts about the means you are choosing to your ends, dictate what descriptions of your intention you must admit" (Anscombe 1982, 23).[10] Donovan also observes that the preservation of the second patient's life through reallocation of the ventilator "has occurred through the action that resulted in the first patient's death," namely, the ventilator's withdrawal, with the upshot that "[t]he good effect has occurred through

---

the patient to make the ventilator available for someone else just is killing the patient or at least highly risking his death; the bad effect is wholly spatiotemporally contained within the intended act (Sulmasy 2007, 119).

[8] Eberl does not make this claim, but Daly (2020, 54) does. He cites a 1980 statement from the Congregation for the Doctrine of the Faith observing that the benefits of a life-sustaining treatment may be judged by physicians and patients to be disproportionate to the burdens, such that the treatment cannot be imposed and may rightfully be refused. Such a refusal amounts to acceptance of the human condition rather than suicide—but this case is manifestly different from that of a patient who is benefiting from a treatment only to have it removed for someone else's use.

[9] See, for much more on double effect reasoning, Prusak (2016, chapters 2 and 3).

[10] Compare Murphy (2013, 200), "the intention with which the agent acts is constrained by the structure or design or nature of the act-type"; and Jensen (2010, 468), "causal structures are the material conditions out of which intentions are made."

the means of the bad effect, violating the rule of double effect" (Eberl and Donovan 2020, n.p.). Admittedly, the first, extubated patient's unexpected survival would not prevent the second patient's survival by the reallocated ventilator, but that bit of moral luck (the medical team is not responsible for killing the first patient) does not change the fact that the plan was to try to save one patient by killing or at least highly risking the death of another.

Finally, Donovan disputes that the proportionality condition is satisfied: no good at all accrues to the first patient; for him, the withdrawal of the ventilator is all loss. Bishop and Eberl (2021b, 2169) comment that they invoke the rule of double effect "only to justify the act of extubating a patient with poor prognosis [*poorer* prognosis?] to make a ventilator *available* for reallocation." But making a ventilator available by withdrawing it from a patient hardly satisfies the proportionality condition: making a ventilator available, considered simply on its own, is not such a great good that it is worth causing someone to die as an allegedly secondary effect.

10.2.3 Just War?

There is more to say about whether causing the patient to die figures as a means to making the ventilator available (it does, for reasons indicated above), but I believe it is clear enough that double effect reasoning does not "justify the … act of withdrawal" (Bishop and Eberl 2021b, 2169). Though not conceding this point, Bishop and Eberl turn to what they call "just war reasoning" as a second strategy for justifying unilateral withdrawal of life-sustaining treatment for purposes of reallocation. Like the rule of double effect, just war reasoning, they propose, "defines a set of criteria that render it permissible for an individual to perform an action that would normally be impermissible"—more precisely, that might appear to be impermissible—"for the sake of proportionate … benefit" (Bishop and Eberl 2021a, 2166).[11] The key analogy they invoke is between, on the one hand, a general ordering "a strategic withdrawal of troops defending a city to reallocate them to where there is a greater chance of victory," and, on the other hand, a triage committee ordering "the unilateral withdrawal of a ventilator from one patient to reallocate it to another who has greater expected survivability" (Bishop and Eberl 2021a). If it is permissible for the general to do what he does in the state of war, the reasoning goes, it is likewise permissible for the triage committee to do what it does in the "state of exception" generated by a pandemic or the like.

Sulmasy and Maldonado rightly question whether the struggle against a novel coronavirus is analogous to a state of war and note further that, even if the analogy holds, not everything goes in war: for example, they submit, "the 'War on Terrorism' does not justify torture" (Sulmasy and Maldonado 2021b, 2170). The general's

[11] They write "social benefit," but that is more accurate for just war reasoning than for double effect reasoning. (The proportionate benefit need not be "social" for a secondary effect to be justifiable.)

withdrawal of troops from one city in order to reallocate them to another certainly seems justifiable, despite the foreseeable loss of life in the first city, but it is not justified through double effect reasoning: the withdrawal and the reallocation are again two distinct actions, and it makes little sense to say that making the troops *available*, all on its own, is such a great good that it is worth doing despite the secondary effect of sacrificing a city. To introduce a further observation beyond the texts under discussion, the good that the general is seeking is winning the war; that is the goal that is guiding his strategy. In other words, he is intent not just on saving some other city; he is after victory in a greater struggle, and it is the presumed moral importance of achieving this victory (assuming that the cause is just and that his country's conduct of the war does not vitiate its claim to have right on its side) that justifies the wrenching strategic decision in question. It follows that, for the analogy between the general's order and the triage committee's order to hold, it would have to be the case that a similar overriding goal is likewise guiding the triage committee. That is, it would have to be the case that the triage committee is after more than only saving a different patient: withdrawing the ventilator from one patient and reallocating it to another would have to be a move in the wider war against SARS-CoV-2. But it is not, and the analogy collapses.

The two schools argue further about the practicality of reallocation (Sulmasy and Maldonado 2021a, 2167; Bishop and Eberl 2021b, 2169), and they also disagree about whether allowing or disallowing reallocation would be more unjustly discriminatory toward underserved and historically poorly served populations (Sulmasy and Maldonado 2021a, 2168; Bishop and Eberl 2021b, 2169). But these differences, though not unimportant, do not go to the heart of what separates the schools. It is interesting in this regard that Bishop and Eberl (2021a, 2166) propose that the reason why just war reasoning "may not seem immediately applicable to the health care context" is that "the last half-century of American medicine has witnessed increasingly abundant resources such that physicians typically have not had to consider larger consequences and thus are privileged to focus more narrowly on the patients under their care." Yet "[t]his situation," they claim, "is merely an artifact of the blessings of a wealthy nation." The implication seems to be that the state of war against sickness and disease—in which human life is presumably poor, nasty, brutish, and short—is the natural state, to which the "state of exception" generated by the COVID-19 pandemic threatened to return us.[12] Further, from this point of view, the traditional patient-physician relationship, in which the physician is understood to be committed to act for the good of the patient and is bound by obligations of fidelity to do so, itself appears to be a fragile, context-dependent artifact, not "the moral foundation of healthcare" that Sulmasy and Maldonado (2021a, 2168) proclaim it to be, or at least not a foundation that can withstand hell or high water. Here,

[12] It could be countered that this story misrepresents the economic dynamics of U.S. health care over the last fifty years; see, for a prescient discussion, Feldstein (1986). For example, arguably, the pandemic was made worse because of staff shortages that hospital systems had previously instituted in the interests of the bottom line. See in this regard Robbins et al. (2022).

then, is a deep difference indeed, and it returns us at last to our question: Who does the physician work for? That is, whose agent is she, or should she be?

10.3 From "Whose Agent?" to "Whose Patient?"

10.3.1 Against Dual Agency

As we have seen, the Georgetown school is critical of the Saint Louis school's two proposed strategies of justifying unilateral withdrawal and reallocation of life-sustaining treatment under crisis standards of care: appeal to double effect reasoning, and appeal to just war reasoning. I share that assessment: in my judgment, neither appeal is compelling; each is seriously flawed. Yet the Georgetown school's most ardent objection to the Saint Louis school's thinking has nothing to do with its faulty logic. Instead, that objection is: "Reallocation policies undermine the patient-physician relationship" (Sulmasy and Maldonado 2021a, 2167). More fully:

> Patients who are struggling just to stay alive ought not to be forced to bear the additional fear that their caregivers might be looking for signs that would justify taking [life-sustaining treatment] away from them. One could not imagine anything more damaging to patient trust. Even if reallocation were practical and might result in a few more lives being saved, it would not be worth sacrificing this moral bedrock. Reallocation makes physicians dual agents (Sulmasy and Maldonado 2021a, 2168).[13]

It also thus contradicts a fundamental tenet of the Georgetown school: "The physician cannot be a double agent; he or she serves primarily the patient or primarily the self, the hospital, the economic or fiscal policy, or the law" (Pellegrino and Thomasma 1993, 44).

Georgetown is often associated with the four principles—respect for autonomy, beneficence, non-maleficence, and justice—formulated by Tom L. Beauchamp and James F. Childress in their book *Principles of Biomedical Ethics*, first published in 1979, when they were both at Georgetown, and now in its eighth edition with a ninth apparently on its way. The Georgetown school in the present controversy, however, has a different derivation: its founding figure is Edmund D. Pellegrino, who taught at Georgetown for more than 30 years and served as the founding director of its Center for Clinical Bioethics. Daniel Sulmasy was a student of Pellegrino's at Georgetown, and he has defended and advanced his teacher's thinking.

According to Sulmasy, Pellegrino's "overarching view was that once one understands what medicine is as a human practice, and understands its purposes, then,

---

[13] It is worth noting that this quotation suggests a different grounding for Sulmasy and Sugarman's (1994) argument against the equivalence thesis: namely, the patient-physician relationship, not Nozick's principle of "original acquisition of holdings." It is similarly noteworthy that, in a paper defending the soundness of double effect reasoning, Sulmasy and Pellegrino (1999) appear concerned first and foremost that undermining it would have adverse consequences for the patient-physician relationship, particularly at end of life.

and only then, can one derive a set of moral expectations for practitioners and establish norms for the relationship between medicine and the society at large" (Sulmasy 2014, 106). In brief, as a human practice, "medicine is first and foremost a relationship" between a vulnerable patient and a healing professional (paradigmatically, a physician) who promises to help and thereby incurs a fiduciary obligation to be worthy of the trust that the patient has accorded her (Sulmasy 2014, 107).[14] The "central question" guiding the physician must then be: "[W]hat is the right and good healing act for this patient in these circumstances?" (Sulmasy 2014, 109).[15] Against this background, Sulmasy notes that Pellegrino

> was deeply troubled by bureaucratic interference in the physician-patient relationship, and particularly opposed to attempts to induce physicians though financial incentives into becoming agents of society, rationing health care resources in the setting of the patient-physician relationship. Such arrangements undermined the trust he held essential to that relationship, and made physicians into dual-agents (Sulmasy 2014, 109).

It should be amply evident from this chapter that Sulmasy is likewise troubled by any arrangement that would undermine the trust between patient and physician and shift physicians' "emphasis from individual patients to improving the health of the entire population" (Sulmasy 2006, 81). To the contrary, Sulmasy declares that, "[i]n any physicians' oath worth its salt, physicians should swear to commit themselves to the service of their patients—putting patients' interests first" (Sulmasy 2006, 105).

10.3.2 Between Individual and Social Obligations

Pushing back against being made a dual agent with divided loyalties under the exigencies of managed care, though, is one thing. Pushing back against being made a dual agent under triage procedures in a public health crisis might well seem like something importantly different. In the first case, what is at stake is, largely, saving money. Pellegrino (1997, 328) was scathing about the supposed imperative "to forestall the national economic disaster of spending too much on health care" when we spend gobs of money on the sports industry, entertainment, recreation, and what he termed "self-indulgence." In the second case, what is at stake is, at least theoretically, saving more lives. Here the question of the practicality—Sulmasy and Maldonado say the *im*practicality in the chaos of a pandemic or the like—of continually reviewing and updating patients' prognoses is relevant. Sulmasy and Maldonado (2021a, 2167) also note that "medical prognostication is notoriously imprecise at the individual level," such that there is little reason to trust the

[14] Compare Pellegrino (1987, 7): "At the center of medical morality is the healing relationship." Or Pellegrino and Thomasma (1993, xii): "the moral essence of a health profession is the special relationship that sickness and the response to illness creates between healer and patient."

[15] "[T]he good of the patient provides the architectonic of the relationship" (Pellegrino and Thomasma 1993, 53). Accordingly, "Justice is a requisite duty because what we owe to the patient is fidelity to the trust we elicited when we offered to help" (Pellegrino and Thomasma 1993).

judgment that one patient has a 40% chance of survival while another has a 60% chance. But grant that there is reason to trust the judgment that one patient has only a 20% chance of survival while another has an 80% chance. The first patient is using life-sustaining treatment that the other patient needs and cannot otherwise get. What then?

We know already the two schools' answers, but both schools need to say more. Bishop and Eberl (2021a, 2165) cogently ask:

> [H]ow should [crisis standards of care] balance physicians' ethical obligation to provide effective life-sustaining treatments to patients under their care with the wider obligation to exercise prudential stewardship over scarce medical resources to maximize survivability for the most people?

But they do not really speak to this question, or anyway they do not wrestle with it. For them, it seems "the wider obligation" to society takes precedence. But just how wide is this wider obligation, and what is its basis? Gesturing to the common good does not suffice (see Bishop and Eberl 2021b, 2169).[16] Further, is the obligation really to *maximize* survivability? Why? The Georgetown school likewise does not really speak to this question or wrestle with it. Instead, for this school, it is clear that "physicians' ethical obligation ... to patients under their care" takes precedence. Yet Pellegrino (1987, 13–14) acknowledged that "[a] serious question for conscientious physicians is how to deal with conflicts between what is good for the patient and what is good for society—between individual and social obligations." He also allowed that, "when resources are limited, there are morally sound and morally unsound ways of confronting the dilemmas of clinical choice. Under certain conditions rationing can be morally justified so long as these conditions are set by ethics and not economics" (Pellegrino 1997, 323). If only he had said more here!

10.4 A Third and Final Thought Experiment

To begin to consider what else might be said, imagine once more that you are a physician, though this time in a Catholic hospital or hospital system. The ethics department has asked you and other physicians and health care professionals to help review the system's triage procedures under crisis standards of care. Some of your colleagues are utilitarians (it is not nearly the case, after all, that everyone working in the system is Catholic, and some Catholics may be utilitarians despite magisterial teaching (see John Paul II 1993, n. 74)); but a purely utilitarian framework seeking to maximize overall net benefit, without heeding the oft-proclaimed "separateness of persons," is a non-starter with the powers-that-be (compare Bishop and Eberl 2021a, 2165). You are skeptical of a policy permitting unilateral withdrawal and reallocation of life-sustaining treatments from one patient to another. First, you have

[16] Compare, too, Pellegrino and Thomasma (1993, 93): "the root of justice lies in the individual's good, not only in the common good, in commutative as well as distributive justice."

questions about the practicality of such a policy in the chaos of a pandemic. Second, you do not find the non-utilitarian arguments purporting to justify such a policy compelling. To your mind, neither appeal to double effect reasoning, nor appeal to just war reasoning is sound. Third, you are troubled by what such a policy would mean for and do to the patient-physician relationship, which in your view is sacrosanct—and what is more, is a means for you to participate in your own way in the healing ministry of Jesus.

But you see where the other side is coming from. If, per hypothesis, there is one patient who has only a 20% chance of survival while another has an 80% chance, and the first patient is using life-sustaining treatment that the other needs and cannot otherwise get, do you want to stand on principle and watch the second patient die? Or is it right and just for you and your colleagues to intervene in such a case, and try to save the patient who seems more likely to live? Time-limited trials, specifying the progress a patient must make in order to stay on life-sustaining treatment, might be a compromise, but then it needs to be asked how high or how low to set the bar that patients must clear. (A 20% chance of winning the lottery is a very high chance given the stakes!) In any event, yes, the physician must be the patient's "advocate with the system" (Pellegrino 1997, 325), and not "*primarily* an advocate for society rather than for his patient" (Pellegrino 1997, 324), that is, "not simply an instrument of social, institutional, or fiscal policies" (Pellegrino and Thomasma 1993, 69). But who is your patient here? In a pandemic or other public health crisis, it might make more sense to say: Neither is *your* patient, or least yours alone; both are the *hospital's*, or at least the hospital's as well as yours. Arguably, the questions to ask then become: What are the hospital's obligations in this situation? Who is the hospital working for? Whose agent is it, or should it be?

A step back is in order. In *After Virtue*, Alasdair MacIntyre warns of the corrupting power of institutions on practices like that of medicine. By 'practice,' MacIntyre (1984, 187) means "any coherent and complex form of socially established cooperative human activity through which goods internal to that form of activity are realized in the course of trying to achieve those standards of excellence which are appropriate to, and partially definitive of, that form of activity."[17] Pellegrino conceived of medicine in those terms (Sulmasy 2014, 108): for him, the internal good it realizes is the good of the patient, which encompasses not only his biomedical good, but also what he understands to be the good choice for himself in his circumstances, his understanding of what is good for human beings, and his understanding of what is good for him as a spiritual being, if he thinks about himself that way (Pellegrino and Thomasma 1993, 58). It is possible to realize that fourfold good only in and through the patient-physician relationship, which is thus precisely the

[17] And the sentence even goes on: "with the result that human powers to achieve excellence, and human conceptions of the ends and goods involved, are systematically extended" (MacIntyre 1984). MacIntyre gives as immediate examples the games of football (fútbol?) and chess, architecture, farming, the enquiries of physics, chemistry, and biology, the work of historians, and painting and music.

form in which medicine, as such, is practiced.[18] Ironically, MacIntyre does not define what he means by an 'institution,' but it seems he has in mind familiar organizations, such as hospitals, universities, and firms, consisting of complex systems of relations distributing powers and opportunities, as well as responsibilities and restrictions. MacIntyre (1984, 194) claims that "[i]nstitutions are characteristically and necessarily concerned with [so-called] external goods," such as money, power, and status. He also claims, however, that institutions are necessary for practices: institutions are the "bearers" of practices; "no practices can survive for any length of time unsustained by institutions" (MacIntyre 1984), for the practitioners of medicine, history, farming, or music themselves need money and other material goods both to do what they do (practices cost money) and to support themselves (practitioners must eat). The upshot is that practices and institutions stand in an "intimate" relationship, and "the ideals and the creativity of the practice are always vulnerable to the acquisitiveness of the institution," which is wont to turn the practice to its purposes (MacIntyre 1984).

You see that risk in health care—it is everywhere visible nowadays for people who have the eyes to see it—and you understand why Pellegrino was so exercised about the corrupting potential of managed care and the like on the practice of medicine. But neither MacIntyre nor Pellegrino seems to think of institutions as capable of and even responsible for fostering and advancing critical thought about the practices of which they are the bearers.[19] Yet that surely is what the hospital system is asking you, through its ethics department, to do in this moment of crisis when its triage procedures must be reviewed and established. The system is asking you how it should act responsibly toward all the patients who will present themselves at it for care. Arguably at least, they will be *its* patients as well as *yours*, and the question is accordingly whether the system has a social responsibility not only to uphold your practice, but also to ask you to suspend it in this exceptional context. Normally, the fact that a patient is yours gives you a reason to be his advocate in the system. Perhaps the fact that, under crisis standards of care, a patient is as much the hospital's as yours is a so-called context-undermining reason (Cullity 2013). That is, perhaps it means that, in this exceptional context, the fact that the patient is yours is *no longer* a reason to be his advocate in the way you normally would. Perhaps, instead, the fact that the hospital system has a wider social obligation gives you reason to change your mind and agree to a policy authorizing unilateral withdrawal and reallocation of life-sustaining treatment under a limited, closely monitored set of circumstances. You will have to give this possibility further thought. (You will need to know more, for example, about this putative wider obligation. Just how wide is it, and what is its basis? And is it really to *maximize* survivability? Why?)

There is, though, one more thought for you to keep in mind: according to the political theorist Giorgio Agamben, states of exception, once invoked, tend to make

[18] See Pellegrino (1997, 12): "It is in the moment when physician and patient together decide what should be done that medicine as medicine comes into being."

[19] Compare Cordell (2011).

themselves the new rule (Agamben 2005, 7). More precisely, the changes they justify do not go away once the state of exception has passed, but persist. It could well be, then, that you should think twice before agreeing to the suspension of the normative patient-physician relationship under crisis standards of care. If it is already being eroded under conventional standards of care, its suspension under crisis standards of care might just be the final blow.

References

Agamben, Giorgio. 2005. *State of exception*. Trans. Kevin Attell. Chicago: University of Chicago Press.

Anscombe, Elizabeth. 1982. Action, intention and 'double effect'. *Proceedings of the American Catholic Philosophical Association* 56:12–25.

Bierer, Barbara E., and Robert D. Truog. 2023. The unresolved challenge of triage. *JAMA Network Open* 6 (8): e2329676.

Bishop, Jeffrey P., and Jason T. Eberl. 2021a. Is it ethically permissible to unilaterally withdraw life-sustaining treatments during crisis standards of care? Yes. *Chest* 159 (6): 2165–2166.

Bishop, Jeffrey P., and Jason T. Eberl. 2021b. Rebuttal from Drs Bishop and Eberl. *Chest* 159 (6): 2169–2170.

Bledsoe, Thomas A., Janet A. Jokela, Noel N. Deep, and Lois Snyder Sulmasy. 2020. Universal do-not-resuscitate orders, social worth, and life-years: Opposing discriminatory approaches to the allocation of resources during the COVID-19 pandemic and other health system catastrophes. *Annals of Internal Medicine* 173 (3): 230–232.

Cahill, Lisa Sowle. 1981. Teleology, utilitarianism, and Christian ethics. *Theological Studies* 42 (4): 601–629.

Cordell, Sean. 2011. Virtuous persons and social roles. *Journal of Social Philosophy* 42 (3): 254–272.

Cullity, Garrett. 2013. The context-undermining of practical reasons. *Ethics* 124 (1): 8–34.

Daly, Daniel J. 2020. Guidelines for rationing treatment during the COVID-19 crisis: A Catholic approach. *Health Progress* 101 (3): 50–56.

Darwall, Stephen. 2006. *The second-person standpoint: Morality, respect, and accountability*. Cambridge, MA: Harvard University Press.

DeMartino, Erin S., Jackson S. Ennis, Susan M. Wolf, and Daniel P. Sulmasy. 2024. Learning from COVID-19 triage schemes to face the next public health emergency. *Journal of the American Geriatrics Society* 72 (4): 1298–1301.

Eberl, Jason T. 2020. Ethics as usual? Unilateral withdrawal of treatment in a state of exception. *American Journal of Bioethics* 20 (7): 210–211.

Eberl, Jason T., and G. Kevin Donovan. 2020. Is it ethical to unilaterally withdraw life-sustaining treatment in triage circumstances? Health Progress 101(2). https://www.chausa.org/publications/health-progress/archive/article/pandemic-coverage/is-it-ethical-to-unilaterally-withdraw-life-sustaining-treatment-in-triage-circumstances. Accessed 18 November 2024.

Elbaum, Alan. 2020. Black lives in a pandemic: Implications of systemic injustice for end-of-life care. *Hastings Center Report* 50 (3): 58–60.

Ennis, Jackson S., Kirsten A. Riggan, Nicholas V. Nguyen, Daniel B. Kramer, Alexander K. Smith, Daniel P. Sulmasy, Jon C. Tilburt, Susan M. Wolf, and Erin S. DeMartino. 2023. Triage procedures for critical care resource allocation during scarcity. *JAMA Network Open* 6 (8): e2329688.

Feldstein, Paul J. 1986. The emergence of market competition in the U.S. health care system: Its causes, likely structure, and implications. *Health Policy* 6:1–20.

Gaurke, Mary Katherine, Bernard G. Prusak, Kyeong Yun Jeong, Emily Scire, and Daniel P. Sulmasy. 2021. Life-years and rationing in the COVID-19 pandemic: A critical analysis. *Hastings Center Report* 51 (5): 18–29.

Iserson, Kenneth V., and John C. Moskop. 2007a. Triage in medicine, part I: Concept, history, and types. *Annals of Emergency Medicine* 49 (3): 275–281.

Iserson, Kenneth V., and John C. Moskop. 2007b. Triage in medicine, part II: Underlying values and principles. *Annals of Emergency Medicine* 49 (3): 282–287.

Jecker, Nancy S. 2022. Too old to save? COVID-19 and age-based allocation of lifesaving medical care. *Bioethics* 36 (7): 802–808.

Jensen, Steven J. 2010. Getting inside the acting person. *International Philosophical Quarterly* 50 (4): 461–471.

John Paul II. 1993. *Veritatis splendor*. https://www.vatican.va/content/john-paul-ii/en/encyclicals/documents/hf_jp-ii_enc_06081993_veritatis-splendor.html. Accessed 20 November 2024.

MacAskill, William. 2016. *Doing good better: How effective altruism can help you help others, do work that matters, and make smarter choices about giving back*. New York: Penguin.

MacIntyre, Alasdair. 1984. *After virtue: A study in moral theory*. 2nd ed. Notre Dame, IN: University of Notre Dame Press.

Moskop, John C. 1998. A moral analysis of military medicine. *Military Medicine* 163 (2): 76–79.

Murphy, James G. 2013. The principle of double effect: Act-types and intentions. *International Philosophical Quarterly* 53 (2): 189–205.

Nozick, Robert. 1974. *Anarchy, state, and utopia*. New York: Basic Books.

Pellegrino, Edmund D. 1987. Toward a reconstruction of medical morality. *Journal of Medical Humanities and Bioethics* 8 (1): 7–18.

Pellegrino, Edmund D. 1997. Managed care at the bedside: How do we look in the moral mirror? *Kennedy Institute of Ethics Journal* 7 (4): 321–329.

Pellegrino, Edmund D., and David C. Thomasma. 1993. *The virtues in medical practice*. Oxford: Oxford University Press.

Prusak, Bernard G. 2016. *Catholic moral philosophy in practice and theory: An introduction*. New York: Paulist Press.

Prusak, Bernard G. 2020. All hospitals want to save lives. Should they also be seeking to save 'life years'? *America: The Jesuit Review*, May 28. https://www.americamagazine.org/faith/2020/05/28/all-hospitals-want-save-lives-should-they-also-be-seeking-save-life-years. Accessed 18 November 2024.

Robbins, Rebecca, Katie Thomas, and Jessica Silver-Greenberg. 2022. How a sprawling hospital chain ignited its own staffing crisis. *New York Times*, December 15. https://www.nytimes.com/2022/12/15/business/hospital-staffing-ascension.html. Accessed 18 November 2024.

Savulescu, Julian, Ingmar Persson, and Dominic Wilkinson. 2020. Utilitarianism and the pandemic. *Bioethics* 34 (6): 620–632.

Scire, Emily, Kyeong Yun Jeong, Mary Katherine Gaurke, Bernard G. Prusak, and Daniel P. Sulmasy. 2022. Rationing with respect to age during a pandemic: A comparative analysis of state pandemic preparedness plans. *Chest* 161 (2): 504–513.

Sulmasy, Daniel P. 2006. *The rebirth of the clinic: An introduction to spirituality in health care*. Washington, D.C.: Georgetown University Press.

Sulmasy, Daniel P. 2007. 'Reinventing' the rule of double effect. In *The Oxford handbook of bioethics*, ed. Bonnie Steinbock, 114–149. Oxford: Oxford University Press.

Sulmasy, Daniel P. 2014. Edmund Pellegrino's philosophy and ethics of medicine: An overview. *Kennedy Institute of Ethics Journal* 24 (2): 105–112.

Sulmasy, Daniel P. 2020a. The care our elders deserve: A medical ethicist on the danger of rationing in the age of coronavirus. *New York Daily Post*, March 18. https://www.nydailynews.com/2020/03/18/the-care-our-elders-deserve-a-medical-ethicist-on-the-danger-of-rationing-in-the-age-of-coronavirus/. Accessed 18 November 2024.

Sulmasy, Daniel P. 2020b. Respirators, our rights, right and wrong: Medical ethics in an age of coronavirus. *New York Daily Post*, March 22. https://www.nydailynews.com/2020/03/22/

respirators-our-rights-right-and-wrong-medical-ethics-in-an-age-of-coronavirus/. Accessed 18 November 2024.
Sulmasy, Daniel P., and Fabien Maldonado. 2021a. Is it ethically permissible to unilaterally withdraw life-sustaining treatments during crisis standards of care? No. *Chest* 159 (6): 2167–2169.
Sulmasy, Daniel P., and Fabien Maldonado. 2021b. Rebuttal from Drs Sulmasy and Maldonado. *Chest* 159 (6): 2170.
Sulmasy, Daniel P., and Edmund D. Pellegrino. 1999. The rule of double effect: Clearing up the double talk. *Archives of Internal Medicine* 159 (6): 545–550.
Sulmasy, Daniel P., and Jeremy Sugarman. 1994. Are withholding and withdrawing therapy always morally equivalent? *Journal of Medical Ethics* 20 (4): 218–222.
Truog, Robert D. 2021. Ventilator allocation protocols: Sophisticated bioethics for an unworkable strategy. *Hastings Center Report* 51 (5): 56–57.

Chapter 11
What Is "Ethics as Usual" Under Crisis Standards of Care?

Jason T. Eberl

11.1 Introduction

The COVID-19 pandemic prompted an unprecedented need for health care institutions to develop and implement triage policies—typically referred to as "crisis standards of care" (CSCs)—for allocating scarce resources if they experience a surge of patients needing life-sustaining treatment (Antommaria et al. 2020). In the absence of definitive guidance from the Church's magisterium concerning specific requirements of such policies, there is space for moral disagreement concerning triage criteria and allowable practices. One area in which Catholic bioethicists have disagreed concerns whether non-futile life-sustaining treatment may be unilaterally withdrawn from one patient, who has a relatively poor expected outcome but is currently benefitting from said treatment, to be made available for reallocation to another patient predicted to have a better chance of survival to discharge; the qualifier "unilaterally" refers to the treatment being withdrawn without, and perhaps even against, the explicit consent of the patient or their surrogate decision-maker.

As in all moral matters, certain guiding principles ought to be operative; as Daniel Sulmasy correctly noted at the outset of the pandemic, it is precisely during such crises that we must maintain "ethics as usual" (Minda 2020; cf. Donovan 2020). For example, respect for each person's *intrinsic dignity* requires that health care professionals not intentionally cause a patient's death by either commission or unjustifiable omission of *ordinary/proportionate* forms of care (USCCB 2018, nn. 23, 56 & 70). Another relevant principle, particularly in the context of scarcity, is the *responsible stewardship* of resources, which includes just allocation of life-sustaining treatments while maintaining a preferential option for the poor and vulnerable (USCCB 2018, nn. 3 & 6). Nevertheless, extraordinary crisis situations

J. T. Eberl (✉)
Albert Gnaegi Center for Bioethics, Saint Louis University, St. Louis, MO, USA
e-mail: jason.eberl@slu.edu

J. T. Eberl (ed.), *Emerging Issues in Catholic Bioethics*, Philosophy and Medicine 153, https://doi.org/10.1007/978-3-032-08862-8_11

requiring life-and-death decisions under conditions of scarcity arguably create a "state of exception" in which conflicts among social, political, and ethical norms—without altering or suspending such norms altogether—may be resolved in abnormal ways when such conflicts inevitably arise (Agamben 2005).

Under non-triage conditions, it is generally held in both secular and Catholic bioethics that, if one has sufficient ethical justification to withhold a life-sustaining treatment, one has equivalent justification to withdraw an already implemented treatment (Beauchamp and Childress 2009, 155–158).[1] Standard justifications include autonomous refusal by the patient or their surrogate, disproportionate balance of the anticipated burdens or risks and expected benefits of the treatment, or physiological futility. In triage situations, however, it is widely held that one may *unilaterally withhold* a scarce life-sustaining treatment from one patient who may not sufficiently benefit from it and allocate it to another in greater need, or who is in equivalent need but is reasonably predicted to have a better chance of benefitting from the treatment. In cases of *unilateral withdrawal*, none of the aforementioned conditions are met and ethical consideration expands from the immediate good of the patient currently receiving treatment to include the good of other patients. The question thus explored in this chapter is whether ethics as usual justifies unilateral withdrawal of non-futile life-sustaining treatment in triage situations.

11.2 Stewardship of Limited Resources within a State of Exception

As noted above, it is a generally—though not universally—held bioethical truism that, if one has sufficient justification to withhold a treatment, one has *equivalent justification* to withdraw said treatment. Hence, justification to unilaterally withhold a treatment should also justify unilateral withdrawal of the same treatment. Yet, this equivalence thesis may not hold in triage situations as the patient who initially receives a treatment arguably has a moral claim—based on Robert Nozick's "principle of original acquisition of holdings"—to continued treatment unless the standard conditions of autonomous refusal, disproportionate burden, or physiological futility are met (Nozick 1975, 150–155; Sulmasy and Sugarman 1994, 219). While ethics as usual allows for withdrawal under these standard conditions, unilateral withdrawal may violate a patient's claim-right to continued treatment.

I have contended previously that the COVID-19 pandemic constituted a "state of exception" in which ordinary social, ethical, and political norms, while still valid, may be applied differently in specific concrete circumstances when two or more norms come into conflict (Eberl 2020). The existence of such a state of exception in

[1] This purported equivalence has been challenged by Donovan (1994, 27–28) and Sulmasy and Sugarman (1994). For responses in support of the equivalence thesis, see Harris (1994) and Wilkinson and Savulescu (2014).

which standard moral norms need to be applied in nuanced, atypical ways may sometimes be unavoidable. To understand how the notion of a state of exception coheres with the Catholic moral tradition, consider Thomas Aquinas's treatment of theft and just war theory. Theft is, of course, condemned by the Seventh Commandment. Yet, the possibility of theft is premised upon the existence of *private property*, the legitimate possession of which is acknowledged by Aquinas (1948, IIa–IIae, q. 66, aa. 1–2) and Catholic Social Teaching (Pontifical Council for Justice and Peace 2004, n. 176). The right to private property is not absolute, however, and Aquinas contends that in circumstances of "manifest and urgent" need "all things are common property" (Aquinas 1948, IIa–IIae, q. 66, a. 7; cf. Pontifical Council for Justice and Peace 2004, n. 177). In a state of exception in which all property has become common, appropriating another's surplus wealth to ameliorate one's dire need does not formally constitute an act of theft. Thus, even if a patient has a *prima facie* claim-right—akin to a property right—to continued treatment once started, situations involving dire need on the part of others call this apparent property-like right into question. But what if one is themselves in dire need of said treatment? Admittedly, Aquinas's case involves depriving another of their surplus of whatever material resource is at issue; yet, the general principle stands that one's claim-right to a resource of which others are in dire need may be abrogated in relevant circumstances.

Another state of exception is war. While there is a general prohibition on killing other human beings, if the conditions of a "just war" are met, a soldier who kills another either to defend their own life, to safeguard the lives of others, or to advance the war's justified aims is not guilty of murder (Aquinas 1948, IIa-IIae, q. 40, a. 1; Walzer 2015; Winright 2018). This does not mean that anything goes once war is justifiably declared—the operative concept of *jus in bello* (Jackson 2011)—but it does mean that actions not normally permitted outside of war may be permitted under rigorous criteria.

In a context in which CSCs must be implemented in order to exercise responsible stewardship of available resources, an act that would otherwise constitute an unjustifiable omission of care—unilaterally withdrawing non-futile life-sustaining treatment—may become justifiable insofar as continuation of such treatment may be disproportionate with respect to the burden placed on the *community* experiencing a surge of patients requiring treatment, which is a legitimate criterion in drawing the ordinary/extraordinary care distinction, albeit not one typically utilized in non-crisis situations (NCBC 2020, n. 5.b.ii).

11.3 Double-Effect Reasoning[2]

Declaring a state of exception, particularly in a public health context, may seem to justify adopting a *utilitarian* framework in which unilateral withdrawal is justified by the goal of maximizing the greatest overall net benefit, typically expressed in terms of maximizing "quality-adjusted life years" (Savulescu et al. 2020). Doing so would not preserve ethics as usual, however, as utilitarianism is inimical to Catholic morality, especially in the clinical context (John Paul II 1993, n. 74; Prusak 2020; Gaurke et al. 2021). Yet, it would be mistaken to ignore altogether the *social consequences* of either one's actions as an individual moral agent or the collective actions of a health care institution in devising and implementing CSCs (Cahill 1981). Eschewing utilitarianism, a possible justification of unilateral withdrawal and reallocation is by appeal to the *rule of double-effect* (RDE).[3]

11.3.1 Argument for the Applicability of RDE

RDE is widely utilized to justify various interventions within Catholic health care that involve both positive and negative effects. Paradigmatic cases include the administration of palliative medication that may hasten death or the removal of a gravid cancerous uterus resulting in fetal demise. The rule has held since Aquinas's (1948, IIa-IIae, q. 64, a. 7) writings on a person's right to self-defense, in which he affirms that one may permissibly perform an act aimed at an intended good end that also has a foreseen and unavoidable side-effect that is outside of one's intention. The principle has evolved to include the following standard criteria:

- There is a good end which one intends to attain by the action performed;
- The action performed is *per se* morally permissible;
- The foreseen negative effect is not intended either as an end or as a means by which the good end is brought about;
- The good end attained is proportionate to the negative effect caused by one's action.[4]

To establish a baseline, let us first see how these criteria are satisfied in a non-triage case of treatment withdrawal—specifically, mechanical ventilation—due to its having been refused by a patient in an advance directive or their surrogate. First, one is aiming at the good of respecting either the patient's autonomous choice or their best

---

[2] This section is derived from Eberl and Donovan (2020). For a recent defense of the applicability of RDE to the case at hand, see Leever (2024).

[3] Following Sulmasy (2007, 115), I will refer to the '*rule* of double-effect' as opposed to referring to it as a 'principle,' 'doctrine,' or 'dogma' as others tend to label it.

[4] For historical and contemporary formulations and discussions of double-effect, see Mangan (1949), Boyle (1980), Woodward (2001), Cavanaugh (2006), and Masek (2018).

interests based on the surrogate's assessment. Second, the act of extubating a living patient from a ventilator is *per se* morally permissible; otherwise, one could not extubate a patient who had recovered from an acute respiratory episode and was able to be discharged. Third, the negative effect that results from extubation—the patient's death—while foreseen and unavoidable, is neither sought as an end nor is it a means toward attaining the end of acting in accord with the patient's autonomous choice or best interests; if the patient were to miraculously start breathing on their own, it would not be counted as a failure. Finally, the good of acting in accord with the patient's choice or interest in not living perpetually hooked up to a ventilator, as assessed by themselves or their surrogate, is arguably proportionate to however many years of continued biological life they could have experienced if ventilation were continued.

Now, let us consider the case of unilateral withdrawal under circumstances requiring triage decisions. First, the intended good end is to make a vital resource available to save a person's life, which immediately points to a fundamental difference between the two cases. In the baseline case, the intended good end and foreseen negative effect both accrue to the same person; in the triage case, the negative effect is borne by one person while the good is enjoyed by another. Consider, though, Aquinas's original case, in which one may kill an unjust aggressor in self-defense; there, too, one person benefits at another's expense. Admittedly, Aquinas's example involves an *unjust aggressor* who is directly threatening one's life, which differs from the triage scenario. However, the fact that the person killed might be construed as *meriting* death by virtue of their guilt is immaterial insofar as a) Aquinas (1948, IIa-IIae, q. 64, aa. 2–3) holds that only a sovereign authority can legitimately kill someone guilty of a capital offense, and b) it is arguable that one may kill even an *innocent* material threat in self-defense (Camosy 2015, 63–66; Mumford 2013, 163–167). The first criterion of RDE is thus met insofar as one may pursue the good end of saving *another's* life even if so doing entails the foreseen negative effect of someone's death; it need not be the case that the good end and negative effect both accrue to the same individual.

The second criterion is satisfied insofar as extubation is *per se* morally permissible, as in the non-triage case. Here, one might claim that the *act* in question is not merely "extubation" but the "reallocation" of the ventilator. This claim points to the question of what constitutes a *directly intended* end from a *more remote* end toward which what one directly intends is aimed. We do not need to address this question here, though. For, even if the act is construed as "reallocation," such an act would be *per se* morally permissible: If a father were autonomously to refuse a ventilator so that it could be reallocated to save his daughter, it would be considered a heroic act of self-sacrifice, not a directly intended suicide. Neither extubation nor reallocation is the issue at hand, but rather that such an act is done *unilaterally*—that is, without or against the patient's or surrogate's explicit consent. Should we thus consider the directly intended act to be "extubation and reallocation without consent"? Again, considered in and of itself, neither act is essentially bad, for extubation (with or without reallocation) could occur for reasons of disproportionate burden or physiological futility regardless of the patient/surrogate's consent.

The third criterion is clearly met insofar as the foreseen death of the patient who is extubated is neither intended for its own sake, nor is it a means of bringing about the good of saving the patient to whom the ventilator is reallocated. If, against expectations, the first patient were to continue breathing on their own post-extubation, and perhaps even eventually recover and be discharged, it would be a desirable outcome. Finally, the proportionality criterion is satisfied insofar as what is gained and lost are two equally valuable lives of persons with intrinsic dignity; one person's life is not being favored over the other due to morally irrelevant criteria.[5] The only criterion is whether there are other patients in need of the ventilator who would be more likely to benefit physiologically from this scarce resource.

RDE thus justifies unilateral withdrawal for the following reasons. First, the directly intended good is to make a life-sustaining resource available for others and not to kill the patient from whom it is removed, although their death is foreseen. Furthermore, extubation, not the patient's death, is the necessary means for the directly intended reallocation to occur and is *per se* morally permissible. Finally, the extubated patient's foreseen death is proportionate to saving the life of any subsequently intubated patient.

11.3.2 Objection to the Applicability of RDE

G. Kevin Donovan articulates three reasons why he thinks RDE does not justify unilateral withdrawal of life-sustaining treatment. First, the conditions of RDE are not met insofar as "it is recognized that the withdrawal from the first patient must result in their death, which means we must also allow the intention of their death. With no proportional good accruing to the first patient, this is no longer a neutral act. Moreover, the paired action of reallocation must then admit that the preservation of the second patient's life has occurred through the action that resulted in the first patient's death. The good effect has occurred through the means of the bad effect, violating the rule of double effect" (Eberl and Donovan 2020).

This objection points to a foundational debate within the literature regarding RDE concerning how one defines the *directly intended act* distinct from any consequential effects that, to use Aquinas's (1948, IIa-IIae, q. 64, a. 7) original term, are *praeter intentionem*—i.e., "outside of the intention." The present case makes this task even more difficult as there are apparently multiple acts involved: first withdrawing the ventilator from one patient and then reallocating it to another patient. As Sulmasy (2007; cf. Sulmasy and Maldonado 2021a) rightly notes, double-effect cases properly involve two effects following from a *single* action. I maintain, however, that RDE is applicable to the case at hand as it is not being employed to justify

[5] This is especially true insofar as, in actual clinical situations, there would not be a specific identifiable patient to whom the ventilator is reallocated; rather, given the time needed to disinfect and reset the ventilator once withdrawn, it would be reallocated to a new patient selected by intake triage criteria during a surge.

two intentional actions: withdrawal (extubation of one patient) followed by reallocation (intubation of another patient). Rather, RDE is invoked only to justify the act of extubating a patient with a sufficiently poor prognosis to make a ventilator *available* for reallocation as a common good resource. In practice, it is not the case that a ventilator is withdrawn from one patient and immediately allocated to another, such that the individual good of one patient is prioritized over another; the ventilator must first be cleaned and reset for future use. RDE is invoked to justify only the first act of withdrawal, not the second act of reallocation (Bishop and Eberl 2021b).

Sulmasy and Maldonado (2021b, 2170) object that "under [this] construal of the scope of the RDE, one could say that one only intended to help a needy family when one stole an automobile from a wealthy owner, and that one did not intend to deprive the owner of her property, provided the former needed it more." The two situations are not analogous, however. In the case of ventilator withdrawal, the directly intended act (extubation) is *per se* morally permissible; the question at issue is whether extubation is morally permissible *when done unilaterally under triage circumstances*. Theft, however, is *per se* morally impermissible and so would not be justified by RDE under any circumstances. If it is case involving "manifest and urgent need," *per* Aquinas (1948, IIa-IIae, q. 66, a. 7), the act of appropriating the automobile would not formally constitute "theft" and thus may be morally permissible and thereby justifiably directly intended: not as theft but as appropriation. Since the appropriation could still involve a negative effect on the wealthy person from whom the automobile is taken, the other conditions of RDE would still need to be met in order for the act to be justified. Further, even if all the conditions of RDE were met and the act of appropriation *per se* justified, there may be *prudential* reasons not to allow such an act resulting in just civil laws prohibiting even morally justified appropriation of another's material goods for the sake of preserving social harmony.

Delineating a single action does not resolve Donovan's objection, though, as his primary point is that the extubated patient's death is the *means* by which the subsequently intubated patient's life is saved. If so, we cannot avoid *intending* the extubated patient's death; as Aquinas (1948, IIa-IIae, q. 8, a. 2) notes, when one intends an end, one also intends any necessary means toward attaining that end. Stated another way,

> To make the ventilator available for reallocation as a common good resource, the ventilator was withdrawn from patient A who had a remote chance of recovery. The ventilator was later reallocated to patient B because this patient had a good chance of survival. The negative outcome, which is the death of patient A, although not directly, essentially becomes the means to bring about the intended good effect, which is the survival of patient B. (Yip et al. 2021; cf. Tham et al. 2021).

This construal of the act performed misses the fundamental point of RDE. Even if an outcome of one's directly intended act is foreseen to occur, whether probabilistically or with certainty, it does not follow that that outcome falls under the agent's direct intention. In Aquinas's original example, if I perform an act with the directly intended end of saving my life, it does not necessarily follow that I also intend the foreseen death of my assailant. For it is not my assailant's death that saves my life, but rather the act I perform that stops their attack; whether that act results in the assailant's death or not is immaterial to attaining the intended end of saving my life. Consider a version of the case in which, after disabling my assailant, I call 911 and emergency help arrives in time to save them from the disabling injury I inflicted upon them; my life is saved regardless of whether my assailant lives or dies afterward. It is, of course, possible that I *do* intend my assailant's death: after disabling them I do not to call 911 and allow them to die from the injury I inflicted. But that scenario would *not* be justified by RDE.

It is thus not determinative of an agent's direct intention if the patient from whom life-sustaining treatment is unilaterally withdrawn will expectedly die as a result. Rather, the patient dies of their underlying condition and the directly intended outcome is to make the withdrawn treatment available for another patient; the first patient's death is not a *means* toward attaining this outcome. If, unexpectedly, the first patient continues breathing after life-sustaining treatment is withdrawn, that would both constitute a desirable outcome and in no way impact the directly intended outcome of having the withdrawn treatment be available for reallocation to another patient. While it *could* be the case that the physician who withdraws life-sustaining treatment does intend the patient's death—as was apparent in the case of Michael Hickson, from whom life-sustaining treatment was withheld due to a discriminatory "quality of life" assessment (Shapiro 2020)—such a case would not be justified by RDE.

Donovan's second reason against the applicability of RDE to the situation at hand is the fact that there are typically multiple agents involved: a physician (or care team collectively) and an institutional triage committee. This makes the application of RDE problematic in two ways: (1) actions justified by RDE should involve only a single moral agent (Sulmasy and Pellegrino 1999); and (2) "If any action or any intention of the multiple agents is considered immoral, delegating the decision to a triage committee cannot make it moral or excuse a physician from complicity through cooperation with such an act" (Eberl and Donovan 2020).

It is indeed true that we need to consider the distinct acts of each set of moral agents; for simplicity, I refer to 'the triage committee' as one collective agent and 'the physician' as the other. As I will show in the next section, the triage committee may be viewed as analogous to a military general who must make a strategic decision regarding the optimal allocation of the troops under their command, the troops being analogous to the health care resources for which the health care institution—with the triage committee as its agent—is morally obliged to exercise responsible stewardship. Just as RDE justifies the general's strategic decision to reallocate troops from one city to protect another without intending the first city's capture or destruction by the enemy, it also justifies the strategic reallocation of life-sustaining treatment from an intubated patient without the triage committee intending the patient's death. If the triage committee's reallocation decision is thereby morally justified, then it does not involve illicit cooperation for the physician, as a distinct moral agent, to adhere to the triage committee's decision and extubate the patient.[6]

Finally, Donovan contends that the proportionality criterion is not satisfied as there are more values at stake than human life alone. Emphasizing trust and loyalty as essential virtues physicians ought to cultivate (Pellegrino and Thomasma 1993, Chap. 5; Trotter 1997), Donovan charges that unilateral withdrawal and reallocation "would inevitably damage the trust relationship between the physician and the patient" (Eberl and Donovan 2020). Sulmasy and Maldonado (2021a, 2168) illustrate the point quite forcefully:

> Patients who are struggling just to stay alive ought not to be forced to bear the additional fear that their caregivers might be looking for signs that would justify taking [life-sustaining treatment] away from them. One could not imagine anything more damaging to patient trust. Even if reallocation were practical and might result in a few more lives being saved, it would not be worth sacrificing this moral bedrock. Reallocation makes physicians dual agents.

This contention follows from Sulmasy's (2006, 105) sound assertion that, "[i]n any physicians' oath worth its salt, physicians should swear to commit themselves to the service of their patients—putting patients' interests first."

I agree wholeheartedly that trust and loyalty are essential virtues for physicians and other health care professionals, and that engendering and preserving a relationship of fidelity between clinicians and patients is of immense value. I take this to the be most powerful counterpoint to my position, warranting careful consideration and the implementation of crucial safeguards. As a point of consideration, it is worth observing that the clinical staff within a hospital arguably forms a relationship of care with every person who enters its doors in need of treatment. Yet, that relationship manifests in different degrees and spheres of responsibility. For instance, in a non-crisis situation in a hospital emergency department, a triage nurse typically assesses each patient's needs and prioritizes them for treatment based on severity of condition and then, all things being equal, on a first-come/first-serve basis. It is only after the patient is admitted "behind the curtain" that a physician with the rest of the associated care team takes responsibility for the patient. This example shows how a *hospital* may bear a degree of responsibility for a patient's care, executed by the triage nurse, while the treating physician bears a different degree of responsibility, manifested only once the triage nurse has admitted the patient for treatment—the bearer of responsibility will shift again if the patient is admitted into intensive care or some other hospital unit.

With this precedent in mind, consider the analogous case of a hospital operating under CSCs experiencing a surge of patients. Once again, the triage nurse will conduct their assessment—perhaps using a modified set of criteria—thereby acting as the hospital's agent exercising responsible stewardship of the institution's limited resources (space, personnel, equipment, etc.). Once admitted to the emergency

[6] For further discussion of the conditions of licit versus illicit cooperation, see Cavanaugh (2007) and Flannery (2019, 2021).

department or other unit, the care team led by a physician takes over responsibility and—in agreement with Donovan and Sulmasy—should have as their *primary* focus the health of each individual patient under their care. Yet, the situation has introduced another, non-standard level of responsibility—the hospital's triage committee—whose sphere of responsibility not only includes the patients currently under treatment but also anticipated future patients—some of whom may already be awaiting treatment outside the gate being kept by the triage nurse. Like the triage nurse, the triage committee is also charged with assessing patients—based on data provided by the care team—and exercising responsible stewardship of the hospital's resources, particularly if there is a specific form of treatment (e.g., mechanical ventilation) with limited availability required by current and anticipated patients.

Just as the physician-led care team is not responsible if a patient is denied treatment, or treatment is delayed, by the triage nurse's assessment and decision, they are not responsible if a treatment resource is taken away from them by the triage committee. Hence, physicians are not forced to function as "dual agents" beyond reporting vital facts about the patient's condition to the triage committee, who are the agents responsible for allocating/reallocating resources for the physician to utilize. Of course, this conclusion trades on the soundness of the moral equivalence between withholding and withdrawing treatment, which Sulmasy and others have challenged (Sulmasy and Sugarman 1994), but we cannot beg the question against the equivalence thesis simply by denying the validity of its application in this context.

Given the significant value at stake—namely, trust in the care relationship—extreme caution must be employed in the formulation of CSCs, the assessment criteria utilized in triage decisions, the decision to implement CSCs, and their employment in practice.[7] There will undoubtedly be damage to the trust between patients/families and the health care institution/system that unilaterally orders the withdrawal of life-sustaining treatment, but such a loss of trust could also result from a health care institution/system failing to provide life-sustaining treatment to a patient expected to benefit from it. The best we could hope for is to shield individual physicians and care teams from having to make such triage decisions themselves, allowing them to focus on care for the patients under their charge to the best of their ability with the resources that are allotted to them by the institution exercising its proper stewardship responsibility.[8]

[7] A key consideration here is the specific triage criteria utilized. Throughout the COVID-19 pandemic, the primary triage criterion typically employed was the SOFA (Sequential Organ Failure Assessment) score (Antommaria et al. 2020); however, the use of the SOFA score for triage decisions is not without criticism (Wynia and Sottile 2020), especially in the context of the COVID-19 pandemic (Raschke et al. 2021). In fact, use of the SOFA score may have inadvertently discriminated against non-Hispanic Black patients (Tolchin, Hull, and Kraschel (2021); Tolchin et al. (2021); Roy et al. 2021).

[8] This line of reasoning parallels that of Kaczor (2017) in arguing that the locus of responsibility for deciding whether conjoined twins Jodie and Marie ought to be surgically separated was justifiably removed from their parents, who should be oriented toward preserving the life and health of both of their daughters, to the judiciary.

11.4 Just War Reasoning[9]

A closely related mode of ethical analysis, one born in the state of exception and considers social consequences, is *just war reasoning* (JWR). Like RDE, JWR defines a set of criteria that render it permissible for an individual to perform an action that would normally be impermissible for the sake of proportionate, in this case social, benefit. JWR, when its criteria are met, justifies everything from the political authority of one state declaring war against another (*jus ad bellum*) to the strategic commands of generals and admirals, the tactical decisions of battalion leaders and ship captains, and the individual actions of soldiers and sailors in defending their own lives and prosecuting the war's aims (*jus in bello*).[10] Furthermore, JWR, when combined with RDE, permits the lives of non-combatants to be jeopardized so long as they are not intentionally killed; consider the difference between "strategically" bombing an enemy munitions factory within an urban area and "terror" bombing an urban area to force the enemy into submission, JWR/RDE permitting the former but not the latter (Cavanaugh 2006, Chap. 5.4).

JWR may not seem immediately applicable to the health care context, but that is because the last half-century of American medicine has witnessed increasingly abundant resources such that physicians have typically not had to consider larger social consequences, and thus are privileged to focus more narrowly on the patients under their care. Practicing physicians in the US and other technologically advanced nations usually do not have to weigh the social consequences of their actions at the bedside. This situation is merely an artifact of the blessings of a wealthy nation, whereas the state of exception forces physicians in industrialized nations to reapply the norms of the non-crisis situation.

Consider this example: JWR justifies a general to order a strategic withdrawal of troops defending a city to reallocate them to where there is greater chance of victory. The general foresees, but does not intend, that the non-combatants residing in the city will be killed by the enemy. Likewise, in the state of exception created by a pandemic, a triage committee orders the unilateral withdrawal of a ventilator from one patient to make it available for reallocation to another who has greater expected survivability. The triage committee—and the physician bound by their decision—foresees, but does not intend, that the patient from whom the ventilator is withdrawn will die. In the state of exception, it may be permissible to remove a ventilator from one patient and deploy it for the benefit of another more likely to survive.

[9] This section is derived from Bishop and Eberl (2021a).

[10] An important related concept is *jus post bellum*, which refers the ethical nature of what is done to end a war and after a war concludes: e.g., ensuring political stability, securing democracy, rebuilding infrastructure, promoting reconciliation (Bass 2004). In the present context, *jus post bellum* would analogously ground the ethical obligation to (a) provide palliative and spiritual care once life-sustaining treatment is withdrawn (Sulmasy 2020), (b) continually reevaluate CSCs and the criteria justifying unilateral withdrawal (DeMartino et al. 2024), and (c) take practical steps to avoid future infectious disease outbreaks or a scarcity of resources necessitating future employment of CSCs.

As noted above, by identifying the locus of responsibility for the stewardship of a health care institution's resources with a triage committee separate from the care team oriented toward the health of the patient under their charge, we can hopefully ameliorate to some extent the potential ensuing damage to trust between patients/families and physicians. Consider the captain of a platoon who has promised a city's residents that the platoon will defend the city at all costs; however, the captain is under the authority of the general who may order the platoon redeployed to defend another city with a greater chance of success. While the city's residents may appropriately feel betrayed by the general's orders, they cannot reasonably hold the captain to their promise under the circumstances—the captain is absolved of their responsibility for defending the first city unless the general's order were demonstrably unjust. Of course, a health care institution's administrative hierarchy is not as strict as a military hierarchy, but it is a hierarchy nonetheless and physicians can only be held responsible for their use (or non-use) of the treatment resources left at their disposal, under CSCs, by the triage committee.

The appropriateness of analogizing medicine with war may be questioned (Sulmasy and Maldonado 2021b) and certainly it must be done cautiously so as to avoid implying that "anything goes" once a war or pandemic crisis is declared; recall the concept of *jus in bello*, referring to the moral assessment of actions done within a justly declared war. Nevertheless, I contend that it is an apt analogy insofar as, like the military, medicine, and the health care system in which medicine is practiced, is a not a private but a *social good* meant for, among other things, protecting people from disease just as the military is meant for protecting people from the enemy (Craig 2014).

11.5 Social Implications

Among the reasons why unilateral withdrawal of life-sustaining treatment under CSCs may not be morally permissible is the potential for discrimination against persons with disabilities or other disadvantaged groups (Sulmasy and Maldonado 2021a). Certainly, a disabled, long-term ventilator-dependent person should not have their ventilator reallocated insofar as it has arguably become constitutive of their physiological state of being and their psychological self-identity (Reynolds, Guidry-Grimes, and Savin 2021). Additionally, reallocation decisions should not be made on the basis of "quality of life" considerations or prospects for long-term survival, only *survivability to discharge* in order to avoid discrimination against persons with disabilities or chronic health conditions (Ennis et al. 2023). Finally, any other form of discrimination based on age (Jecker 2022), race (Elbaum 2020), or any other non-medical consideration should be avoided both explicitly, with respect to the formulation of CSCs, and implicitly, with respect to their application.[11]

Nevertheless, in considering the *common good* and the obligation of Catholic health care institutions to exercise responsible stewardship of resources, we should note the social implications of disallowing unilateral withdrawal insofar as it would entail a default criterion of "first-come/first-serve."[12] Such a criterion, while seemingly fair *prima facie*, may disadvantage those lacking ready access to health care services, especially those without their own means of transportation, live in rural areas, or otherwise experience access disparities (Tolchin, Hull, and Kraschel 2021). When a situation requiring implementation of CSCs is unavoidably forced upon a Catholic health care institution, due consideration should be given to the *health access disparities* that disproportionately burden the poor and vulnerable for whom we should have a "preferential option" (CDC 2020). Such disparities may result in members of economically and socially disadvantaged groups not being able to access appropriate health care prior to those who are more well-positioned. A first-come/first-serve policy is also arguably unjust with respect to those who practice measures to safeguard themselves from infection but become ill later on or persons who, for whatever reason, experience a longer incubation period between infection and the onset of illness. There may also be justification to reallocate resources based on the need to prioritize health care workers, first responders, and others fulfilling essential social roles.

11.6 Conclusion

I have argued that in a state of exception brought on by a pandemic, no one is morally due life-sustaining treatment when reallocation, even if by unilateral withdrawal, is based on clinically reasonable and sufficiently justified criteria.[13] Whether the state of exception obtains, however, depends on what resources are available and how they are already allocated. Such is the responsibility of political and public health authorities, as well as health care system administrators, who are duty-bound to exert every morally sustainable effort prior to and amidst an outbreak to stave off surges of patients that would create conditions of "manifest and urgent need" due to a scarcity of resources. Just as JWR requires that all peaceful alternatives be

[11] A particularly egregious example of evidently intersecting disability and racial discrimination is the case of Michael Hickson, a Black man with quadriplegia who was denied ventilatory support when he contracted COVID-19, with his purported lack of "quality of life" the explicit justification given by the physician to his wife (Shapiro 2020).

[12] In a survey of triage policies early in the COVID-19 pandemic, it was found that, while 23% utilized a first-come/first-serve criterion within their overall allocation framework, no policy utilized it as the sole or exclusive criterion (Antommaria et al. 2020).

[13] Excepting persons who have previous long-term dependency on such treatment due to chronic illness or disability.

exhausted before war is declared, CSCs should be implemented only once all other options to stave off an infectious disease outbreak, prevent such an outbreak from developing into a pandemic, and ameliorate the scarcity of requisite medical resources have been exhausted: "Rationing must always be a last resort. When resources are scarce, we should offer alternatives that are equivalent or nearly equivalent, share resources with others, and transfer needy patients to other facilities that can provide the resources before engaging in triage" (Sulmasy 2020).[14]

While society has an obligation to take all feasible prudential measures to prevent a state of exception in which CSCs must be implemented, and unilateral withdrawal of non-futile life-sustaining treatment should always be a last resort, ruling out the potential moral justification of such an act unwarrantedly constrains Catholic health care institutions' obligation to exercise responsible stewardship over available resources, as well as to consider in the development and implementation of their triage policies not only the intrinsic dignity of each individual patient, but also the needs of the common good, especially those of socially disadvantaged persons. Of course, when triaging is forced upon a Catholic health care institution and unilateral withdrawal of life-sustaining treatment becomes an ethically unavoidable option, as Sulmasy (2020) affirms, "we must never stop caring for those we can't save. All dying persons deserve pain control and palliative care" along with spiritual care for both patients and their families.

Ethics as usual requires us not only to consider whether the purported ethical equivalence of withholding and withdrawing life-sustaining treatment can be sustained when such actions are performed unilaterally. It also requires us to consider whether other salient ethical principles, such as distributive justice and preferential consideration for those who experience disparities in health care access, might be violated by disallowing such equivalence in circumstances involving scarce resources. Ethics as usual requires us to take differing circumstances into account such that, as a matter of general principle, what would violate an individual's *prima facie* rights in one situation does not do so in another—just as Aquinas contends what looks like theft actually is not in situations of dire need in which all things have become common property.

One lesson of COVID-19 for Catholic bioethicists is that, without reducing moral theory *tout court* to "situation ethics" (Fletcher 1966), the general moral principles that ethics as usual comprises may have built-in context-dependency. General moral principles, even if they are categorically valid, are best understood as "rules of thumb" since no such principle can cover all contexts. This is why practical wisdom is an "indispensable virtue" for health care professionals (Pellegrino and Thomasma 1993, Chap. 7). One must be practiced at knowing when context alters the applicability of a given principle. In the state of exception, it is not the case that general moral principles are no longer valid, but rather it is recognized that such principles sometimes do not fit certain circumstances. In the context of a pandemic,

---

[14] Fins (2023) documents first-hand the impact of a lack of preparedness, including a lack of carefully formulated CSCs, on clinical decision-making during the height of the pandemic during March–April 2020 in New York City.

it is up to moral agents—whether clinicians at the bedside, members of triage committees, or public health authorities—to prudentially decide whether certain general moral principles cohere with the particularities of the exceptional state at hand.

References

Agamben, G. 2005. *State of exception*. Trans. Kevin Attell. Chicago: University of Chicago Press.
Antommaria, A. H. M., T. S. Gibb, A. L. McGuire, et al. 2020. Ventilator triage policies during the COVID-19 pandemic at U.S. hospitals associated with members of the Association of Bioethics Program Directors. *Annals of Internal Medicine*. https://doi.org/10.7326/M20-1738.
Aquinas, T. 1948. *Summa theologiae*. Trans. English Dominican Fathers. New York: Benziger.
Bass, Gary J. 2004. Jus post bellum. *Philosophy and Public Affairs* 32 (4): 384–412.
Beauchamp, T. L., and J. F. Childress. 2009. *Principles of biomedical ethics*. 6th ed. New York: Oxford University Press.
Bishop, J. P., and J. T. Eberl. 2021a. Point: Is it ethically permissible to unilaterally withdraw life-sustaining treatments during crisis standards of care? Yes. *Chest* 159 (6): 2165–2166.
Bishop, J. P., and J. T. Eberl. 2021b. Rebuttal from Drs Bishop and Eberl. *Chest* 159 (6): 2169–2170.
Boyle, Joseph. 1980. Toward understanding the principle of double effect. *Ethics* 90:527–538.
Cahill, Lisa Sowle. 1981. Teleology, utilitarianism, and Christian ethics. *Theological Studies* 42 (4): 601–629.
Camosy, Charles C. 2015. *Beyond the abortion wars: A way forward for a new generation*. Grand Rapids: Eerdmans.
Cavanaugh, T. A. 2006. *Double-effect reasoning: Doing good and avoiding evil*. New York: Oxford University Press.
Cavanaugh, T. A. 2007. Cooperation: Material and formal. In *Encyclopedia of Catholic social thought, social ccience, and social policy*, ed. M. Coulter, S. M. Krason, R. S. Myers, and J. A. Varacalli. Lanham: Scarecrow Press.
Centers for Disease Control (CDC). 2020. COVID-19 racial and ethnic health disparities. https://www.cdc.gov/coronavirus/2019-ncov/community/health-equity/racial-ethnic-disparities/index.html. Accessed 10 December 2020.
Craig, David M. 2014. *Health care as a social good: Religious values and American democracy*. Washington, DC: Georgetown University Press.
DeMartino, Erin S., Jackson S. Ennis, Susan M. Wolf, and Daniel P. Sulmasy. 2024. Learning from COVID-19 triage schemes to face the next public health emergency. *Journal of the American Geriatrics Society* 72 (4): 1298–1301.
Donovan, G. Kevin. 1994. *Nutrition/hydration and PVS: Ethical considerations in the provision of life-sustaining food and water for the permanently unconscious patient*. MA Thesis,. University of Oklahoma.
Donovan, G. Kevin. 2020. Catholic ethics and the challenge of COVID-19 - Part one: DNR orders for COVID-19 patients and reallocation of ventilators. *Catholic Health Association and Pellegrino Center for Clinical Bioethics webinar*. https://www.youtube.com/watch?v=8oezcKaQ6MY&feature=youtu.be. Accessed 16 April 2020.
Eberl, J. T. 2020. Ethics as usual? Unilateral withdrawal of treatment in a state of exception. *American Journal of Bioethics* 20 (7): 210–211.
Eberl, J. T., and G. K. Donovan. 2020. Is it ethical to unilaterally withdraw life-sustaining treatment in triage circumstances? *Health Progress* 101 (2): 2167–2169.
Elbaum, Alan. 2020. Black lives in a pandemic: Implications of systemic injustice for end-of-life care. *Hastings Center Report* 50 (3): 58–60.

Ennis, Jackson S., Kirsten A. Riggan, Nicholas V. Nguyen, Daniel B. Kramer, Alexander K. Smith, Daniel P. Sulmasy, Jon C. Tilburt, Susan M. Wolf, and Erin S. DeMartino. 2023. Triage procedures for critical care resource allocation during scarcity. *JAMA Network Open* 6 (8): e2329688.

Fins, Joseph J. 2023. COVID-19 and clinical ethics: Reflections on New York's 2020 spring surge. In *COVID-19 and the law: Disruption, impact and legacy*, ed. I. Glenn Cohen, Abbe R. Gluck, Katherine Kraschel, and Carmel Shachar, 17–27. New York: Cambridge University Press.

Flannery, Kevin L. 2019. *Cooperation with evil: Thomistic tools of analysis*. Washington, D.C.: Catholic University of America Press.

Flannery, Kevin L. 2021. Avoiding illicit cooperation with evil: Alphonsus Liguori, Thomas Aquinas, and contemporary issues. *National Catholic Bioethics Quarterly* 21 (2): 231–246.

Fletcher, J. 1966. *Situation ethics: The new morality*. Louisville, KY: Westminster John Knox Press.

Gaurke, Mary Katherine, Bernard G. Prusak, Kyeong Yun Jeong, Emily Scire, and Daniel P. Sulmasy. 2021. Life-years and rationing in the COVID-19 pandemic: A critical analysis. *Hastings Center Report* 51 (5): 18–29.

Harris, John. 1994. Are withholding and withdrawing therapy always morally equivalent? A reply to Sulmasy and Sugarman. *Journal of Medical Ethics* 20 (4): 223–224.

Jackson, David Michael. 2011. Jus in bello. In *Encyclopedia of global justice*, ed. Deen K. Chatterjee, 584–585. Dordrecht: Springer.

Jecker, Nancy S. 2022. Too old to save? COVID-19 and age-based allocation of lifesaving medical care. *Bioethics* 36 (7): 802–808.

John Paul, II. 1993. *Veritatis splendor*. Vatican City: Libreria Editrice Vaticana.

Kaczor, Christopher. 2017. Is it ethically permissible to separate conjoined twins? Murder, mutilation, and consent. In *Contemporary controversies in Catholic bioethics*, ed. Jason T. Eberl, 123–135. Dordrecht: Springer.

Leever, Martin G. 2024. On the morality of reallocating life-sustaining interventions in times of scarcity. *Christian Bioethics* 30 (3): 178–188.

Mangan, Joseph. 1949. An historical analysis of the principle of double effect. *Theological Studies* 10:41–61.

Masek, Larry. 2018. *Intention, character, and double effect*. Notre Dame, IN: University of Notre Dame Press.

Minda, J. 2020. Sulmasy says it should be 'ethics as usual,' even in extraordinary times. *Catholic Health World*. https://www.chausa.org/publications/catholic-health-world/archive/article/pandemic-coverage/sulmasy-says-it-should-be-'ethics-as-usual-'-even-in-extraordinary-times. Accessed 14 April 2020.

Mumford, James. 2013. *Ethics at the beginning of life: A phenomenological critique*. New York: Oxford University Press.

National Catholic Bioethics Center (NCBC). 2020. Triage protocol guidelines. https://www.ncbcenter.org/resources-and-statements-cms/triage-protocol-guidelines?rq=triage. Accessed 16 April 2020.

Nozick, R. 1975. *Anarchy, state, and utopia*. New York: Basic Books.

Pellegrino, E. D., and D. C. Thomasma. 1993. *The virtues in medical practice*. New York: Oxford University Press.

Pontifical Council for Justice and Peace. 2004. *Compendium of the social doctrine of the Church*. Vatican City: Libreria Editrice Vaticana.

Prusak, Bernard G. 2020. All hospitals want to save lives: Should they also be seeking to save 'life years'? *America: The Jesuit Review*, 28 May. https://www.americamagazine.org/faith/2020/05/28/all-hospitals-want-save-lives-should-they-also-be-seeking-save-life-years. Accessed 18 November 2024.

Raschke, Robert A., Sumit Agarwal, C. Pooja Rangan, William Heise, and Steven C. Curry. 2021. Discriminant accuracy of the SOFA score for determining the probable mortality of patients with COVID-19 pneumonia requiring mechanical ventilation. *JAMA* 325 (14): 1469–1470.

Reynolds, Joel Michael, Laura Guidry-Grimes, and Katie Savin. 2021. Against personal ventilator reallocation. *Cambridge Quarterly of Healthcare Ethics* 30 (2): 272–284.

Roy, S., M. Showstark, B. Tolchin, N. Kashyap, J. Bonito, M. C. Salazar, J. L. Herbst, K. A. Nash, M. J. N. Tiako, K. Jubanyik, N. Kim, D. Galusha, K. H. Wang, and C. Oladele. 2021. The potential impact of triage protocols on racial disparities in clinical outcomes among COVID-positive patients in a large academic healthcare system. *PLoS ONE* 16 (9): e0256763.

Savulescu, J., I. Persson, and D. Wilkinson. 2020. Utilitarianism and the pandemic. *Bioethics* 34 (6): 620–632.

Shapiro, J. 2020. One man's COVID-19 death raises the worst fears of many people with disabilities. NPR.org. https://www.npr.org/2020/07/31/896882268/one-mans-covid-19-death-raises-the-worst-fears-of-many-people-with-disabilities. Accessed 31 July 2020.

Sulmasy, Daniel P. 2006. *The rebirth of the clinic: An introduction to spirituality in health care*. Washington, DC: Georgetown University Press.

Sulmasy, Daniel P. 2007. 'Reinventing' the rule of double effect. In *The Oxford handbook of bioethics*, ed. Bonnie Steinbock, 114–150. New York: Oxford University Press.

Sulmasy, Daniel P. 2020. The care our elders deserve: A medical ethicist on the danger of rationing in the age of coronavirus. *New York Daily Post*. https://www.nydailynews.com/2020/03/18/the-care-our-elders-deserve-a-medical-ethicist-on-the-danger-of-rationing-in-the-age-of-coronavirus/. Accessed 18 March 2020.

Sulmasy, Daniel P., and Fabien Maldonado. 2021a. Is it ethically permissible to unilaterally withdraw life-sustaining treatments for reallocation during crisis standards of care? No. *Chest* 159 (6): 2167–2169.

Sulmasy, Daniel P., and Fabien Maldonado. 2021b. Rebuttal from Drs Sulmasy and Maldonado. *Chest* 159 (6): 2170.

Sulmasy, Daniel P., and Edmund Pellegrino. 1999. The rule of double effect: Clearing up the double talk. *Archives of Internal Medicine* 159:545–550.

Sulmasy, Daniel P., and Jeremy Sugarman. 1994. Are withholding and withdrawing therapy always morally equivalent? *Journal of Medical Ethics* 20 (4): 218–222.

Tham, Joseph, Louis Melahn, and Michael Baggot. 2021. Withdrawing critical care from patients in a triage situation. *Medicine, Health Care and Philosophy* 24:205–211.

Tolchin, B., S. C. Hull, and K. Kraschel. 2021. Triage and justice in an unjust pandemic: Ethical allocation of scarce medical resources in the setting of racial and socioeconomic disparities. *Journal of Medical Ethics* 47:200–202.

Tolchin, B., C. Oladele, D. Galusha, N. Kashyap, M. Showstark, J. Bonito, M. C. Salazar, J. L. Herbst, S. Martino, N. Kim, K. A. Nash, M. J. N. Tiako, S. Roy, R. V. Greeno, and K. Jubanyik. 2021. Racial disparities in the SOFA score among patients hospitalized with COVID-19. *PLoS One* 16 (9): e0257608.

Trotter, Griffin. 1997. *The loyal physician: Roycean ethics and the practice of medicine*. Nashville: Vanderbilt University Press.

US Conference of Catholic Bishops (USCCB). 2018. *Ethical and religious directives for Catholic health care services*. 6th ed. Washington, DC: USCCB.

Walzer, Michael. 2015. *Just and unjust wars: A moral argument with historical illustrations*. 5th ed. New York: Basic Books.

Wilkinson, D., and J. Savulescu. 2014. A costly separation between withdrawing and withholding treatment in intensive care. *Bioethics* 28 (3): 127–137.

Winright, Tobias. 2018. Why I shall continue to use and teach just war theory. *Expositions* 12 (1): 142–161.

Woodward, P. A., ed. 2001. *The doctrine of double effect: Philosophers debate a controversial principle*. Notre Dame: University of Notre Dame Press.

Wynia, Matthew K., and Peter D. Sottile. 2020. Ethical triage demands a better triage survivability score. *American Journal of Bioethics* 20 (7): 75–77.

Yip, Yu-Yeung, Jonathan Chun-Hei Cheung, and Koon Ngai Lam. 2021. Rule of double-effect and utilitarianism: Intensivist perspectives. *Chest* 160 (2): e250–e251.

Chapter 12
Catholic Vaccine Ethics: Reexamining the Role of Virtues, Moral Principles, and Magisterial Teachings

John F. Brehany

12.1 Introduction

Catholic Church leaders have long supported immunization in principle as well as most vaccines offered to promote public health among children and adults (Tornielli 2021). Prior to 2000, there were few signs of significant ethical controversy among moral theologians or ethicists about vaccines. Since then, two major disputes have arisen: (1) whether or not, on ethical grounds, parents or patients should utilize or refuse, or Catholic health care providers should offer or refuse to provide, certain vaccines because abortion-derived cell lines (ADCL) have been utilized in producing or testing them; and (2) whether or not, on ethical grounds, persons may refuse a novel vaccine mandated in the course of a legally declared pandemic. It appears that the majority of bishops and health care ethicists have supported mandated vaccination in the course of these controversies and have encouraged Catholics to accept vaccination first and to resolve ethical dilemmas by means other than avoidance.[1]

[1] Many dioceses in the United States refuse religious exemptions for prospective students at Catholic schools who have not received all immunizations set by state department of health policy or state law. For example, all dioceses in Texas have been following this policy since 2009: https://txcatholic.org/wp-content/uploads/2024/07/TCCB-ED-immunization-policy-2024-only.pdf. TheArchdioceseofPhiladelphia has had a similar policy in place for over 10 years: https://hfrcs.org/documents/Vaccination%20Requirements%20AoP.pdf. With regard to COVID-19 mandates and religious exemptions, the Archdioceses of Chicago, Los Angeles, New York, Philadelphia, San Diego, St. Paul and Minneapolis, and Seattle, and the dioceses of El Paso, Honolulu, and Lexington, Kentucky issued statements refusing to endorse, or participate in validating, religious exemptions to COVID-19 vaccines: https://web.archive.org/web/20210827012137/https://www.catholicnews.

J. F. Brehany (✉)
The National Catholic Bioethics Center, Broomall, PA, USA
e-mail: jbrehany@ncbcenter.org

J. T. Eberl (ed.), *Emerging Issues in Catholic Bioethics*, Philosophy and Medicine 153, https://doi.org/10.1007/978-3-032-08862-8_12

Representatives of the majority appear to rely on the history of the Church's acceptance and encouragement of vaccines as well as standard social narratives and practices—namely, that vaccines are safe and effective and that public health standards (such as the childhood vaccine schedule) and processes (federal oversight of vaccine development) are reliable. Within this framework, vaccination is presented principally in terms of duty, especially in the face of doubts or disagreements regarding established standards and practices. Recently, some scholars have been contending there are no valid reasons to avoid or deny this duty (Eberl 2021; Eberl and Winright 2021; Lysaught 2021a, 2021b). After mulling these issues for years, starting with issues surrounding ADCL-dependent vaccines and more so during arguments over society-wide vaccine mandates starting in 2021, I have come to wonder whether this approach is correct or sustainable not only regarding controversies that arose during the COVID-19 pandemic but, even, perhaps, regarding a number of established vaccines and practices. In addition to more carefully examining certain ethical principles and arguments, some important developments over the last forty years should be acknowledged and, ultimately, assessed.

First, the number of diseases addressed via immunization and vaccine products have been growing dramatically over the last thirty years[2] and are set to increase even further in the foreseeable future (Medicines in Development 2020, 5).[3] The nature of the immunizing products is changing as well. mRNA vaccines are being developed for many infectious diseases and disease states, including cancer, food, and environmental allergies, heart conditions and stroke, and neurodevelopmental disorders (Penn Medicine n.d.; Karam and Daoud 2022). The implementation of this new technology raises a number of practical and ethical issues (Banoun 2023). Moreover, these changes have been occurring, in the US, within the legal and financial framework established by the Vaccine Injury Act of 1986, which removed liability for vaccine injuries from pharmaceutical companies and obligated the federal government to adjudicate claims and fund restitution for damages. This framework arguably undercuts incentives to ensure vaccine safety and creates incentives to develop more vaccines. And, while there has been some progress in replacing ADCL-dependent vaccine products (e.g., Merck's Zostavax was replaced by GlaxoSmithKline's Shingrix beginning in 2018), support for use of ADCL in biologics R&D is continuing, if not growing (Dumont et al. 2016).

com/vaccine-mandates-open-door-for-bishops-to-discern-an-appropriate-action/. Eberl and Winright (2021) argue that there is "no sufficient moral reason for Catholics" to claim religious exemptions to various mandatory vaccination policies during the COVID-19 pandemic. Lysaught (2021c) criticizes arguments justifying religious objections to COVID-19 vaccine mandates and defends the legitimacy of vaccine mandates in general.

[2]In 1963, the US childhood vaccine schedule contained three vaccine shots addressing five diseases. In 1986, it included eleven shots for seven diseases. By 2023, the schedule recommended sixty-six to sixty-eight doses for seventeen diseases for children between birth to age 18. (CDC 2025a; IOM 2013).

[3]The report lists over 150 vaccines in development, not counting vaccines for cancer and COVID-19.

Second, the development and implementation of the COVID-19 vaccine has posed distinct ethical issues concerning the novelty and speed of the virus and the vaccine development. Third, relevant information and questions have been raised about the process of testing and approving vaccines.[4] And finally, governments, public health agencies, and institutions have begun exerting more pressure than ever on individuals to accept vaccine products.[5]

These developments militate in favor of reexamining the terms within which ethical duties regarding vaccines have been framed within Catholic health care ethics, both to better address old disputes and to prepare to address new ones. The goal of this chapter is not to "settle" the two disputes noted at the beginning of this chapter—e.g., by systematically reviewing the arguments and data supporting both sides. Rather, I hope to outline a broader ethical framework than I have yet seen articulated which will contest the essential stance of the majority noted above. This broader framework will also, however, provide a more adequate framework within which individuals can make and defend a wide range of personal health care decisions, scholars can debate a more complete list of considerations, and Christians can more effectively witness to various moral goods and harms inherent in routine and emergency immunization programs.[6]

Given limitations of space, this chapter will focus on providing a more adequate framework than a duty- and mandate-based approach for vaccine ethics for individual decision-makers by examining three issues: (1) key issues in prudential decision-making; (2) the interpretation of three major moral principles in the Catholic moral tradition: charity, stewardship, and the common good; and (3) magisterial teachings.

[4] There are multiple publications and details on this topic, some of which are contested by establishment figures in academe, medicine, and government. For a recent acknowledgement of past shortcomings in studying, following up upon, and reporting safety data, see (Salmon et al. 2024).

[5] California (2015) and the New York (2019) have removed all religious and ethical considerations as a basis for objecting to vaccines, leaving only medical reasons as a basis for exemption. During the COVID-19 pandemic, unprecedented pressures were exerted by the federal government, employers, and universities—including Catholic dioceses and institutions—to induce individuals to accept novel vaccines developed and approved under Emergency Use Authorization.

[6] For example, some scholars might argue that the issue of ADCL-linked vaccines is moot, since the CDF has formally recognized at least twice, in (CDF 2008) and (CDF 2020), that individual patients or parents may choose to utilize them. However, the CDF has addressed only the permissibility of individual use, rather than a moral obligation to use. Moreover (CDF 2008, para 35) enjoined strict limits and duties on researchers and healthcare providers to avoid use of "biological material of illicit origin," noted the responsibility of decision-makers in organizations which make use of ADCL-linked research and therapeutics, and highlighted the need to "mobilize consciences in favor or life." Focusing exclusively on the issue of "permissibility" fails to adequately account for other important ethical considerations and principles.

12.2 Key Issues in Prudential Decision-Making About Vaccines

12.2.1 *Observations on the Unique Nature of Vaccines*

This section focuses mainly on the indispensable role of informed consent and the virtue of prudence in ethical decision-making about vaccines, which will be covered after some brief observations on the nature of vaccines. Limitations of space permit addressing only two other ethical considerations subsequently, conscience and moral certitude.

The first observation is that vaccines have unique features among medical interventions. Historically, they have been preventive rather than curative, designed to protect a person from infection or from suffering the burdens of a fully developed disease.[7] Moreover, vaccines can benefit not only the person receiving them, but also others, who may be protected from transmission of infectious diseases directly or via a "herd effect" and even the broader society, if the incidence or the severity of illness can be lessened.

A second observation is that, although they may seem well-established in theory and practice, by their nature vaccines are a *medical* intervention, and a potentially significant one at that. While vaccines may appear to be minimally burdensome, similar to blood draws and taking vital signs, immunization involves a medical intervention (often injections) that seeks to alter a complex system of persons' bodies for the long term.[8] The goal of vaccines is to manipulate or train a person's immune system for years, if not decades. Like other medical interventions, therefore, vaccines should be evaluated by the moral agents responsible for accepting them, and these agents should carefully assess a vaccine's promised benefits and potential burdens, including adverse events and other safety risks. The nature of vaccines as medical interventions has important implications for ethical discernment about their obligatory character, including but not limited to the obligation of moral agents to accept vaccines for the sake of their own health, for the sake of others' health, or for mandates imposed by a government or other institution.

Finally, because vaccines are medical interventions, they are subject to the requirements of informed consent. This creates mutual obligations for both the healthcare professionals dispensing, and the moral agents responsible for accepting, vaccination. This important consideration can be overlooked, in part because of vaccines' established place in public health initiatives concerning children and in part

---

[7] This chapter is focused on preventive diseases rather than the curative vaccines currently under development (Medicines in Development 2020, 2).

[8] The implications of such an alteration can be seen in a recent article about pertussis vaccines. James Cherry (2019), a world expert on diphtheria, tetanus, and pertussis vaccines, recounts how the effort to create a safer vaccine has rendered all children who received the DTaP vaccines more susceptible to pertussis throughout their lifetimes, and admitted that there is no easy way to decrease this increased lifetime susceptibility.

when the focus is on the social benefits of vaccination. Indeed, some hold that informed consent is not required from subjects before vaccination. The California Department of Health (2024) has asserted that "There are no Federal or California State requirements for informed consent specifically relating to immunization." The Centers for Disease Control (CDC 2019) has admitted that Vaccine Information Statements (VIS) given to patients or parents prior to vaccination are not "informed consent" forms but "provide enough information that anyone reading them should be adequately informed."[9] While these statements may reflect common practice or long-established opinions in government agencies, they are ethically flawed and may be contested or resisted by appropriate measures.

12.2.2 Key Elements of Decision-Making: Prudence, Conscience, and Moral Certitude

Decision-making about vaccines is unavoidably complex. There are a number of diseases or conditions (caused by bacterial or viral agents) for which vaccine products are available. Multiple risk profiles exist for these diseases based on age, health status, prior infection, and geographic location. Moreover, vaccines frequently are provided, not in single doses, but in multiple-antigen products (e.g., MMR-V; Pediarix). This can make it difficult to distinguish the effects associated with a given antigen[10] and create practical and ethical dilemmas when one antigen is either particularly ineffective or problematic.[11] Finally, vaccines have been and continue to be contested by individuals and entire communities. The advent of the internet and social media have only made arguments about vaccines more complex (Larson et al. 2021). In addition to reams of complex, clinical information and opposing points of view, people frequently can be subject to strong emotional reactions. All these factors make promotion of the virtue of prudence indispensable.

[9] Additionally, the CDC states, "Some states have informed consent laws, covering either procedural requirements (e.g., whether consent may be oral or must be written) or substantive requirements (e.g., types of information required). Check your state medical consent law to determine if there are any specific informed consent requirement relating to immunization. VISs may be used for informed consent as long as they conform to the appropriate state laws."

[10] The negative and positive implications of this practice are only beginning to receive serious attention (Stabell Benn et al. 2020).

[11] Regarding the mumps antigen in the MMR, see United States ex rel. Krahling v. Merck & Co., 44 F. Supp. 3d 581 (E.D. Pa. 2014): https://casetext.com/case/united-states-ex-rel-krahling-v-merck-amp-co.

12.2.2.1 Prudence

Prudence is the virtue that enables human beings to consistently understand and effectively choose the good(s) available in concrete human action. While prudence is an act of, and perfects, the practical reason, it can be developed and exercised only in concert with the other cardinal virtues in recognizing the truth of what constitutes good and fulfilling moral action (Cessario 1991, 76ff).

To make a prudent decision about whether to accept a vaccine, a parent or patient should ask: What will immunization achieve regarding a particular disease and in terms of overall health? Is the vaccine effective (and how) for the person receiving it and for the sake of others or the common good? Is the vaccine safe and known to be safe? What adverse events are associated with it and how could these impact one's health and well-being? After appropriately weighing efficacy and safety, are there any other factors to consider regarding a particular vaccine product, for example, are there valid alternatives to vaccination or is a particular vaccine dependent on ADCL in a material manner? These questions should be asked on a case by case basis; ideally, one disease and one vaccine at a time. However, as just noted above, frequently this will not be possible because a number of vaccines are given in multiple-vaccine products.

Efficacy and Benefit

It is hard to overstate support for the benefits of vaccines and immunization among physicians, public health officials, and policy makers (Greenwood 2014). Many vaccines on the childhood schedule have significantly reduced the incidence of reported cases of many illnesses and arguably some morbidity and mortality associated with them. Most people accept vaccines on this basis. During the Covid-19 pandemic, US-based vaccines for the SARS-CoV-2 virus were promoted, recommended, and mandated as effective for preventing disease, hospitalization, transmission, and death. Given all this, are there any grounds for prudential reservations or decisions that do not fully align with public standards or mandates based on benefit? At least three come to mind. For the sake of simplicity I will outline these with regard to Hepatitis B immunization.

First, some people may have legitimate objections some immunizations, at least at some times. For example, in the case of immunization for Hepatitis B, the current practice is to provide a first dose on the first day of a baby's life. Yet Hepatitis B is a viral infection spread by contact with bodily fluids of an infected individual, most commonly by sexual contact and intravenous drug use. The vast majority of Hepatitis B infections occur in individuals who engage in high-risk sexual behavior and drug use, who live with an infected individual, or who live in communities where there is a high prevalence of Hepatitis B. Thus, many people may prudentially decide against their infant(s) receiving this vaccine on their first day of life.

Second, some people may have valid alternative strategies for protecting or responding to threats to their own health or that of their dependents. Hepatitis B can be avoided by most people through responsible behavior, Hepatitis B infections can resolve on their own or can be treated with effective medications, obviating the need for Hepatitis B immunization at all.

Third, evidence may arise of primary or secondary vaccine failure which may lead some people, after careful discernment, to forgo use, or repeated use, of some vaccine products. For example, it has been well-documented that the mumps component of Merck's MMR vaccine product was not as efficacious as promised over the course of decades.[12] And, while the benefits of measles vaccination are largely stable over time, there have been reports of breakthrough infections and transmissions among MMR-vaccinated individuals.[13]

The efficacy of COVID-19 vaccine products also should be carefully assessed in the course of prudential decision-making. On the one hand, many have argued strongly for the efficacy of the mRNA vaccines licensed for use in the US and for their relative importance for populations at risk, including the elderly (CDC 2025c). Yet, enough questions and concerns have been raised about efficacy to justify careful prudential assessment and a diverse approaches to acceptance of COVID-19 vaccines. First, it became increasingly evident over time that the efficacy of the mRNA vaccines waned rapidly, making boosters necessary. It also became clear very quickly that COVID-19 vaccines did not and could not prevent transmission (Brown et al. 2021; Kampf 2021). Not even the original boosters were effective for the long-term. Indeed, some evidence suggest that the more boosters or total injections patients received, the more susceptible they become to COVID-19 infection and hospitalization (Shrestha et al. 2023; Shrestha et al. 2024; Uversky et al. 2023). Yet there are important differences in the risk profile of patients threatened by SARS-CoV-2 (Pezzullo et al. 2023). Finally, those approving and promoting COVID-19 mRNA vaccines focused principally on benefits of relative risk reduction as opposed to absolute risk reduction. Relative risk measures the percentage reduction in the risk of an outcome (e.g., disease) in a vaccinated group compared to an unvaccinated (control) group relative to the risk in the unvaccinated group. It tends to yield larger, more impressive numbers, connoting a vaccine's efficacy. Absolute risk reduction captures the actual reduction in risk for an individual due to vaccination. The latter is more important for individuals and society. In the end, the absolute risk reduction afforded by COVID-19 vaccines was much lower and less clinically significant than people were led to believe (Brown 2021).

[12] Information is available in a whistleblower case, United States ex rel. Krahling v. Merck & Co., 44 F. Supp. 3d 581 (E.D. Pa. 2014): https://casetext.com/case/united-states-ex-rel-krahling-v-merck-amp-co. The case ultimately was dismissed but key facts about efforts to correct or to hide (in)efficacy of the mumps vaccine were not in dispute.

[13] For example, see the discussion in (LeBaron et al. 2007) and (Rosen 2014).

Adverse Events and Safety

The safety of a vaccine is an essential consideration for prudential assessment. To be accepted in the course of prudential decision-making, a vaccine should be safe and *known* to be safe. Since the safety of vaccines is regularly and rigorously defended (American Academy of Pediatrics 2022), I will briefly note some questions about safety related to some vaccines on the childhood schedule, using the Hepatitis B and MMR vaccines as examples, some specific questions about testing of vaccines, and safety issues concerning the mRNA vaccine products developed for SARS-CoV-2 to illustrate the real and relevant questions that could factor into prudential assessment. Some issues about "known safety" will be addressed further below in the context of discussing moral certitude.

The Hepatitis B vaccine has been associated with common side effects (fever, diarrhea, and fatigue/weakness) as well as serious side effects (seizure, at a rate of 1/1300 children vaccinated). Other serious adverse events have been associated with the Hepatitis B vaccine, including neurological disorders and autoimmune diseases. The Institute of Medicine has recognized that "the evidence is inadequate to accept or reject a causal relationship" between the hepatitis B vaccine and a wide range of conditions, including fibromyalgia, Guillain-Barré Syndrome, Type-1 diabetes, and arthritis (IOM 2012, 435ff). Reports of some risks and adverse events, as well as the lack of definitive answers on a connection between the Hepatitis B vaccine and such events, can and should be carefully evaluated in the course of prudential decision-making, especially when there also may be questions about efficacy and overall benefit.

The safety of the vaccine for measles long given in the US, the MMR II vaccine, has evoked both strong concerns and defenses. Details of the debate cannot be done justice here. However, one statistic from a recent study on an alternative MMR vaccine product can help to shed light on some potential risks. Recently, a study was conducted to assess the relative safety and efficacy of the GlaxoSmithKline MMR vaccine, Priorix (MMR-RIT), a proposed alternative to Merck's MMR II (Klein et al. 2020). A supplement to this study listed the nature and percentage of serious adverse events associated with the two MMR vaccines. Just one set of Adverse Events of Special Interest (AESIs), "new onset chronic diseases"—including autoimmune disorders, asthma, type I diabetes, vasculitis, celiac disease, conditions associated with sub-acute or chronic thrombocytopenia, and allergies—occurred at a rate of 3.4% for MMR-RIT and 3.7% for MMR-II among recipients. The authors concluded the vaccines were similarly safe, but parents legitimately could have concerns about the nature and frequency of these and other adverse events.

COVID-19 mRNA Vaccines

The safety profile of COVID-19 mRNA vaccines is as contested and challenging to discuss as issues of efficacy. However, there appear to be more legitimate questions about safety than efficacy. It can be difficult to demonstrate the causality and frequency of harm caused by COVID-19 vaccines in particular. One reason has to do with the manner of defining "vaccinated status." In most, if not all, Western nations, patients were considered fully vaccinated only 14 days after the second of Pfizer and Moderna's two-dose regimen. Thus, if adverse events, deaths, or infections occurred within this timeframe, those suffering from them would be labeled "unvaccinated" (Fenton and Neil 2024, 81ff, 202–211). It is taking time for adverse events to become manifest, researched, and reported upon in part because of delays in pharmaceutical companies' public reporting (Thorp et al. 2023), but efforts in research and publication are beginning to show some results.[14] While some adverse events from COVID-19 vaccines may have been caused by specific batches of the product (Schmeling et al. 2023), other major problems are now known to affect specific populations. For example, COVID-19 vaccines appear to cause elevated rates of myocarditis in young men.[15]

The information outlined above highlights the need for people to exercise the virtue of prudence in making free and informed decisions about immunization for themselves and others. People should consult with trusted medical professionals and carefully evaluate reliable sources of information and well-founded arguments before making an informed choice. Moreover, it is only within the overall context of prudential discernment of issues of efficacy, safety, and overall benefit that people should address other ethical issues related to immunization such as ADCL-dependent vaccine products. Given the multiple considerations and questions, moral agents may validly decide to defer or decline some established and some COVID-19 vaccine products. Overly broad and coercive efforts to mandate acceptance of all vaccine products are not ethically sustainable.

12.2.2.2 Conscience

The Church teaches that conscience plays an indispensable role in human action (*Catechism of the Catholic Church* 1992, 1776–1794). Conscientious judgments and objections can be significant not only existentially, but also legally and socially for those claiming them.[16] Given the Catholic Church's historical support for immunization in principle and practice, it might appear that there is no support within the Catholic faith or moral tradition for conscientious objection to vaccination. Indeed, during the COVID-19 pandemic, some scholars argued that, at best, conscientious objections might be subjectively sincere but objectively unfounded or mistaken and hence not worthy of protection or respect (Eberl 2021; Eberl 2022).

However, while acknowledging there can be significant confusion about appropriate formation and exercise of conscientious judgment in general, and as related

[14] For an overview of multiple issues being brought forward, see Faksova et al. (2024).

[15] This issue was brought to the attention of the CDC in February 2021, but was not publicly acknowledged until June 2021 (Kracalik et al. 2022).

[16] The complexities of the legal status of Catholics' objections to some vaccines has been well addressed in Anders (2020).

to issues of vaccination, in particular, I contend that Catholics can reach valid judgments in conscience that accepting a given vaccine, including a novel vaccine for COVID-19, is morally wrong for them—that is, such judgments can be subjectively and objectively grounded, not merely the result of emotion or ignorance, and consistent with the Catholic moral tradition and magisterial teachings.

First, as noted above, people can legitimately make prudential decisions that some vaccines are not sufficiently efficacious and safe to warrant acceptance. Having reached such decisions, the threat to their informed judgment and agency over their bodily health, posed by outside pressure or a formal mandate, can be viewed as a form of moral evil. Apart from issues of informed consent, there are ethical goods or harms that can arise in the context of vaccination and, in particular, mandated vaccination, that persons can legitimately strive to address: some may wish to avoid involvement with ADCL-dependent vaccines consistent with their duty to protect their own health and the common good, and view pressure to become entangled with such vaccines as a moral threat. Some parents may see inappropriately broad federal or state mandates as a principled threat to their legitimate responsibility for the health and well-being of their children. And some people may conclude that, given recent and potential future trends, it is morally wrong for governments, public health agencies, or nongovernmental organizations to hold ultimate responsibility for their own immunity to disease.[17] Such moral goods or harms are complex. Protecting and advancing them certainly demands exercising the virtue of prudence. However, the presence of mandates or coercion can raise what might be viewed as a matter of practical judgment or acceptance of societal expectations to the level of ethical concern and conscientious judgment.

Reaching and acting upon such conscientious judgments can be consistent with the objective moral framework of the Catholic moral tradition. First, Aquinas recognized that people do and can legitimately focus on distinct elements of good or harm present in human actions and thus come to different decisions or conclusions (Aquinas 1948, Ia–IIae, q. 13, a. 6).[18] Second, Aquinas recognized that the content of the natural law, both in terms of what is required and what can be known by moral agents, can differ the further one departs from general principles and immediate conclusions therefrom and descends into matters of detail (Aquinas 1948, Ia–IIae, q. 94, a. 4). It is difficult to overstate the number of relevant details to be considered in matters of immunization that thus create space for a diversity of different assessments and decisions. Third, diversity in judgments about conserving human life and

[17] The significance of artificially managed immunity was acknowledged in a Plenary Address to the World Health Organization ("we're in a unique position in human history where we've shifted the human population to vaccine-induced dependency and vaccine-induced immunity and that's on the great assumption the populations would cooperate and for many years people lined up …") (Larson 2019); transcript available from the author.

[18] I recognize that article 6 addresses a basis for human freedom and not the ethical propriety of a particular choice. However, the complex interplay of goods and harms in earthly choices can and must be addressed appropriately by moral agents. There is a range of legitimate options as evidenced by Aquinas's comments on natural law, infra, and based on what has been said already about informed consent.

health is consistent with the magisterial teachings of the Church in part, as will be outlined below, because there is no authoritative teaching of the Church *obligating* Catholics to receive any and all vaccines at all times. Finally, it is not true that Catholics may conscientiously object *only* to actions violating formally defined doctrines of the Catholic Church. The moral life is more complex than that.[19]

12.2.2.3 Moral Certitude

In reaching a judgment of conscience about the moral quality of an act, an agent should have moral certitude that what he is doing is ethically right or not ethically wrong. It is widely recognized by reputable moral theologians that an agent should not act without appropriate certitude about the moral quality of a proposed action.[20] For a variety of reasons, the system of moral theology and methods of resolving doubts of conscience that arose in the post-Tridentine era largely collapsed beginning in the 1960s (Grisez 1983, 304ff). However, the need for moral certitude still exists.[21]

One should have moral certitude, not merely about whether one's actions could be deemed sinful in a laxist, rigorist, or probabilist framework (McHugh and Callan 1958, 245–78), but about whether actions are truly good and fulfilling for oneself and for others in community. With regard to the issue at hand, it is critical that decisions about immunization be made with sufficient moral certitude that a vaccine product in question is efficacious, safe, and a good choice all things considered. Lack of moral certitude can justify deferring a choice until certitude can be attained. Factors that may obviate moral certitude include lack of sufficient information (e.g., safety testing has not been sufficiently conducted or reported) or inability to trust certain authorities.

There are multiple reasons, both with regard to some established vaccines and novel mRNA vaccines, why moral agents could lack the information or trust necessary to act with moral certitude in accepting some vaccine products. With regard to lack of sufficient information, it appears that, with the exception of one vaccine (Gardasil 9), no vaccine on the childhood vaccine schedule has been tested, and its safety profile verified, in long-term, randomized, non-reactive placebo-controlled trials despite the fact that such testing is standard for new drugs approved by the FDA.[22] Second, it appears that testing for some vaccine products on the childhood

---

[19] In this regard, the recent beatification of Franz Jägerstätter, who resisted service in the Wehrmacht even against the urging of his own bishop, should be of significant interest (Bischof 2019).

[20] The scriptural basis for this concern is Romans 14:23. Regarding its applicability to issues in bioethics, see Austriaco (2021, 45–48).

[21] For a contemporary, scholarly account of the types of certitude and their relevance for moral action, see Cataldo (2017, 200–205).

[22] Full documentation of this claim is possible, but too detailed for the scope of this chapter. Essential confirmation can be found in *Informed Consent Action Network v. United States Department of Health and Human Services*, No. 18-cv-03215-JMF (S.D.N.Y. 2018), which

schedule may have been insufficient to demonstrate safety on a long-term, population basis. Using Hepatitis B vaccine products, approved by the FDA for use on the first day of life again as an example, for RECOMBIVAX HB, only 147 infants and children were enrolled in a clinical study, and these were followed for five days after each of three shots for a total of 437 doses (FDA 1986). The other vaccine product, ENGERIX-B's approval was based on more clinical studies (36) subjects (5071) and doses (13,495), but the subjects were followed for only 4 days after each dose (FDA 1989). The small number of subjects means that adverse effects that occur at a rate of ≥ 5000+ would not be detected in the trials. The absence of long-term follow-up means that if certain adverse events—such as new onset chronic diseases—resulted from the vaccine, their causal origin could be obscured. Finally, according to an Institute of Medicine (IOM) committee established to study the safety of the entire immunization schedule, "studies designed to examine the long-term effects of the cumulative number of vaccines or other aspects of the immunization schedule have not been conducted" (IOM 2013, 6).[23] Even renowned vaccinologist Stanley Plotkin has acknowledged this concern in a recent article: "[Vaccine] Postauthorization studies are needed to fully characterize the safety profile of a new vaccine, since prelicensure clinical trials have limited sample sizes, follow-up durations, and population heterogeneity" (Salmon et al. 2024). Such factors can undermine the ability of moral agents to have confidence in the safety of some specific vaccines or in the cumulative effects of the childhood vaccine schedule.

Significant obstacles to trust of government authorities were exacerbated during the COVID-19 pandemic, not only regarding public policy responses (e.g., lockdowns and masking) but also novel mRNA vaccines. Trust was undermined in part because important authorities, such as CDC Director Rachelle Walensky and President Biden, made significant factually incorrect statements about the efficacy of COVID-19 vaccines and in part because neither Pfizer nor Moderna released full copies of their research data when emergency use authorization was given (Lybrand 2021).[24] Worse, the government helped to obstruct the release of important information about the safety and efficacy of mRNA vaccines.[25] After Pfizer refused to

---

includes the text of a HHS letter acknowledging it had no records responsive to a legal demand for evidence of long-term, randomized, non-reactive placebo-controlled trials for vaccines on the childhood schedule. There is an extant principled ethical argument against doing randomized, placebo-controlled studies on vaccine safety; namely, that such studies would unjustly deprive some children of the benefits of some vaccinations. While it is an open question whether that argument is convincing, it does not change the fact that these studies have not been done, and thus key objective safety data are not available.

[23] Recommendations to conduct future studies can be found at pp. 12–14.

[24] Biden made exaggerated claims on at least two occasions in 2021. See, e.g., https://www.politifact.com/factchecks/2021/dec/22/joe-biden/biden-says-vaccinated-people-cant-spread-covid-19-/.

[25] The FDA tried to delay full release of data from the Pfizer research trials for up to 75 years. Lawsuits and Freedom of Information Act requests were necessary to dislodge other data critically important for public claims and individual informed consent; see, e.g., Deese (2022).

release complete copies of trial data, the FDA petitioned a federal court to seal the records for up to 75 years. Only when the documents were released and analyzed later in 2022 did the full picture of adverse events begin to emerge. Similarly, the CDC refused to release data from its V-Safe app until ordered to do so by federal judges.[26] Even publicly reported data was not accurate, or accurately characterized, and needed to be recalculated (Fraiman et al. 2022). For these and other reasons, novel COVID-19 mRNA vaccines could not have been known to be safe at the time they were being aggressively recommended and mandated. Finally, since they were authorized for use only on an emergency basis, novel COVID-19 mRNA vaccines were *experimental*, at least until the formal end of their research trials. The Moderna trial ended in October 2022; Pfizer's trial officially ended in February 2024. Indeed, one can make a strong case that Moderna and Pfizer's novel mRNA products remain experimental morally speaking. In either case, uncertainty about their long-term efficacy and safety was and is reasonable.

The actions just described and others undermined trust in governmental officials and agencies among health care professionals and the public (Oliver 2021). If parents and patients through no fault of their own are unable to achieve moral certitude about whether some vaccine products are sufficiently efficacious or safe or both, they are entitled, if not advised, to defer or refuse medical interventions until they can attain appropriate moral certitude.

12.3 Fundamental Ethical Principles: Charity, Stewardship, and Common Good

12.3.1 Charity and the Order of Love

During the implementation and mandating of COVID-19 vaccines, the topic of the importance and obligations of charity was raised. For example, in a video message produced in conjunction with the Ad Council, Pope Francis (among other bishops and cardinals) urged people to get vaccinated, calling it an "act of love" (Catholic News Agency 2021). The chairmen of the US Conference of Catholic Bishops Committees on Doctrine and Pro-Life Activities described COVID-19 vaccination as an act of love in two statements, (USCCB 2020, 2021) as have other Catholic ethicists (e.g., McTavish and Eberl 2022). Based on these statements and on the presumed benefits of vaccination in general, Catholic advocates of accepting and mandating COVID-19 vaccines have argued that Catholics have an obligation in charity to accept vaccination or that accepting vaccination is best understood as an act of love (Lysaught 2021a). It can be questioned, however, whether the statements or presumptions noted above can withstand careful scrutiny.

---

[26] *Public Health and Medical Professionals for Transparency v. Food and Drug Administration*, No. 4:21-cv-1058-P (W.D. Tex. 2022).

Charity is the greatest of the theological virtues and a unique type of love, as explained by St. Thomas Aquinas (1948, IIa–IIae, qq. 23–46).[27] Charity, an infused virtue which enables us to love God above all things and our neighbors as ourselves, should be an integral part of every human act, including decisions about vaccination for ourselves and our dependents. Yet, the move from the unique nature of charity as revealed in the life, mission, and word of Jesus Christ to accepting particular therapeutic products or interventions should be made with care. While an act of charity requires a specific intention, intention (or motivation) alone does not make a human act charitable *per se*. Aquinas (1948, IIa–IIae, q. 23, a. 1) notes that beneficence, or well-wishing alone, does not suffice for the friendship of charity. Apart from the mutual exchange of love that is required for friendship, for the issue at hand it can be argued that, beyond intention, the good to be served by or advanced in charity must be real to be truly or effectively charitable. Prevention of transmission was at the heart of the claims for the charitable nature of vaccination by bishops and COVID-19 vaccine advocates. But this means that COVID-19 vaccination could be an *effective* act of charity only if it meaningfully prevented transmission. Yet, despite some initial claims by government officials and the media, it became undeniably clear that the mRNA vaccines distributed in the US did not prevent infection with or transmission of the SARS-CoV-2 virus[28] and what protection was available waned rapidly (see, e.g., Nordström et al. 2022). Beyond the issue of transmission, a claim to an obligation in charity also should not be made in the abstract, without taking appropriate account of burdens, side effects, or needs based on risk profile. The person best positioned to take these into account is the patient or moral agent, in the context of informed consent, optimally given with the assistance of the virtues of both prudence and charity.

Beyond charity itself, principles inherent in the order of love must be appropriately respected in decisions about, and demands to accept, some vaccine products. Aquinas (1948, IIa–IIae, q. 26, a. 5 *ad* 3) argues that one should love his neighbor's soul more than one's own body, but that one need not imperil his own body for a neighbor's welfare unless there is a specific obligation to do so and, if he does, that belongs to the perfection of charity. It can be debated what a constitutes a "specific obligation" in general. In retrospect, it is highly debatable that COVID-19 vaccination could so qualify. But describing a choice as the "perfection of charity" presumes that the decision is voluntary. This accords with the generally recognized

[27] Charity is a form of friendship founded on God's communication of love to us in Jesus Christ (Aquinas 1948, IIa-–Iae, q. 23, a. 1).

[28] Indeed, the ineffectiveness of mRNA vaccines in preventing transmission could not be denied after an incident in Massachusetts in July 2021 (Brown, C. 2021). Of note, the prevention of transmission was not a primary or secondary outcome measure in the Pfizer-BioNTech (https://clinicaltrials.gov/study/NCT04368728) or Moderna (https://clinicaltrials.gov/study/NCT04470427) clinical trials. It is true that, later in the pandemic, some scientists have argued that COVID-19 vaccines can prevent transmission in some respects (Oordt-Speets et al. 2023). However, retrospective, observational, and other *post hoc* studies can never provide the certainty that was promised, but not delivered, in the COVID-19 clinical trials and subsequent governmental representations (Lybrand 2021).

standard that someone is not obligated to endanger his own life or health to benefit others (Wechsler 2013). Rather, such decisions are left to free and informed consent in the context of undergoing medical treatments. One has an important duty in charity to take good care of his own bodily life and health. There are many factors to weigh in this regard both for established and for novel vaccines (see, e.g., Quinn et al. 2025).

Aquinas (1948, IIa–IIae, q. 26, a. 8) further raises the question of whether one should preferentially love some relatives more than others. The principle articulated here bears upon the issue of whether parents' choice to decline a vaccine on the childhood vaccination schedule is necessarily selfish or flawed. Spaeder (2016, 249) suggests there is one correct answer for parents' obligation to protect the health of their children and educate them to care for the vulnerable: "to vaccinate a young child in accordance with the carefully constructed, standard vaccine schedule." However, while caring for the bodies and souls of children is a duty, it is debatable whether this duty demands acceptance of the childhood vaccine schedule *tout court*. Parents should engage in prudential assessment, motivated by charity, in caring for their children, considering all available evidence and consulting competent, trusted authorities. But there are enough questions about the safety or overall benefit of some vaccines for parents to legitimately conclude that some vaccine products are not in the best interest of their children. For example, the CDC (2025b) currently recommends COVID-19 vaccines for children ages six months and older. However, parents may well have legitimate practical and ethical reasons to decline to follow such recommendations. It appears that no European nation recommends COVID-19 immunization for children as young as six months but rather have diverse recommendations based on age and vaccine products (Van Kessel 2023). And reputable clinicians and scholars have argued that there are good reasons to question CDC recommendations, both for children and young adults (Malhotra 2022a, 2022b; Prasad 2022; Bardosh et al. 2024). For practical and principled reasons, parents legitimately may take particular care to protect the well-being of their children, guided by charity and by prudential assessment.[29]

[29] Beyond legitimate parental discernment regarding issues of immediate benefit or adverse events *vis-à-vis* immunization, Hart (2020) asks a significant question that demands further reflection: "should children be impeded in developing natural immunity to Covid-19?" There may well be room to apply Feinberg's (1980) concept of a child's right to an open future to COVID-19 vaccines, given some evidence of long-term impact on the immune system of some COVID-19 mRNA vaccines (Uversky et al. 2023) and even to some established vaccines, given in light of Cherry (2019).

12.3.2 Stewardship and the Principles of Ordinary and Extraordinary Means

Defining duties regarding love of one's own life and health in the context of medical interventions was developed in the Catholic moral tradition in terms of the principles of ordinary and extraordinary means (Cronin 2011). Proper effort, sacrifice, and employment of medical arts for the sake of the body, its life or health, instantiate a proper love of self.

Eberl and McTavish (2022) have argued that vaccination to protect from the SARS-CoV-2 virus counts as an ordinary means and thus is morally obligatory. And at first glance this may appear plausible, not only for COVID-19 vaccines but also for established vaccines. Yet there are reasons why the principles of ordinary means cannot be applied so apodictically as to declare a general moral duty. These reasons stem from nuances in the nature and application of these principles in the Catholic moral tradition and from a lack of authoritative Church teaching designating vaccines as ordinary means.

First, the entire framework of the principles of ordinary and extraordinary means was premised on serious threats to life and health for which interventions (natural and artificial; pharmaceutical and surgical) need be considered. What was viewed as essential is that people take appropriate measures to protect life and health, not accept one specific type of intervention. Yet some diseases for which there are vaccines (e.g., Hepatitis B) can be avoided without immunization.

Second and relatedly, as Cronin (2011, 158ff) explains, only relative norms for ordinary means can exist. While extraordinary means of conserving life are not ever obligatory, there is no absolute norm according to which certain means of conserving life are *per se* ordinary for all.[30] Scholars in the Catholic moral tradition,[31] and Pope Pius XII (1957/2009), have recognized a zone of latitude or discretion for individuals based on details or judgments relative to them.[32] The Congregation for the Doctrine of the Faith (CDF 1980, Sect. 4) addressed this issue when it stated, "Therefore one cannot impose on anyone the obligation to have recourse to a technique which is already in use but which carries a risk or is burdensome." And this teaching has been incorporated into the Ethical and Religious Directives for Catholic Health Care Services (USCCB 2018, dir. 32). As noted above, in making a prudential decision, a person may conclude that, after sound reasoning and in good conscience, a particular vaccine product has insufficient benefit or unjustified risks.

---

[30] Cronin is arguably the expert on the Church's moral tradition in this area.

[31] In addition to Cronin, in general, for the legitimacy of subjective factors warranting special consideration such as *vehemens horror*, see Lisson (1994).

[32] Indeed, Cronin (2011, 172 & 204) specifically discussed vaccines, listing them among interventions that may qualify as ordinary means; but he went on to note that, "However, even in these treatments, the relative physical, psychological, and economic condition of an individual can change the case."

Finally, Cronin (2011, 118–119) highlights a significant and eminently applicable distinction in this area—between an "ordinary medical procedure," i.e., a procedure commonly and routinely performed and recognized as "standard"—and an ordinary means. Vaccination certainly is an ordinary medical *procedure* but arguably not as an *ordinary means per se*. What ultimately determines whether a medical intervention, including a specific vaccine or vaccine product, counts as an ordinary or extraordinary means—absent an intervention by the magisterium—is the patient's informed judgment. This assessment should be made under the guidance of reason, with due control of emotions and of bias.

12.3.3 The Common Good: Nature and Claims about Duty to Vaccinate

Appeals to the common good in the context of vaccine ethics and mandates are ubiquitous and often treated as dispositive, whether in terms of an aspirational good to be achieved ("promote and respect the common good") or in terms of the moral duties owed to neighbors and fellow citizens. While these appeals are understandable, they should be carefully framed to ensure they are consistent with the full understanding of the common good in the Catholic moral tradition and that flawed or incomplete accounts of the common good are not misused to violate human dignity. There is ample space for qualifying information and recommendations in this regard.

First, the concept of the common good is both ancient and complex, much deeper than what is conveyed in the oft-cited quotation, "the sum total of social conditions which allow people, either as groups or as individuals, to reach their fulfillment more fully and more easily" (Vatican II 1965, n. 26; *Catechism* 1992, n. 1906). This brief definition encapsulates a decades-long effort to apply the Church's perennial understanding of the common good in the modern world, but does not fully capture the content of the Church's understanding (Lewis 2019; Hittinger 2021). Overidentifying the content of the common good with this brief formulation can encourage undue focus on material, measurable elements of social conditions, to the detriment of deeper truths.

One significant truth that easily can be lost is the clear Church teaching that the demands of the social order must respect the good of the human person:

> the common good presupposes respect for the person as such. In the name of the common good, public authorities are bound to respect the fundamental and inalienable rights of the human person... In particular, the common good resides in the conditions for the exercise of the natural freedoms indispensable for the development of the human vocation, such as "the right to act according to a sound norm of conscience and to safeguard ... privacy, and rightful freedom also in matters of religion" (Vatican II 1965, n. 26; *Catechism* 1992, n. 1907).

To these examples of "conditions for the exercise of natural freedoms" can be added respect for free speech, the rule of law, and informed consent in all matters medical, all of which were threatened during the COVID-19 pandemic.

There are many ways that people can contribute to the common good, from civic engagement to paying just taxes and, perhaps above all, by entering into lifelong, faithful marriages and building strong families. Even in the course of a pandemic, there are many ways that people should have contributed to the common good, from visiting the sick to maintaining essential goods and services. All such efforts are necessary and worthy of support and encouragement. Thus, it is unfortunate that some advocates of vaccine mandates so hastily defined obligations to the common good in terms of accepting novel vaccines and obeying mandates.

Such insistence on vaccination *vis-à-vis* the common good ultimately was ill-founded for two reasons. First, it is arguably inconsistent with the Church's understanding of the proper relationship between persons and the common good to mandate or coerce utilization of a novel and experimental medical intervention. Consent for vaccination may well be pursued consistent with informed consent and, perhaps, facilitated through public funding and other measures. Compelling or coercing compliance for what constituted, in principle, an extraordinary means was an ethical bridge too far.[33] Second, a mandate or duty to accept vaccination for the common good, i.e., to protect others, can only be sustained if the vaccine in question actually prevents transmission. Hopes were expressed and claims were made that the COVID-19 vaccines prevented transmission. However, as noted above, the ability of the mRNA vaccines to prevent transmission was not even formally researched, much less established.

Indeed, even categorical claims about the common good *vis-à-vis* established vaccines, including some on the childhood vaccine schedule, should be avoided, or least made with great caution, since they may not be sustainable on factual grounds. First, not all required vaccines protect against contagious diseases (e.g., tetanus). Second, some established vaccines lose their ability to prevent transmission over time.[34] Third, it is not clear that the FDA approval process always requires proof of a vaccine's ability to prevent transmission.[35] And finally, the nature and feasibility of "herd immunity" is not always clearly applicable or feasible in all cases (Bullen et al. 2023). Of course, immunization is a worthy goal. Vaccines should be carefully tested, monitored, and improved and accepted, when appropriate, with free and informed consent. But what should be avoided, even during social crises, are unfounded demands to accept questionable medical interventions, in particular when made on the basis of an incomplete understanding of the common good.

[33] I understand that there is a longstanding police power attributed to states to maintain public health and that the Supreme Court has affirmed this power in the past. I am disputing whether this is indeed ethical; others have begun questioning whether it is constitutional (Mariner et al. 2005).

[34] For example, measles (Tranter et al. 2024) and pertussis (Esposito et al. 2019).

[35] For example, the approval of the MMR II vaccine does not appear to include prevention of transmission as an endpoint or criterion of success (FDA 2022, 8).

12.4 Magisterial Teachings

Two sets of arguments are frequently employed by Catholic vaccine advocates to establish a moral duty to vaccinate in general or to accept the COVID-19 vaccines in particular: (1) there are statements by ecclesiastical figures and bodies expressing support for vaccines; and (2) the Church has never formally condemned vaccination *per se*, even ADCL-dependent vaccines, but has long supported the concept and practice of immunization (Tornielli 2021). Lysaught (2021), for example, appeals to statements from Congregation for the Doctrine of the Faith and the "Vatican COVID-19 Commission" in collaboration with the Pontifical Academy for Life, statements of members of the USCCB, and multiple statements of Pope Francis. While it is not possible to parse them here, it is clear from the sources and nature of the statements cited by Lysaught that none, individually or taken together, establish a binding moral duty for Catholics to accept any or all forms of vaccination.

First, some key sources cited by Lysaught have no magisterial authority, that is, no authority to teach in a binding manner. Neither the Vatican COVID-19 Commission, the Pontifical Academy for Life, nor the Chairmen of USCCB Committees have any teaching authority. Second, none of the statements made by these bodies teach that there is a binding duty to accept vaccination. They either note that there is no bar to receiving or accepting vaccines based on their dependence upon ADCL or characterize the acceptance of vaccination as a good act, even an "act of love." But none taught specifically and with authority that there is an obligation to accept a specific vaccine. Third, magisterial authorities, such as the Congregation for the Doctrine of the Faith, have addressed the issue of vaccination. For example, the CDF (2008, 2020) has taught that it could be *permissible* to accept ADCL-dependent vaccines, not that accepting vaccines was *obligatory*.

In this regard, it is important to recall that the CDF has played a critical role in issuing authoritative moral guidance on a moral duty regarding ordinary means. For instance, the CDF (2007a) has clarified that nutrition and hydration, even if artificially administered, constitute in principle an ordinary and proportionate means of conserving human life. A Commentary to the CDF's Response clarified the Church's magisterial and moral tradition over the preceding fifty years of debate (CDF 2007b). However, there is no moral or magisterial tradition even remotely similar supporting a moral duty to vaccinate, or to accept novel, experimental vaccines. Establishing such a tradition will require no less time, scholarly debate, and discernment than did the matter of nutrition and hydration before any authoritative teaching can be issued.[36]

Fourth, Pope Francis did make multiple statements about, and in favor of, COVID-19 vaccines. And it is true that "religious submission of mind and will must

[36] In the course of this debate, the recent trend toward mRNA-based "vaccine" products itself will require special focus. These are not traditional vaccines but rather "genetic therapies" which will require even greater discernment of clinical and ethical considerations surrounding the nature and goals of genetic engineering.

be shown in a special way to the authentic magisterium of the Roman Pontiff, even when he is not speaking ex cathedra … according to his manifest mind and will … [which] may be known either from the character of the documents, from his frequent repetition of the same doctrine, or from his manner of speaking" (Vatican II 1964, n. 25).[37] However, it not clear from the character of the documents and his manner of speaking that Pope Francis was exercising an "authentic magisterium" on this matter—i.e., intending to teach a binding moral duty for all to accept a particular vaccine product for a particular disease. In terms of "documents," Pope Francis made statements in interviews with the press, in speeches to diplomats, and in a public service announcement. None of these constitute the type of documents associated with authoritative teachings of the Church, such as encyclicals, apostolic exhortations, or even allocutions like those utilized by Pope Pius XII (Carlen 1990, xi–xiv).

In his manner of speaking, Pope Francis expressed opinions ("I believe that morally everyone must take the vaccine"),[38] made assertions ("Getting the vaccines authorized by the respective authorities is an act of love"),[39] alluded to the suitability of vaccines to address our "responsibility to care for ourself and our health, and … respect for the health of those around us" (Francis 2022)[40] and addressed other topics, such as access to vaccines. It was clear that Pope Francis was convinced of the safety and efficacy of COVID-19 vaccines and believed that people should take them. Still, he did not utilize formulas and documents associated with binding moral teachings (see, e.g., John Paul II 1993, n. 62; Paul VI 1968, nn. 11, 14, and 15). While Catholics have a duty to respect the noninfallible magisterium of a pope, it is the responsibility of popes to teach clearly. If a pope truly wishes to teach about detailed and novel moral obligations then, after ensuring that a proper foundation exists in Scripture and Tradition, he should avail himself of the formulas and tools of the Catholic moral and magisterial tradition.

---

[37] "[T]hat is, it must be shown in such a way that his supreme magisterium is acknowledged with reverence, the judgments made by him are sincerely adhered to, according to his manifest mind and will. His mind and will in the matter may be known either from the character of the documents, from his frequent repetition of the same doctrine, or from his manner of speaking" (Vatican II 1964, n. 25).

[38] Interview on Canale 5/TG24 (Jan. 10, 2021).

[39] Ad Council (August 17, 2021). Unity Across the Americas | COVID-19 Vaccine Education: https://youtu.be/zY5rwTnJF0U?si=4DqHuYXgIOiq8MTM.

[40] This speech contains perhaps the Pope's most extensive comments about use of the COVID-19 vaccines.

12.5 Conclusion

Immunology is a well-established scientific theory and there are decades of apparently successful vaccination programs. However, as well there are complex clinical and ethical issues surrounding some vaccine products and public policies thereon; some are based on unresolved issues from the past, some on the rapid expansion of new vaccine products and technologies, and some derive from the inherent complexity of the diseases, persons, and products involved in vaccination. And so, while there is significant support within the Catholic moral tradition for immunization, this tradition cannot be reduced to a simple duty to accept mandates and the narratives of public health and civic authorities to the detriment of prudential discernment, conscientious conviction, and informed consent.

This chapter suggests a valid path to navigate these issues, beginning with the recognition that vaccines are medical interventions about which people must make free and informed decisions. It continues by enabling and encouraging Catholics in particular to deploy a virtue-based approach, centered around the virtues of prudence and charity, and to thoughtfully draw upon the rich resources of the Catholic moral tradition as well. Within this framework, people can more appropriately and effectively address other important moral goods, such as advocating for higher standards of safety, more individualized choices, and ending the role of ADCL in vaccine products and biologics research and development.

The magisterium has accepted the validity of immunization in general and has clearly confirmed the permissibility of utilizing vaccine products dependent upon ADCL. But much more scholarship and reflection are required before a magisterial authority can issue binding moral teaching that Catholics must use some or all vaccine products approved for use in their locale. And, after a sufficient process of discernment, it may well be recognized that no binding teaching should, or shall, be issued.

References

American Academy of Pediatrics. 2022. Vaccine safety: Examine the evidence. https://www.healthychildren.org/english/safety-prevention/immunizations/pages/vaccine-studies-examine-the-evidence.aspx. Accessed 6 September 2022.

Anders, C. C. 2020. Individual and institutional religious exemptions from vaccines federal law and catholic teaching. *National Catholic Bioethics Quarterly* 20 (3): 501–523.

Aquinas, T. 1948. *Summa theologiae*. Trans. English Dominican Fathers. New York: Benziger.

Austriaco, N. 2021. *Biomedicine and beatitude: An introduction to catholic bioethics*. 2nd ed. Washington, DC: Catholic University of America Press.

Banoun, H. 2023. mRNA: Vaccine or gene therapy? The safety regulatory issues. *International Journal of Molecular Sciences* 24 (13): 10514.

Bardosh, K., et al. 2024. COVID-19 vaccine boosters for young adults: A risk benefit assessment and ethical analysis of mandate policies at universities. *Journal of Medical Ethics* 50:126–138.

Bischof, G. 2019. The story of Austrian Catholic Resister Franz Jägerstätter. *The National WWII Museum.* https://www.nationalww2museum.org/war/articles/story-austrian-catholic-resister-franz-jagerstatter.

Brown, R. B. 2021. Outcome reporting bias in COVID-19 mRNA vaccine clinical trials. *Medicina* 57:199.

Brown, C., et al. 2021. Outbreak of SARS-CoV-2 infections, including COVID-19 vaccine breakthrough infections, associated with large public gatherings - Barnstable County, Massachusetts, July 2021. *MMWR* 70 (31): 1059–1062.

Bullen, M., G. S. Heriot, and E. Jamrozik. 2023. Herd immunity, vaccination, and moral obligation. *Journal of Medical Ethics* 49 (9): 636–641.

California Department of Health. 2024. Laws and regulations. https://www.cdph.ca.gov/Programs/CID/DCDC/Pages/Immunization/laws.aspx.

Carlen, Claudia. 1990. *Papal pronouncements: A guide*. Vol. 1. Ypsilanti, MI: The Pierian Press.

Cataldo, P. 2017. Moral certitude in the use of levonorgestrel for the treatment of sexual assault survivors. In *Contemporary controversies in catholic bioethics*, ed. J. T. Eberl. Dordrecht: Springer.

Catechism of the Catholic Church. 1992. https://www.vatican.va/archive/ENG0015/_INDEX.HTM.

Catholic News Agency. 2021. Pope Francis promotes COVID-19 vaccine in the Americas. https://youtu.be/IamLikEn-mM. Accessed 18 August 2021.

CDC. 2019. Vaccine information statement|frequently asked questions|VIS|CDC. https://www.cdc.gov/vaccines/hcp/vis/about/vis-faqs.html. Accessed 15 November 2022.

CDF. 1980. *Declaration on Euthanasia*. Vatican City: Libreria Editrice Vaticana.

CDF. 2007a. Response to certain questions of the United States conference of catholic bishops concerning artificial nutrition and hydration. https://www.vatican.va/roman_curia/congregations/cfaith/documents/rc_con_cfaith_doc_20070801_risposte-usa_en.html.

CDF. 2007b. Commentary on response to certain questions of the United States conference of catholic bishops concerning artificial nutrition and hydration. https://www.vatican.va/roman_curia/congregations/cfaith/documents/rc_con_cfaith_doc_20070801_nota-commento_en.html.

CDF. 2008. Instruction dignitas personae, on certain bioethical questions. https://www.vatican.va/roman_curia/congregations/cfaith/documents/rc_con_cfaith_doc_20081208_dignitas-personae_en.html.

CDF. 2020. Note on the morality of using some anti-Covid-19 vaccines. https://www.vatican.va/roman_curia/congregations/cfaith/documents/rc_con_cfaith_doc_20201221_nota-vaccini-anticovid_en.html.

CDC. 2025a. *Child* and adolescent immunization schedule by age. https://www.cdc.gov/vaccines/hcp/imz-schedules/child-adolescent-age.html. Accessed 29 May 2025.

CDC. 2025b. Staying up to date with COVID-19 vaccines. https://www.cdc.gov/covid/vaccines/stay-up-to-date.html. Accessed 7 January 2025.

CDC. 2025c. Covid-19: Benefits of getting vaccinated. https://www.cdc.gov/covid/vaccines/benefits.html. Accessed 13 January 2025.

Cessario, Romanus. 1991. *The moral virtues and theological ethics*. Notre Dame: University of Notre Dame Press.

Cherry, J. 2019. The 112-year Odyssey of pertussis and pertussis vaccines—Mistakes made and implications for the future. *Journal of the Pediatric Infectious Diseases Society* 8 (4): 334–341.

Cronin, D. 2011. *Ordinary and extraordinary means of conserving life (15th Anniversary Edition)*. Philadelphia: National Catholic Bioethics Center.

Deese, K. 2022. Judge scraps 75-year FDA timeline to release Pfizer vaccine safety data, giving agency eight months. *Washington Examiner*. https://www.washingtonexaminer.com/news/2381224/judge-scraps-75-year-fda-timeline-to-release-pfizer-vaccine-safety-data-giving-agency-eight-months/. Accessed 7 January 2022.

Dumont, J., et al. 2016. Human cell lines for biopharmaceutical manufacturing: History, status, and future perspectives. *Critical Reviews in Biotechnology* 36 (6): 1110–1122.

Eberl, Jason T. 2021. Vaccine mandates are comingCatholics have no moral reason to oppose them. *America: The Jesuit Review*. https://www.americamagazine.org/faith/2021/08/10/covid-vaccine-mandate-exemptions-voluntary-ignorance-241196. Accessed 21 August 2021.

Eberl, Jason T. 2022. Is there a moral obligation to be vaccinated for COVID-19? *Health Care Ethics USA* (Winter/Spring): 36–39.

Eberl, Jason T., and Tobias Winright. (2021, March). Catholics have no grounds to claim exemption from COVID vaccine mandates. https://www.ncronline.org/news/coronavirus/catholics-have-no-grounds-claim-exemption-covid-vaccine-mandates.

Esposito, S., et al. 2019. Pertussis prevention: reasons for resurgence, and differences in the current acellular pertussis vaccines. *Frontiers in Immunology* 10:1344.

Faksova, K., et al. 2024. COVID-19 vaccines and adverse events of special interest: A multinational Global Vaccine Data Network (GVDN) cohort study of 99 million vaccinated individuals. *Vaccine* 42:2200–2211.

FDA. 1986. RECOMBIVAX HB package insert. – https://www.fda.gov/media/74274/download

FDA. 1989. ENGERIX-B: Package insert. https://www.fda.gov/media/119403/download

FDA. 2022. Statistical review-Measles, mumps, and rubella virus vaccine STN: 101069/5846, Section 6.1.8. https://www.fda.gov/media/166052/download?attachment.

Feinberg, J. 1980. The child's right to an open future. In *Whose child? Children's rights, parental authority, and state power*, ed. W. Aiken and H. LaFollette, 124–153. Lanham, MD: Rowman and Littlefield.

Fenton, N., and M. Neil. 2024. *Fighting Goliath: Exposing the flawed science + statistics behind the Covid-19 event*. New York: Sovereign Publishing.

Fraiman, J., et al. 2022. Serious adverse events of special interest following mRNA COVID-19 vaccination in randomized trials in adults. *Vaccine* 40:5798–5805.

Francis. 2022. Address to the members of the diplomatic corps accredited to the Holy See. https://www.vatican.va/content/francesco/en/speeches/2022/january/documents/20220110-corpo-diplomatico.html. Accessed 10 January 2022.

Greenwood, B. 2014. The contribution of vaccination to global health: Past, present and future. *Philosophical Transactions of the Royal Society B* 369:20130433.

Grisez, G. 1983. *The way of the Lord Jesus volume one: Christian moral principles*. Chicago: Franciscan Herald Press.

Hart, E. M. 2020. Is it ethical to impede access to natural immunity? The case of SARS-CoV2? *British Medical Journal* 368:m1089.

Hittinger, R. 2021. Some reflections on the common good in modern catholic social doctrine. Thomistic Institute. https://youtu.be/1L8RVF6IKns.

IOM. 2012. *Adverse effects of vaccines evidence and causality*. Washington, DC: Academic.

IOM. 2013. *The childhood immunization and safety: Shareholder concerns, scientific evidence, and future studies*. Washington, DC: Academic.

John Paul, II. 1993. *Veritatis splendor*. Vatican City: Libreria Editrice Vaticana.

Kampf, G. 2021. The epidemiological relevance of the COVID-19-vaccinated population is increasing. *The Lancet Regional Health - Europe* 11:100272.

Karam, Mia, and Georges Daoud. 2022. mRNA vaccines: Past, present, future. *Asian Journal of Pharmaceutical Sciences* 17 (4): 491–522.

Klein, Nicola P., et al. 2020. Immunogenicity and safety of a measles-mumps-rubella vaccine administered as a first dose to children aged 12 to 15 months: A phase III, randomized, non-inferiority, lot-to-lot consistency study. *Journal of the Pediatric Infectious Diseases Society* 9 (2): 194–201.

Kracalik, I., et al. 2022. Myocarditis outcomes after mRNA COVID-19 vaccination investigators and the CDC COVID-19 response team. outcomes at least 90 days since onset of myocarditis after mRNA COVID-19 vaccination in adolescents and young adults in the USA: A follow- up surveillance study. *The Lancet Child and Adolescent Health* 6 (11): 788–798.

Larson, H. 2019. Vaccine safety in the next decade: Why we need new modes of trust building. *WHO Global Vaccine Safety Summit*. https://www.who.int/news-room/events/detail/2019/12/02/default-calendar/global-vaccine-safety-summit. Accessed 12 March 2019.

Larson, Heidi J., Isabelle Sahinovic, Madhava Ram Balakrishnan, and Clarissa Simas. 2021. Vaccine safety in the next decade: Why we need new modes of trust building. *BMJ Global Health* 6 (Supp. 2): e003908.

LeBaron, Charles W., et al. 2007. Persistence of measles antibodies after 2 doses of measles vaccine in a postelimination environment. *Archives of Pediatric Adolescent Medicine* 161 (3): 294–301.

Lewis, V. 2019. Catholic social teaching on the common good. In *Catholic social teaching: A volume of scholarly essays*, ed. G. V. Bradley and E. C. Brugger, 235–266. New York: Cambridge University Press.

Lisson, E. 1994. Moral repugnance as a source in moral analysis. *Linacre Quarterly* 61 (1): 53–65.

Lybrand, H. 2021. Fact check: Four times Walensky's comments were out of step with CDC guidance. CNN.com. https://www.cnn.com/2021/05/21/politics/walensky-comments-cdc-guidance-fact-check/index.html.

Lysaught, M. Therese. 2021a. All vaccines are morally acceptable, says member of Pontifical Academy for Life. *National Catholic Reporter*. https://www.ncronline.org/news/guest-voices/all-vaccines-are-morally-acceptable-says-member-pontifical-academy-life. Accessed 5 March 2021.

Lysaught, M. Therese. 2021b. Catholics seeking 'religious' exemptions to vaccines must follow true Church teachings on conscience. https://www.ncronline.org/news/opinion/catholics-seeking-religious-exemptions-vaccines-must-follow-true-church-teaching. Accessed 21 September 2021.

Lysaught, M. Therese. 2021c. Catholic vaccine Exemption Requests: A resource for organizations. *National Catholic Reporter*. https://www.ncronline.org/files/file_attachments/Catholic%20Vaccine%20Exemption%20Request%20Organizational%20Response%20Template.pdf. Accessed 28 August 2021.

Malhotra, A. 2022a. Curing the pandemic of misinformation on COVID-19 mRNA vaccines through real evidence-based medicine - Part 1. *Journal of Insulin Resistance* 5 (1): a71.

Malhotra, A. 2022b. Curing the pandemic of misinformation on COVID-19 mRNA vaccines through real evidence-based medicine - Part 2. *Journal of Insulin Resistance* 5 (1): a72.

Mariner, W., G. Annas, and L. Glantz. 2005. *Jacobson v Massachusetts*: It's not your great-great-grandfather's public health law. *American Journal of Public Health* 95 (4): 581–590.

McHugh, John A., and Charles J. Callan. 1958. *Moral theology: A complete course*. Vol. I, 245–278. New York: Wagner.

McTavish, J., and J. T. Eberl. 2022. Is COVID-19 vaccination ordinary (morally obligatory) treatment? *National Catholic Bioethics Quarterly* 22 (2): 319–333.

Medicines in Development. 2020. *Report: Vaccines*. https://phrma.org/-/media/Project/PhRMA/PhRMA-Org/PhRMA-Org/PDF/medicines-in-development-for-vaccines-2020.pdf.

Nordström, P., M. Ballin, and A. Nordström. 2022. Risk of infection, hospitalisation, and death up to 9 months after a second dose of COVID-19 vaccine: A retrospective, total population cohort study in Sweden. *The Lancet* 399 (10327): 814–823.

Oliver, D. 2021. Trust in CDC, fda took a beating during pandemic. *WebMD Health News*, 9 June.

Oordt-Speets, A., et al. 2023. Effectiveness of COVID-19 vaccination on transmission: A systematic review. *COVID* 3:1516–1527. https://doi.org/10.3390/covid3100103.

Paul, VI. 1968. *Humanae vitae*. Vatican City: Libreria Editrice Vaticana.

Penn Medicine. n.d.. https://www.pennmedicine.org/mrna.

Pezzullo, A. M., et al. 2023. Age-stratified infection fatality rate of COVID-19 in the non-elderly population. *Environmental Research* 216:114655.

Pius XII. 1957/2009. The prolongation of life: Address to an international congress of anesthesiologists National Catholic Bioethics Quarterly 9(2): 327–332.

Prasad, V. 2022. Covid vaccines shouldn't be 'routine' for kids. *The Free Press*. https://www.thefp.com/p/covid-vaccines-shouldnt-be-routine. Accessed 24 October 2022.

Quinn, G., et al. 2025. What lessons can be learned from the management of the COVID-19 pandemic? *International Journal of Public Health* 70:1607727.

Rosen, Jennifer B. 2014. Outbreak of measles among persons with prior evidence of immunity, New York City, 2011. *Clinical Infectious Diseases* 58 (9):

Salmon, D., et al. 2024. Funding postauthorization vaccine-safety science. *New England Journal of Medicine* 391 (2): 102–105.

Schmeling, M., V. Manniche, and P. Hansen. 2023. Batch-dependent safety of the BNT162b2 mRNA COVID-19 vaccine. *European Journal of Clinical Investigation* 53 (8): e13998.

Shrestha, N. K., P. C. Burke, A. S. Nowacki, and S. M. Gordon. 2023. Risk of coronavirus disease 2019 (COVID-19) among those up-to-date and not up-to-date on COVID-19 vaccination by US CDC criteria. *PLoS One* 18:e0293449.

Shrestha, N. K., P. C. Burke, A. S. Nowacki, and S. M. Gordon. 2024. Effectiveness of the 2023–2024 formulation of the COVID-19 messenger RNA vaccine. *Clinical Infectious Diseases* 79 (2): 405–411.

Spaeder, G. 2016. The moral obligation to vaccinate: Autonomy and the common good. *National Catholic Bioethics Quarterly* 16 (2): 245–254.

Stabell Benn, C., et al. 2020. Vaccinology: Time to change the paradigm? *The Lancet Infectious Diseases* 20 (10): e274–e283.

Thorp, J., et al. 2023. COVID-19 vaccines: The impact on pregnancy outcomes and menstrual function. *Journal of American Physicians and Surgeons* 28 (1): 28–34.

Tornielli, Andrea. 2021. Vaccines for everyone, for the poor: examples from Pius VII and Pius IX. *Vatican News*. https://www.vaticannews.va/en/pope/news/2021-05/vaccines-for-all-history-pius-vii-xi-editorial.html.

Tranter, I., et al. 2024. Onward virus transmission after measles secondary vaccination failure. *Emerging Infectious Diseases* 30 (9): 1–8.

USCCB. 2018. *Ethical and religious directives for catholic health care services*. 6th ed. Washington, DC: USCCB.

USCCB. 2020. Moral considerations regarding the new covid-19 vaccines. https://www.usccb.org/resources/moral-considerations-covid-vaccines.pdf. Accessed 11 December 2020.

USCCB. 2021. U.S. Bishop Chairmen for Doctrine and for pro-life address the use of the Johnson & Johnson covid-19 vaccine. https://www.usccb.org/news/2021/us-bishop-chairmen-doctrine-and-pro-life-address-use-johnson-johnson-covid-19-vaccine. Accessed 2 March 2021.

Uversky, V. N., E. M. Redwan, W. Makis, and A. Rubio-Casillas. 2023. IgG4 antibodies induced by repeated vaccination may generate immune tolerance to the SARS-CoV-2 spike protein. *Vaccine* 11:991.

Van Kessel, R. 2023. Divergent COVID-19 vaccine policies: Policy mapping of ten European countries. *Vaccine* 4:2804–2810.

Vatican, II. 1964. *Lumen gentium*. Vatican City: Libreria Editrice Vaticana.

Vatican, II. 1965. *Gaudium et spes*. Vatican City: Libreria Editrice Vaticana.

Wechsler, M. 2013. Good Samaritan Laws and the duty to help or rescue someone. https://www.thelaw.com/law/good-samaritan-laws-the-duty-to-help-or-rescue-someone.218/.

Chapter 13
Vaccine Mandates Serve the Common Good

Matthew P. Schneider

13.1 Introduction

Eradicating smallpox and greatly reducing what we now consider vaccine-preventable disease is one of the greatest public health achievements of the past 100 years (Amin et al. 2012). This lack of disease is a great common good from which all in society benefit. Infectious diseases have gone from over half of deaths in 1900 to a single digit percentage by 1997, while life expectancy rose significantly (Cohen 2000). Yet this is a health concern around which we need to maintain vigilance as, unless a disease is completely eliminated, herd immunity can be lost if vaccination rates drop. To maintain this common good of less disease, vaccine mandates of various forms are often implemented.

This chapter puts forward a defense of many vaccine mandates as serving the common good. This is not a universal approval of all mandates, as mandates must serve the common good. It begins with some prerequisites, defines mandates, divides mandates into types, speaks about vaccines serving the common good, then notes how mandates can often serve the common good, looks at various factors in different vaccines, and concludes with a note about exemptions.

13.2 Prerequisites

Three pre-requisites are assumed in this chapter. First, this chapter assumes the majority, consensus or mainstream scientific view of all existing or proposed vaccines. This includes all aspects such as effectiveness, safety, etc. If one wants more,

M. P. Schneider (✉)
Holy Apostles College and Seminary, CT, Cromwell, USA
e-mail: mpschneider@legionaries.org

J. T. Eberl (ed.), *Emerging Issues in Catholic Bioethics*, Philosophy and Medicine 153, https://doi.org/10.1007/978-3-032-08862-8_13

there is plenty in peer-reviewed literature by experts in virology, immunology, vaccinology, etc. This includes other facts about diseases, such as if one does not get an infection, they cannot transmit it; thus, by preventing the one vaccinated from being infected, they also help others (CDC 2024).

Second, this chapter also assumes that getting vaccinated against diseases where it is recommended is a morally good act. In fact, it is often morally obligatory as multiple prior publications have noted. However, even if something is clearly the morally right thing to do, that does not directly imply it can be mandated. Carson and Flood (2017a) conclude, "Catholic social teaching entails a duty to vaccinate in order to protect the vulnerable." They base this conclusion on the fact that infectious diseases are a problem requiring collective action, such as needing 95% vaccination to end endemic spread and thus gain herd immunity. This requires all (or almost all) to serve the common good by vaccinating, not just a few individually making a morally right choice. This will be explained further below. Specifically regarding COVID-19 vaccines, McTavish and Eberl (2022) concluded similarly, "to be vaccinated is, in general, a morally good act. It has positive health benefits, is not too burdensome, and is economic. Vaccination can thus be considered ordinary." They come to this conclusion from an analysis of ordinary versus extraordinary forms of treatment. Similar conclusions have been reached by Austriaco (2021), Anderson et al. (2021), the United States Conference of Catholic Bishops (USCCB) (Rhoades and Naumann 2020), the bishops of England and Wales (Mason and Sherrington 2020), and the Vatican (Ladaria 2020).

This is true even if the vaccine is tested or grown on cell lines developed from previously aborted fetuses. The sources cited in the previous paragraph all note this as an objection to which they respond. Several others have written on cooperation with or appropriation of a moral evil more specifically. I have provided a needed clarification about Church teaching, noting that the Church has only approved skipping vaccines *grown* on fetal cell lines—though not requiring Catholics to conscientiously refuse such vaccines—and all references by the Church's magisterium imply that mere *testing* on fetal cell lines is not significant enough for any Catholic to skip any medication, including vaccines (Schneider 2021a). Further Mominee and Bellocq (2023) wrote an extensive analysis and noted, "We consider the position that regards the use of these vaccines as intrinsically immoral to be a minority position." In fact, they quote only bishop Athanasius Schneider and attorney Christopher Ferrara as holding such a view and neither is an expert in moral theology or bioethics. Their main point is to discuss if such vaccines involve appropriation or cooperation, both of which allow one to morally use any recommended vaccine where there is not a more moral option available. Melissa Moschella (2021) also notes similarly, "Using HEK 293 [a fetal cell line] or its products involves no cooperation with current evil, facilitation of future evil, or ratification of past evil, and because the dangers to moral character and the risk of scandal are weak." Anders (2020) goes so far as to indicate, "When significant risk exists, individuals in good conscience may have the duty to vaccinate despite their valid objections to vaccines of illicit origin."

Third, this chapter assumes the one issuing a vaccine mandate is a legitimate authority who has care for the common good of the community. The *Compendium of the Social Doctrine of the Church* notes this about political authority, but it also applies to other situations like the director of a hospital: "Political authority must guarantee an ordered and upright community life … for the attainment of the common good." (Pontifical Council for Justice and Peace 2004) The common good is "the sum of those conditions of social life which allow social groups and their individual members relatively thorough and ready access to their own fulfillment." (Vatican II 1965) Vaccine mandates are a type of civil law and Eberl (2022a) notes, "Aquinas defines civil laws as those made by appropriate authorities, utilizing prudential reason to craft ordinances that serve the *common good*."

13.3 Defining Vaccine Mandates

Vaccine mandates refer to any requirement by a legitimate authority to get vaccinated where not vaccinating involves some negative effect imposed by that authority. These negative effects can be anything from fines, to regular testing, to the inability to do certain things.

Vaccine mandates started with smallpox: the earliest recorded anywhere is a principality in the Italian peninsula, but by 1809 Massachusetts had the first mandate on US soil (Salmon et al. 2006). In 1905, the US Supreme Court decided *Jacobson v Massachusetts* when infectious disease was the leading cause of death, which affirmed vaccine mandates as a legitimate exercise of state police power (Mariner et al. 2005). In 1922, the Supreme Court ruled that requiring proof of smallpox vaccination for school was constitutional (Mariner et al. 2005). "By 1980, all 50 states had laws that linked vaccination with school entrance" (Salmon et al. 2006). All such rules seem to require vaccination in order to do some other thing; Mariner et al. (2005) argue that actually holding people down and forcing vaccination would likely be unconstitutional.

The least restrictive and coercive means available to one in charge of the common good should be used to accomplish the goal in the interest of the common good (Eberl 2022a). Williams (2022) notes, "Mandates are a form of government *coercion*; that is, they limit the autonomy of an individual to make a free personal choice by threatening punishment for non-compliance." Thus, as mandates are a significant restriction, they should only be used more as a last resort and should be more limited if that achieves the goal of higher vaccination rates to reach herd immunity. Mandates also need to be cautious in implementation to avoid having large groups of people vaccinating as a behavior to fulfill the mandate while internally opposed. Attitudes and behaviors need this distinction to ensure that an overwhelming majority still agree in attitude, not just in behavior (Hendrix et al. 2016).

To meet these goals, the Nuffield Council on Bioethics developed an eight-level ladder with respect to any public health intervention: (1) do nothing, (2) provide information, (3) enable choice (e.g.: cycling lanes to enable cycling), (4) guide choices through changing the default policy (e.g.: restaurants must seat customers in a non-smoking section unless they ask for smoking), (5) guide choices through incentives (e.g.: offer cash to parents who have their children vaccinated on time as Australia does [Walkinshaw 2011; Amin et al. 2012]), (6) guide choice through disincentives (e.g.: financial fines for not vaccinating), (7) restrict choice (e.g.: make an onerous process to get a vaccine exemption, or make hospital employment dependent on vaccination), and (8) eliminate choice (e.g.: isolation of infectious disease patients) (Nuffield Council on Bioethics 2007). Obviously, if vaccination levels can be raised by levels 1–5, those should be done before resorting to a mandate (levels 6–8). Salmon et al. (2006) indicate, "Compulsory vaccination should add to, not replace, other strategies to reach and sustain high rates of immunisation."

Some will argue vaccine mandates in principle go against the Congregation [now Dicastery] for the Doctrine of the Faith's (CDF) note issued in December, 2020, regarding the new COVID-19 vaccines: "Practical reason makes evident that vaccination is not, as a rule, a moral obligation and that, therefore, it must be voluntary." (Ladaria 2020) However, this is untrue for several reasons. First, many mandates have a degree of coercion, such as being required for a job or to get education, or to pay a fine if one chooses not to vaccinate without a valid exemption, but "they are not forcing anyone to be vaccinated against their will." (Eberl 2022a) Second, all but the most universal mandates are required only for specific situations one is free to do or not, like working in a hospital or traveling. Having a requirement if one chooses a particular option leaves the person free to voluntarily choose that option or not. For example, a state can have a law requiring cyclists on public roads to wear a helmet, without making it not voluntary: one is still free not to cycle or only cycle off public roads. (This argument does not work as much for universal mandates such as for school attendance.) Third, as Eberl and McTavish (2022) note, "There are two ways in which something could be a *rule*: absolutely or prima facie. Understanding the CDF as asserting an *absolute* rule would contradict Pope Francis's exhortation, 'I believe that morally everyone must take the vaccine. It is the moral choice because it is about your life but also the lives of others.' (McElwee 2021)" This means that vaccination is only voluntary *as a rule* and in cases like a pandemic it is no longer voluntary. And the Vatican note does not exclude the possibility of vaccine mandates (Cataldo 2021). In fact, the Vatican had quite restrictive mandates for COVID-19 vaccines, especially after the omicron variation emerged (Allen 2021).

Vaccine mandates need to try to get all stakeholders in the society on board. Failing to do this can result in opposition unrelated to vaccines. For example, in the mid-1800s, the smallpox vaccine was put forward by the privileged class, and some in the poor and working classes objected, not because of the vaccine, but because of the sense that the privileged class tended to make laws to favor their own class at the expense of the poorer classes (Amin et al. 2012).

13.4 Types of Vaccine Mandates

Various sources subdivide vaccine mandates slightly differently, but for simplicity I will divide them into three broad categories: limited profession or medical need mandates, travel mandates, and general mandates.

For decades, it has been standard practice to require full vaccination for all staff in places like hospitals and nursing homes: this was the only part of the COVID vaccine mandates that held up in court: "Healthcare systems, and HCWs [Health Care Workers], have special obligations to promote patient interests, a duty often thought to have its source in requirements of beneficence." (Gur-Arie et al. 2023) HCWs also have professional codes of ethics related to these obligations (Yeh 2022). This duty is the basis for many vaccine mandates for medical staff. Likewise, all members of the US military are required to be fully vaccinated, and often to have vaccines for conditions people in the US are not routinely vaccinated for like anthrax (against possible biological warfare) and yellow fever (not present in the USA). These are examples of limited profession mandates. Hospital mandates have been effective in increasing vaccination rates and decreasing disease transmission: a 2005–2010 study (long after influenza vaccines were recommended for HCWs) found mandating and encouraging influenza vaccines raised annual influenza vaccination from 29.5–54% in the prior three years to 97.6–98.9% in the following five years. Studies have shown such vaccination prevents illness in HCWs, plus morbidity and mortality among patients (Rakita et al. 2010). The requirement to be fully vaccinated before receiving an organ transplant can also be seen as a limited mandate due to medical need. Organs are scarce and the transplant recipient must take immunosuppressant drugs so is at higher risk of death from vaccine-preventable disease, and thus not vaccinating becomes medical factor in how long the organ might last (Duchini et al. 2003). These mandates are limited to situations where the risk of transmission and the risk of death to oneself (in the case of an organ recipient) or those to whom one might transmit the disease (in the case of HCWs) is higher, while at the same time employment mandates are well known so one can seek different employment if they strongly object to being vaccinated, so these restricted mandates take less away from the common good compared to other mandates as few if any are really forced to do something against their will. As they take least away for it, these mandates need the least amount of common good to be just. Some have even suggested a mandate based on old age or health factors that increase risk to severe outcomes from infection in order to protect the health system from being overwhelmed (Williams 2022). A mandate based on old age could work as a universal mandate, but a mandate based on health factors seems practically impossible due to the obligation to maintain patient confidentiality.

Second, we have vaccine mandates related to travel or immigration. While there is a general "right of people to migrate," (Pius XII 1952) "Regulating immigration according to criteria of equity and balance is one of the indispensable conditions for ensuring that immigrants are integrated into society with the guarantees required by recognition of their human dignity." (Pontifical Council for Justice and Peace 2004)

Such vaccine mandates are meant to ensure the local health care system is not overrun and that diseases from outside do not enter a particular area. Some of these mandates are for specific diseases endemic to one part of the world but not another, while others are to integrate people into the local society by getting them up to date on what is regularly expected in that society. For example, to receive a US Green Card one must demonstrate the recommended full vaccination for 10 diseases or toxins and others may be added in the future (US Citizenship and Immigration Services 2022). These are similar to the first type of mandate in that their coerciveness is limited and does not take away significantly from the common good, so they only require a moderate level of the common good to justify them. These mandates are generally applied to those making free choices, however, given there is a right to immigrate but not to work in a hospital, these mandates require stronger evidence for serving the common good than a hospital staff mandate.

Finally, we have universal or near-universal vaccine mandates. These are mandates like requiring vaccination for school, for (virtually) all employment, or for access to public assistance. These are common in developed nations. School mandates are the most common form of these as "Most of the diseases [vaccinated against] historically have had very high incidence in school-aged children because of the high potential for transmission in the congregate setting." (Malone and Hinman 2007) But as they are the broadest form and they can easily be coercive which takes away from the common good, they need significantly more evidence they serve the common good, a larger common good served, and a broader set of exemptions. As Yeh (2022) notes, "Ordinary citizens are not held to the same moral obligations and professional codes of ethics as are health care workers; hence, the justifications for health care workers might not be applicable to ordinary citizens. However, ordinary citizens might still have some degree of obligation toward their fellow citizens." These mandates also need overwhelming public support as, in a democratic society, they cannot be used to force all to get childhood vaccines but to target the last few percent who are stragglers (Walkinshaw 2011). For example, it would be hard to justify a mandate for COVID or flu vaccines this way, and when the US government tried to do so the court said that a universal mandate like this was beyond the president's power, while it simultaneously said it was in his power to require a mandate for health care facility staff (Gur-Arie et al. 2023; Sherman 2022). This distinction on the court's part at least implicitly recognizes a difference between mandates for health care staff and more universal mandates.

13.5 Vaccines Serve the Common Good

Several arguments have shown each individual Christian has a general moral obligation to vaccinate for the common good. This section will examine three (groups of) authors who come to almost identical conclusions.

Carson and Flood wrote about the duty of individual Catholics to vaccinate for the common good in 2017 (they explicitly leave mandates to the side in their discussion [Carson and Flood 2017b]). They begin by noting that vaccination is a collective action problem, with higher vaccination leading to lower infection across populations, including among the unvaccinated. In addition to the third prerequisite above about the common good, they (2017a) explain, "The common good … makes necessary reference to goods achievable only in and through the social matrix of shared living and shared responsibilities." This goes back to how herd immunity is a common action problem where we all need to get vaccinated to prevent disease in all. They (2017a) explain this further: "The principle of the common good refers to the objective realities of the interdependence of human beings and cooperation necessary to achieve constitutive ends of human nature … The interdependence of human beings and the resulting common good are not localized to a person's own family or even immediate community but rather extend to all people." We have this obligation to build up the common good across the whole of society which means we need to vaccinate for the common good. They see solidarity as an extension of the common good as it expresses our duty towards it. They (2017a) note, "the duties flowing from solidarity both commit Catholics to acting for the benefit of others and are non-optional because solidarity itself is as essential moral virtue to which all Catholics are called." When we have the virtue of solidarity, we will vaccinate for the good of others, not just for ourselves: this leads to the principle of solidarity when we as a society make such decisions together.

Bow (2023) expanded on Carson and Flood to try to answer the "when" question they leave unanswered. He begins by summarizing that they present a wider view of clinical interactions by bringing in the common good. He also connects individual health to the common good: "the health of an individual contributes to the common good in a more fundamental sense. Man's social/interdependent nature means that the good health of each individual benefits not only those proximate (e.g. family, friends, colleagues), but all mankind collectively." If we are healthy, those around us can more easily pursue flourishing: obviously more so for those closer to us. On top of this, the absence of an infectious disease improves others' health. He points out that the duty to vaccinate is not absolute but more general and in principle. After summarizing Carson and Flood, Bow spends most of the article on twelve criteria "to be met for vaccination to realize its full potential to contribute to the common good." (Bow 2023: note a correction to criterion 5 in Schneider 2024) Most vaccine mandates fit all twelve criteria.

Jason Eberl has written on his own and with James McTavish on the common good specifically regarding COVID vaccines. He begins his article on all Christians quoting the National Council of Churches (2021): "As Christians, our faith tradition is clear and consistent about how we are to treat others. We are taught that each person is made in the image and likeness of God and we are to be concerned about their well-being. Christ, as our example, reminds us in both his words and his actions that how we treat the most vulnerable matters to God." This shows a spectrum of support for vaccination as helping ourselves and others made in the *imago Dei*. He notes principles supporting mandates for the common good. The first: "One's liberty may be restricted to prevent risk of *harm to others*, and ample evidence exists of the risks from lack of vaccination." (Eberl 2022b) However, he notes that the least restrictive means should be employed and gives a shorter schema similar to the intervention ladder noted above. He quotes Moschella (2021) near the end of this piece:

> The widespread (or just highly publicized) refusal of pro-lifers to take the vaccine could actually lead to a sort of counter scandal in which the pro-life movement is seen as unreasonably thwarting the crucial effort to achieve herd immunity even though taking the vaccine is not actually immoral. Moreover, this scandal could be increased by the recognition that the vaccines have no greater connection to abortion than do countless other products (like common medications and processed foods) that pro-lifers use on a regular basis without complaint.

This shows how the refusal to vaccinate becomes a counter-witness and harms the pro-life cause.

When writing with McTavish, the focus is on how COVID vaccines are ordinary means of preventive treatment, but ordinary means on their own do not necessitate mandates as, although people have a moral obligation to care for themselves, we do not generally use laws to enforce behaviors unless others might be harmed: you can legally get drunk alone in your basement, but drunk driving is a crime. They indicate vaccination serves the common good, noting the common good in giving reasons why vaccination is both moral and, under certain conditions, may constitute a moral obligation: "To be vaccinated is a clearly a good thing when viewed from both the perspective of safeguarding one's own health and of promoting the *common good*, and all good acts are to be encouraged." (McTavish and Eberl 2022) The encouragement noted here could in theory include vaccine mandates. They also quote the USCCB: "Being vaccinated safely against COVID-19 should be considered an act of love of our neighbor and part of our moral responsibility for the common good." (Rhoades and Naumann 2020) They note that COVID restrictions such as lockdowns and quarantines were in many ways restrictions on the common good and vaccination can free us from such restrictions, thus serving the common good beyond just health.

From these three, it is clear on the personal level that medically recommended vaccines are obligatory in a general sense on each Christian, and this obligation comes from serving the common good of personal and public health.

13.6 Vaccine Mandates and the Common Good

After noting in the prior section that vaccines serve the common good, we need to turn to how mandating such vaccines can serve the common good as well. Malone and Hinman (2007) note, "An important characteristic of most vaccines is that they provide both individual and community protection." We generally mandate things for community protection, not individual good. This section will show this through the following arguments: the common good achieved by widespread vaccination, the tragedy of the commons, solidarity, stragglers, and conditions.

Hendrix et al. (2016) state clearly, "A population that is appropriately vaccinated against highly infectious diseases is a common good to the very society of which its members are a part." But the problem is that this common good requires everyone's assistance to maintain it, in a way analogous to how we all benefit from roads or military spending, which requires all to pay taxes to maintain them. For the common good of herd immunity, "Maintaining this common good requires that all vaccine-eligible individuals be vaccinated." (Hendrix et al. 2016) Those who refuse vaccination for non-medical reasons thus negatively affect the common good: for both those who choose to vaccinate and those who refuse. Cases like the 2014 Disneyland measles outbreak show how the lack of herd immunity can cause serious damage to the common good (Zipprich et al. 2015). When an authority is set to require anything for the common good which has a harm prevention goal, Williams (2022) explains, "It must be shown that the risk of harm [from not doing an action] is sufficiently high, and the cost of removing it proportionately low." Her example is traffic laws that reduce danger to other motorists and pedestrians. She later notes, directly in relation to COVID vaccines, that the "difference in risk–benefit profiles is important, as it affects the level of burden that an individual is being asked to take on as a contribution to the public good of removing the public health threat of COVID-19." The risks from the vaccine needs to be not only lower than the risk from the disease, but that risk also needs to be an order of magnitude or more lower, and it is generally multiple orders of magnitude for universally recommended vaccines; Hotez (2020) demonstrates this with three common vaccines.

Vaccine mandates have been shown to serve the common good: Briss et al.'s (2000) review shows how school-based vaccine mandates improve vaccination coverage and reduce disease. Gur-Arie et al. (2023) indicate, "The argument for population-wide mandates turns principally on the public health benefit of widespread vaccination in this circumstance, and second, the claim that the government's interest in promoting these outcomes meets the justificatory burden." Given the Catholic view of just government authority as one who oversees the public good for the community (see above), it would seem that once it is clearly indicated that this public health benefit is a sufficient public good with minimal negative effects, a vaccine mandate is justified by the leader's role in caring for the common good. Eberl (2022b) argues similarly: "Insofar as vaccination mandates help to create a social condition, namely, herd immunity, that allows people to reach their fulfillment more fully … the state is fulfilling its proper role to promote the common good and respect the dignity of the human person." Yeh (2022) concurs: "The government's duty [is] to pursue the common good in the form of preventing vulnerable populations from being infected." Yeh is more concerned with the health of vulnerable people than the whole population, but that complements the arguments by others for the health of the whole population. Overall, by objective measures, the common good of improved health is achieved.

Garrett Hardin's essay, "The Tragedy of the Commons," presents how common goods can be lost when each individual has an interest contrary to the common good. He begins with "Picture a pasture open to all. It is to be expected that each herdsman will try to keep as many cattle as possible on the commons." This makes sense and so long as there is plenty of common pasture for each herdsman, that works fine, but when the field is limited each herdsman is motivated to add more cattle as the advantage to him is great while the cost of overgrazing is borne by all, so this negative effect affects very little. Hardin (1968) continues: "Therein is the tragedy. Each man is locked into a system that compels him to increase his herd without limit-in a world that is limited." Hardin continues this analogy with several modern examples like how polluting harms each person a tiny bit but is usually quite beneficial for the polluter unless there is regulation to the contrary. The tragedy of the commons happens whenever logically seeking one's personal good harms the common good more, but such harm is only minorly felt by the person seeking their personal good. Malone and Hinman (2007) take up Hardin and note, "A community free of an infectious disease because of a high vaccination rate can be viewed as a common." Once herd immunity, or close to it, is reached, those who do not vaccinate are still at low risk of infectious disease. Vaccines are not risk-free (Schneider 2024) and once the risk of the disease is lower, even the extremely limited risk of serious negative side effects from vaccines provide a reason for some not to immunize.

Yeh (2022) presents the common good more in terms of solidarity. He presents three tiers: "At tier 1, solidaristic practices are expressed informally at an interpersonal level. At tier 2, some group practices appear as informal customs or social atmosphere. At tier 3, the practices are institutionalized in the forms of contract, policy, law, or other solidaristic institutions." Solidarity is preferable as it is not simply an external deontology, but reflects the common value of the people. In vaccines, we are dealing with a group of individuals, such as citizens who together carry costs to assist each other by getting vaccinated, knowing possible adverse events. This creates a "Self-imposed a moral obligation to assist each other. This self-imposed obligation could be considered a preexisting moral obligation for a mandatory vaccination policy." (Yeh 2022) A strong group solidarity in levels 1 and 2 is needed for a practical and moral implementation of one of level 3, which is where a legal mandate would be, imposing solidarity on all not just the majority who take it upon themselves. Otherwise, the imposition of government can take more from the common good than vaccines add.

Hendrix et al. (2016) propose adding solidaristic reasoning into vaccine decisions: "Ethical considerations include whether patients (or their parents) bear a responsibility to consider that their immunization decisions can affect others." An example would be: "A family's four month old infant (too young to have received all recommended vaccinations) might be totally dependent on the vaccination of all family members and others to be protected against VPDs [Vaccine Preventable Disease] such as measles or pertussis" (Amin et al. 2012). Some even suggest that non-vaccinators could be sued for not vaccinating if it results in someone else's infection (Hendrix et al. 2016). Holding those who refuse vaccines accountable would help them realize that their choices negatively affect not just them, but the common good. It is much harder to justify a mandate that only helps the individual as it is rare to use coercive power to force people to do good for themselves; rather, we use it to require doing good for others or at least preventing harm. For example, speed limits on public roads protect others from the dangers of someone speeding but if someone wants to go 200 mph alone on a closed track, nobody else is endangered like they would be on the interstate. Law abiding for the common good is generally a form of solidarity.

The goal of a vaccine mandate, not just normal social pressure, is to get the stragglers to vaccinate. Many of these stragglers demonstrate the free-rider problem: as each vaccination serves the common good, at a certain point even the unvaccinated are also unlikely to catch a disease as it no longer circulates. In a study using simulations, it was found that "the probability of vaccination decreased by 19% point as the proportion of previous round vaccination among others increased from 0 to 1." (Ibuka et al. 2014) Free-riding was observed in all paradigms changing other factors. Ibuka et al. (2014) note, "A change in perceived risk of infection with a disease alters individuals' behavior regarding prevention of the disease," which is manifest in people not getting vaccinated when they can after there is a high vaccination rate as then their risk of the disease seems diminished. The problem of free-riders with vaccination is that if enough become free-riders, herd immunity goes away and the disease starts spreading in the population again such as the Disneyland measles outbreak (Zipprich et al. 2015). Free-riding goes contrary to solidarity and the common good: mandates to reduce free-riding, like vaccine mandates or mandates to have automobiles insured to drive on public roads, protect the common good.

When something like a vaccine is mandated, especially in more universal mandates, several things are needed to put the individual good in conflict with the common good as little as possible. First, mandated vaccines should be free so someone does not fail the mandate for economic reasons. Second, some system needs to be set up to compensate those with severe negative reactions to mandated vaccines. This follows the reciprocity principle as expressed by Upshur (2002): "The reciprocity principle holds that society must be prepared to facilitate individuals and communities in their efforts to discharge their duties." In other words, if we as a society give someone a duty, we need to provide reasonable means to fulfill that duty. Serious adverse reactions are very rare, but are real and should not be ignored. The current National Vaccine Injury Compensation Program funded by $0.75 per dose paid by vaccine manufacturers seems like a reasonable way to do this (Sugarman 2007). It is important to track these adverse reactions insofar as, if they rise above a certain threshold, their harm to the common good would be grounds to cancel the mandate.

In most cases where vaccines are mandated, the mandates would serve the common good. It is evident that widespread vaccination is a common good for society, and looking at it through various lenses (tragedy of the commons, solidarity, and free-riding), the common good is served by such mandates.

13.7 Factors Affecting the Ethics of Specific Mandates

Several factors affect the ethics of a vaccine mandate such as the mode of and probability of disease transmission, non-vaccine methods of prevention, disease danger as in the chance of dying, disease mutation rate, vaccine efficacy, herd immunity, disease timing, dangers of vaccine side effects, and the breadth of the mandate. The breadth of the mandate was covered above; the others will be covered in this section in that order.

Different diseases have different modes of transmission and different probabilities in different transmission modes. A vaccine against a disease transmitted through the air or breath droplets is much more likely to be moral to mandate widely than one only transmitted through sexual or blood contact. Blood contact is possible at a hospital, even if generally avoided, so they can be mandated there, but mandating vaccines where people elsewhere can easily avoid any possible transmission situation is tougher to justify. The more likely a disease is to transmit, the higher the vaccination level needed for herd immunity to stop social transmission, and thus the more it can be mandated to encourage those last few to vaccinate. For example, for measles, the herd immunity threshold is 96%–99% of a given population to be vaccinated (Hendrix et al. 2016). Diseases transmitted mainly by mosquitos or other things in the environment are often candidates for vaccines mandated for travel but not mandates where the environmental risk is not present. For example, yellow fever is mainly mosquito-borne and the species of mosquito is not present in the US, but present in other countries (WHO 2023). A related factor is how likely transmission is via different modes. Sexually transmitted infections can sometimes also be transmitted via other means, but such transmission is less likely (Petca et al. 2020). From a Catholic perspective, chastity is always an effective way to prevent transmission of disease via sexual contact, so a mandate in such a case is hard to justify; however, as most people do have several sexual partners before marriage and more intense non-sexual touching can transmit it, HPV vaccines are still generally recommended but not mandated for teens.

Voluntary means that can prevent transmission risk without a vaccine need to be considered when making a mandate, especially with the possibility of making exemptions for those willing to undergo some other means of prevention. Gur-Arie et al. (2023) note regarding COVID that N-95 masks without a vaccine are as likely or more likely to prevent transmission as a vaccine without such a mask. Based on this, and their understanding of mandates for medical staff (above), they conclude, "HCWs who refuse COVID-19 vaccination do not straightforwardly violate an obligation of beneficence, nor is it clear that considerations of beneficence outweigh the considerations that count against a mandate." Other means like regular testing and other prophylactic measures may also reduce the need for a vaccine, thus making the mandate less enforceable if such means are implemented or given as an optional alternative to vaccination. These means would require identifying vaccinated and unvaccinated individuals to require more prophylactic measures from those who did not vaccinate, and so become something like level 5 on the Nuffield scale (just below mandates) where choices are guided through incentives like not having to have a weekly COVID test *via* an unpleasant nasal swap (Nuffield Council on Bioethics 2007).

The more dangerous a disease is, the stronger the case for a vaccine mandate. This can be looked at on several grounds like deaths and hospitalizations per 100,000 people in an unvaccinated population or per 100,000 infections. Danger can also be related to the non-vaccine modes to treat it. The mortality rate of many diseases has changed significantly based on newer treatments reducing risk significantly. Amin et al. (2012) argue, "Compared to other procedures for addressing the same issue, the vaccination should be the best opportunity for successfully preventing disease as compared to the risk for harm." This would exclude a vaccine mandate for any disease where vaccine efficacy was low or effective treatment was available such that the danger was low to non-existent. It would also affect mandates insofar as treatments for a disease were available even if the treatment only significantly reduced risk and did not eliminate it if other factors went against this mandate serving the common good.

Some vaccinations are needed only once or twice and they last for life, while others are annual. Some of this is due to how our immune systems work in that immunity can wane over time for some conditions like pertussis (McGirr and Fisman 2015) and some is due to how some diseases, particularly RNA viruses (Elena and Sanjuán 2005), rapidly mutate and thus evade existing immunity, making full herd immunity illusive. The longer a vaccine remains effective, the greater the common good achieved by vaccination and thus the more justification mandating it is. This is why universal or near-universal mandates for COVID or flu vaccines seem impractical and are not nearly as likely to serve the common good.

Different vaccines have different levels of efficacy (Lopalco and DeStefano 2015). A vaccine that is more effective obviously has a stronger justification for a mandate as it provides more of the common good of health and enables herd immunity so endemic transmission can likely end. Salmon et al. (2006) note "a reliable supply of safe and effective vaccines" as a pre-requisite for making these vaccines mandatory. Bow (2023) even makes this a pre-requisite for it being mandatory, giving the criterion: "The vaccine effectively enhances the body's immune response to the pathogen."

Varicella presents a particular issue where timing affects severity in an unusual manner: if a child has the disease, it is usually mild chicken pox; but if an adult has it, it is shingles which is much more serious (Harpaz et al. 2008). When a large portion choose to vaccinate such that endemic childhood transmission is reduced from almost everyone getting it to almost zero, the risk of unvaccinated individuals getting it later rather than earlier becomes an issue of concern indicating parents may be required to vaccinate their children against chickenpox for that child's future good.

No medical intervention has zero possible negative side effects. Different vaccines have different side effects with different frequencies. Minor side effects like a sore arm for 24 hours or a slight tiredness are not significant factors. Extremely infrequent (less than 1 in one million) serious interactions also have little effect. However, more common relatively serious side effects can affect the ethics of mandating vaccines. For example, the relatively rare side effect of myocarditis from mRNA COVID vaccines in younger people (Bozkurt et al. 2021) should be a factor when mandating such a vaccine. Vaccines should be sought with even lower incidences of side effects if possible: for example vaccination for pertussis dropped due to reporting on relatively rare (0.12% of vaccinations) side effects, until a vaccine with even lower side effect risk came out (Amin et al. 2012). Salmon et al. (2006) note that the vaccines must be reliably safe to be compulsory, noting this as a government responsibility if they are the ones mandating them. This is important as the government cannot mandate something that is just slightly better for the common good, but the government needs to achieve significantly more of the common good by following the mandate compared to not following it, so in vaccine mandates the good needs to outweigh the good taken by negative side-effects by at least an order of magnitude. For a non-medical example, you need a much larger common good at stake to draft young people into the frontlines than to require a year of volunteer service as frontline fighting is much more dangerous than spending a year working at a corporal work of mercy. Amin et al. (2012) further argue, "If harm does result from the vaccination, the benefit of vaccination to the subject should at least compensate for the harm incurred." As noted above, the current National Vaccine Injury Compensation Program offers a reasonable way of doing this.

This section has looked at various factors that could affect how a vaccine and thus a vaccine mandate affect the common good to make a stronger or weaker argument for vaccine mandates in particular cases. Most currently enforced vaccine mandates in Canada and the US still serve the common good after counting all these factors, but these factors may indicate certain proposed vaccine mandates may or may not serve the common good. Universal mandates for vaccines against RNA viruses do not seem to generally serve the common good, but none are currently in place in these two countries.

13.8 A Note About Vaccine Mandate Exemptions

There are at least five main types of exemptions offered which will be discussed briefly. Then some issues around working with such exemption claims will be covered. These are not separate from the common good as the common good is concerned both for the well-being and development of the group (served by mandates) and respect for persons (which would include respect for those who object) (Anders 2020).

First, there are medical condition exemptions. These can be things like allergies to ingredients, immunosuppressant drugs, reactions to prior vaccines, or conditions that make vaccination futile or particularly dangerous. We should err on the side of granting such exemptions when medical evidence is uncertain as any negative reaction makes it harder for skeptical people to agree to vaccination. For example, in a study on the mandating the flu vaccine in a hospital, staff "had their accommodation requests approved if they had a prior history of Guillain-Barré syndrome, even though the connection between that and influenza vaccine is controversial" (Rakita et al. 2010). However, given that some with pseudoscience objections (see below) will claim this type of objection, they cannot be given on patient testimony alone. This seems like serving the common good as this has a reasonable chance of imposing a significantly higher burden on those persons.

Second, there is infection or antibody testing. These are not exemptions from being protected against a condition but proving that one is protected by a means other than a vaccine. For example, proof of chicken pox usually allows one to skip varicella vaccination, and antibody testing after a first MMR vaccine can be done to determine if a second dose is needed. Different places have different rules about antibody testing: as a child in Canada my mom had me tested for antibodies after the first dose, so I only received one MMR vaccine to fulfill the school requirement, but when I got a US Green Card, that was insufficient, so I received a second dose. Generally, this type of exemption seems valid as the end of mandatory vaccination is achieved via other means so long as antibody testing for that disease has been shown to provide sufficient protection. Antibodies from a prior infection do not provide full protection in other conditions, especially RNA viruses that change frequently, so a reasonable calculation needs to be made of the probability of infection with just prior infection versus prior infection plus vaccination.

Third are philosophical, religious, or conscientious objections. These is the most common and most controversial. In general, such exemptions should be respected, but not always, especially in more narrower mandates as often it is clear full vaccination is expected in the military or a hospital and nobody must work specifically in either of those fields or has an inherent right to that work.

The distinction between religious and philosophical objections is represented in law: 48 US states allow religious objections but only 15 allow philosophical objections (Malone and Hinman 2007). Legally, the lack of non-medical exemption is likely constitutional given the state's "interest in protecting the public against measles [or other serious diseases] is compelling enough to outweigh an individual's religious exemption from measles [or other serious diseases] vaccinations" (Anders 2020). Malone and Hinman (2007) conclude: "No constitutional right exists to either a religious or philosophic exemption to these requirements [vaccine mandates]." The higher number allowing religious exemptions recognizes religion properly as a more fundamental area of the person's self-identity than a personal philosophy. It is still hard to define religious belief: Is it this person's own view of the higher power, or that of some organized religion? No large, organized religion, including faith-healing religions like Christian Science, object in principle to mandatory vaccination (Ostling 2021; Eberl 2022b). However, smaller groups or individual views of the divine may have a belief against vaccination, and persons can have unique interpretations of their tradition contrary to that tradition's official representatives (Flescher 2023). In the Catholic sphere, the Diocese of Bridgeport noted that there was no Catholic reason to skip COVID vaccines and that priests were not to sign any vaccine exemptions, but allowed individual faithful to object on their own based on their conscience (Schneider 2021b).

Often conscientious exemption can be based on a misinformed conscience. Eberl (2022a) gives the example of "Catholics who seek an exemption from a vaccination mandate due to the remote—and, in the case of the Pfizer/BioNTech and Moderna vaccines, merely probable—material connection to past abortions." He concludes that public authorities need not respect these. In fact, when I presented many of these Catholics with a vaccine that they could promote without any such connection, one significant "Catholic" "anti-vaccine" source (LifeSiteNews) wrote a piece admitting they did not like COVID vaccines at all for pseudoscientific reasons (next type) (Owen et al. 2022). As Hendrix et al. (2016) explain, "It is one thing to reject vaccines because families believe it is inconsistent with their sincerely held religious beliefs (more defensible), but it is another to reject vaccines because of a belief in a government conspiracy (less defensible)."

Fourth is the pseudoscience or conspiracy theory objection. Often this is classed with the prior objection and many claiming the prior objection really have this objection when pushed on the issue (as exemplified by the LifeSiteNews example). This is an objection where the person denies overwhelming evidence about the existence of the disease, the effectiveness of the vaccine, the rarity of side effects, or something similar. The one in charge of the common good could include mandatory vaccine education before approving non-medical exemptions in the hope of changing the minds of people like this. Walkinshaw (2011) notes that if people refuse vaccination in Latvia, "Health providers have a duty to explain the health consequences and if the patient hasn't been persuaded to change his mind," then get a signature from the non-vaccinator acknowledging this. This seems like it may help overcome such unfounded objections.

Fifth is the purely emotional objection. This is something like "I do not want to see my child cry at the needle," or the like. This is where vaccines have been victims of their own success where parents only see the brief tears of vaccination but do not know of the cries of pain children who become infected by vaccine preventable diseases. Like the prior one, mandatory education before approving non-medical exemptions may be helpful. This is the least founded of the five and should have the least ground given to it.

Decisions about latter three types of objections, especially the philosophical, religious, or conscientious objection, are important for prudential judgment about applying vaccine mandates. Compulsory vaccination needs a strong majority in support to be effective, and Salmon et al. (2006) note, "To force vaccination on children of parents who have strong convictions against vaccination could create a public backlash and serve to galvanise anti-vaccine groups." Their argument is that accepting a small number of these exemptions limits public backlash, and ultimately provides higher vaccine coverage. They note the early 2000s reaction to MMR vaccines, but similar things can be seen with COVID vaccines where more kids are getting non-medical exemptions from school vaccine mandates (Seither 2023). This correlates with a backlash to attempts at universal or near universal COVID vaccine mandates with large opposition; I have not seen a study show causation, but it seems likely that is at least part of the drop.

Research has found that the difficulty of non-medical exemptions changes the number sought. Gostin (2015) noted "nonmedical exemption rates [were] 2.3 times higher in states with easy administrative policies than those with difficult policies," and this resulted in 50% higher pertussis rates in those states that made non-medical exemptions easy. This makes it important to make those exemptions difficult to get as one who is not willing to endure a more difficult process is more likely a free-rider than an authentic conscientious objector. Flescher notes the problem of being too lenient on these exemptions: "One may simply assert that one's personal interests trump the public good when the two come into conflict. In this manner, a believer exempting oneself from a health-protective vaccine mandate is afforded an opportunity to cloak ideological objections under the guise of religious rationale." (Flescher 2023; cf. Eberl 2022b) This is a concern for personal good overriding the common good and exemptions becoming the "tragedy of the commons," as noted above.

In giving such exemptions and ensuring it is truly a conscientious / religious exemption and not an emotional or pseudoscience one, courts have inconsistently distinguished "sincerely held" and "suddenly held" beliefs (Flescher 2023). This seems like a reasonable distinction to avoid overuse of such exemptions to the detriment of the common good. On a similar vein, Anders (2020) notes the issue of what qualifies as "sincerely held" which must be answered for each case. He notes, "Exemptions [legally] do not qualify as religious if they are motivated by personal fears or based on health concerns." In all cases of exemptions, when there is an outbreak, the unvaccinated given exemptions can be excluded from public places like school when there is an outbreak of that disease they lack vaccination for (Anders 2020). This is only a temporary measure to contain a disease similar to a quarantine. Ultimately, Hendrix et al. (2016) note an unresolvable issue around non-medical exemptions: "Striking a balance between respecting parental rights and autonomy and maximizing the greater good of herd immunity may seem an intractable problem, especially in the current climate of heated vaccine debates." Prudence needs to be applied in each case and there is no short "magic bullet" that will resolve all issues around exemptions from mandates.

This section covered five levels of exemptions from those always given to those with less grounding: medical issues; antibody testing as an alternative; religious, philosophical or conscience; pseudoscience; and emotional. Granting some non-medical exemptions can increase vaccination rates as it decreases the opposition to them, however, they should be made difficult to ensure they are only granted to those with a sincere religious, philosophical or conscience, not simply a free-rider or one with less grounded objection. Exact procedures for how to deal with non-medical exemptions are a question of prudence well beyond the scope of this article.

13.9 Conclusion

Writing about the history of the anti-vaccine movement, Wolfe and Sharpe (2002) explain, "Vaccination is unique among de facto mandatory requirements in the modern era, requiring individuals to accept the injection of a medicine or medicinal agent into their bodies, and it has provoked a spirited opposition. This opposition began with the first vaccinations, has not ceased, and probably never will." This does create a situation where more objections to this can arise than to say speed limits or compulsory elementary education. However, such objections do not indicate that vaccine mandates are contrary to the common good. In fact, they often point to a lack of or poor education on how vaccine mandates serve the common good.

Vaccines clearly serve the common good. Given we vaccinate for others, not just ourselves, vaccine mandates can also serve the common good. This is especially true when such a mandate can achieve herd immunity or the end of endemic disease spread. Most if not all existing mandates in the US and Canada serve the common good to a degree that is justified. However, based on various factors, not every mandate would serve the common good. Serving the common good also allows some limited exemptions to vaccine mandates.

References

Allen EA. 2021. Vatican issues vaccine mandate for all employees. Crux. https://cruxnow.com/vatican/2021/12/vatican-issues-vaccine-mandate-for-all-employees. Accessed 19 Dec 2024.

Amin, A. N. E., M. T. Parra, R. Kim-Farley, and J. E. Fielding. 2012. Ethical issues concerning vaccination requirements. *Public Health Reviews* 34 (1): 1–20. https://doi.org/10.1007/BF03391666.

Anders, C. C. 2020. Individual and institutional religious exemptions from vaccines: Federal law and Catholic teaching. *Natl Cathol Bioeth Q* 20 (3): 501–523. https://doi.org/10.5840/ncbq202020346.

Anderson RT, Austriaco NPG, Condic M, Flannery K, George RP, Snead OC, Tollefsen C, and White TJ. 2021. Statement from pro-life catholic scholars on the moral acceptability of receiving COVID-19 vaccines. In: Ethics Public Policy Cent. http://eppc.org/news/statement-from-pro-life-catholic-scholars-on-the-moral-acceptability-of-receiving-covid-19-vaccines/. Accessed 19 Mar 2021.

Austriaco, N. P. G. 2021. Prudential use of the morally controversial COVID-19 vaccines. *The Linacre Quarterly* 88 (3): 317–320. https://doi.org/10.1177/00243639211017697.

Bow, S. M. A. 2023. When does Catholic social teaching imply a duty to be vaccinated for the common good? *New Bioeth* 29 (4): 304–321. https://doi.org/10.1080/20502877.2023.2261718.

Bozkurt, B., I. Kamat, and P. J. Hotez. 2021. Myocarditis with COVID-19 mRNA vaccines. *Circulation* 144 (6): 471–484. https://doi.org/10.1161/CIRCULATIONAHA.121.056135.

Briss PA, Rodewald LE, Hinman AR, Shefer AM, Strikas RA, Bernier RR, Carande-Kulis VG, Yusuf HR, Ndiaye SM, and Williams SM. 2000. Reviews of evidence regarding interventions to improve vaccination coverage in children, adolescents, and adults. doi:https://doi.org/10.1016/s0749-3797(99)00118-x.

Carson, P. J., and A. T. Flood. 2017a. Catholic social teaching and the duty to vaccinate. *The American Journal of Bioethics* 17 (4): 36–43. https://doi.org/10.1080/15265161.2017.1284914 .

Carson, P. J., and A. T. Flood. 2017b. Response to open peer commentaries on Catholic social teaching and the duty to vaccinate. *The American Journal of Bioethics* 17 (4): W1–W3. https://doi.org/10.1080/15265161.2017.1293753.

Cataldo P. 2021. Why the CDF note on the morality of using some anti-covid-19 vaccines suggests a moral obligation to receive SARS-CoV-2 vaccines. Health Care Ethics USA 29(2):2–6. https://www.chausa.org/publications/health-care-ethics-usa/archive/issue/fall-2021/why-the-cdf-note-on-the-morality-of-using-some-anti-covid-19-vaccines-suggests-a-moral-obligation-to-receive-sars-cov-2-vaccines. Accesses 19 Dec 2024.

Centers for Disease Control [CDC]. 2024. Vaccines & immunizations. U.S. centers for disease control & prevention. https://www.cdc.gov/vaccines/index.html. Accessed 19 Dec 2024.

Cohen, M. L. 2000. Changing patterns of infectious disease. *Nature* 406 (6797): 762–767. https://doi.org/10.1038/35021206.

Duchini, A., J. A. Goss, S. Karpen, and P. J. Pockros. 2003. Vaccinations for adult solid-organ transplant recipients: Current recommendations and protocols. *Clinical Microbiology Reviews* 16 (3): 357–364. https://doi.org/10.1128/cmr.16.3.357-364.2003.

Eberl, J. T. 2022a. Is there a moral obligation to be vaccinated for COVID-19? *Health Care Ethics USA* 30 (1): 33–39.

Eberl, J. T. 2022b. Are Christians morally obligated to be vaccinated for COVID-19? *Rev Expo* 119 (1–2): 64–75. https://doi.org/10.1177/00346373221132201.

Elena, S. F., and R. Sanjuán. 2005. Adaptive value of high mutation rates of RNA viruses: Separating causes from consequences. *Journal of Virology* 79 (18): 11555–11558. https://doi.org/10.1128/jvi.79.18.11555-11558.2005.

Flescher, A. 2023. How well do religious exemptions apply to mandates for COVID-19 vaccines? *Religions* 14 (5): 569. https://doi.org/10.3390/rel14050569.

Gostin, L. O. 2015. Law, ethics, and public health in the vaccination debates: Politics of the measles outbreak. *JAMA* 313 (11): 1099–1100. https://doi.org/10.1001/jama.2015.1518.

Gur-Arie, R., B. Hutler, and J. Bernstein. 2023. The ethics of COVID-19 vaccine mandates for healthcare workers: Public health and clinical perspectives. *Bioethics* 37 (4): 331–342. https://doi.org/10.1111/bioe.13141.

Hardin, G. 1968. The tragedy of the commons. *Science* 162 (3859): 1243–1248. https://doi.org/10.1126/science.162.3859.1243.

Harpaz R, Ortega-Sanchez IR, and Seward JF. 2008. Advisory Committee on Immunization Practices (ACIP) Centers for Disease Control and Prevention (CDC) prevention of herpes zoster: Recommendations of the Advisory Committee on Immunization Practices (ACIP). Morb. Mortal. Wkly. Rep. 57(RR-5):1–30; quiz CE2-4.

Hendrix, K. S., L. A. Sturm, G. D. Zimet, and E. M. Meslin. 2016. Ethics and childhood vaccination policy in the United States. *American Journal of Public Health* 106 (2): 273–278. https://doi.org/10.2105/AJPH.2015.302952.

Hotez P. 2020. You are unvaccinated and got sick. These are your odds. New York Times. https://www.nytimes.com/2020/01/09/opinion/vaccine-hesitancy.html. Accessed 19 Dec 2024.

Ibuka, Y., M. Li, J. Vietri, G. B. Chapman, and A. P. Galvani. 2014. Free-riding behavior in vaccination decisions: An experimental study. *PLoS One* 9 (1): e87164. https://doi.org/10.1371/journal.pone.0087164.

Ladaria L. 2020. Note on the morality of using some anti-Covid-19 vaccines. https://www.vatican.va/roman_curia/congregations/cfaith/documents/rc_con_cfaith_doc_20201221_nota-vaccini-anticovid_en.html. Accessed 19 Dec 2024.

Lopalco, P. L., and F. DeStefano. 2015. The complementary roles of phase 3 trials and post-licensure surveillance in the evaluation of new vaccines. *Vaccine* 33 (13): 1541–1548. https://doi.org/10.1016/j.vaccine.2014.10.047.

Malone, K. M., and A. R. Hinman. 2007. Vaccination mandates: The public health imperative and individual rights. In *Law in public health practice*, ed. R. A. Goodman, R. E. Hoffman, W. Lopez, G. W. Matthews, M. Rothstein, and K. Foster, 0. Oxford University Press.

Mariner, W. K., G. J. Annas, and L. H. Glantz. 2005. Jacobson v Massachusetts: It's not your great-great-grandfather's public health law. *American Journal of Public Health* 95 (4): 581–590. https://doi.org/10.2105/AJPH.2004.055160.

Mason P, and Sherrington J. 2020. The Catholic position on vaccination. https://www.cbcew.org.uk/home/our-work/health-social-care/the-catholic-position-on-vaccination/. Accessed 19 Dec 2024.

McElwee J. 2021. Pope Francis suggests people have moral obligation to take coronavirus vaccine. Natl. Cathol. Report. https://www.ncronline.org/vatican/pope-francis-suggests-people-have-moral-obligation-take-coronavirus-vaccine. Accessed 19 Dec 2024.

McGirr, A., and D. N. Fisman. 2015. Duration of pertussis immunity after DTaP immunization: A meta-analysis. *Pediatrics* 135 (2): 331–343. https://doi.org/10.1542/peds.2014-1729.

McTavish, J., and J. T. Eberl. 2022. Is COVID-19 vaccination ordinary (morally obligatory) treatment? *Natl Cathol Bioeth Q* 22 (2): 319–333. https://doi.org/10.5840/ncbq202222226.

Mominee, J. J., and A. Bellocq. 2023. Cooperation with past evil? : A Defense of the magisterial teaching on vaccines with a connection to abortion. *Ann Theol* 37 (1): 11–63. https://doi.org/10.17421/ATH371202301.

Moschella, M. 2021. Dignitas personae, HEK 293, and the COVID vaccines. *Natl Cathol Bioeth Q* 21 (1): 107–121. https://doi.org/10.5840/ncbq202121110.

National Council of Churches. 2021. Fact sheet: Christian approach to vaccine hesitancy. https://nationalcouncilofchurches.us/fact-sheet-christian-approach-to-vaccine-hesitancy/. Accessed 19 Dec 2024.

Nuffield Council on Bioethics. 2007. Public health: ethical issues. Nuffield Council on Bioethics, London. https://cdn.nuffieldbioethics.org/wp-content/uploads/Public-health-ethical-issues.pdf. Accessed 19 Dec 2024.

Ostling R. 2021. Big question right now: What religious groups oppose vaccination, even during epidemics? In: GetReligion. https://www.getreligion.org/getreligion/2021/9/20/big-question-right-now-what-religious-groups-oppose-vaccination-even-during-epidemics. Accessed 13 June 2024.

Owen H, Ambrosetti T, and Ambrosetti V (2022) Judge just judgment: A plea for an end to theological hit pieces. In: LifeSiteNews. https://www.lifesitenews.com/opinion/judge-just-judgment-a-plea-for-an-end-to-theological-hit-pieces/. Accessed 1 June 2024.

Petca, A., A. Borislavschi, M. E. Zvanca, R.-C. Petca, F. Sandru, and M. C. Dumitrascu. 2020. Non-sexual HPV transmission and role of vaccination for a better future (review). *Experimental and Therapeutic Medicine* 20 (6): 186. https://doi.org/10.3892/etm.2020.9316.

Pius XII (1952) *Exsul familia nazarethana*. https://www.papalencyclicals.net/pius12/p12exsul.htm.

Pontifical Council for Justice and Peace. 2004. Compendium of the social doctrine of the church. Libreria Editrice Vaticana / United States Conference of Catholic Bishops, Vatican City / Washington, DC.

Rakita, R. M., B. A. Hagar, P. Crome, and J. K. Lammert. 2010. Mandatory influenza vaccination of healthcare workers: A 5-year study. *Infection Control and Hospital Epidemiology* 31 (9): 881–888. https://doi.org/10.1086/656210.

Rhoades KC, and Naumann JF. 2020. Moral considerations regarding the new COVID-19 vaccines. https://www.usccb.org/moral-considerations-covid-vaccines. Accessed 19 Dec 2024.

Salmon, D. A., S. P. Teret, C. R. MacIntyre, D. Salisbury, M. A. Burgess, and N. A. Halsey. 2006. Compulsory vaccination and conscientious or philosophical exemptions: Past, present, and future. *The Lancet* 367 (9508): 436–442. https://doi.org/10.1016/S0140-6736(06)68144-0.

Schneider MP. 2021a. Testing on fetal cell lines: What does the church teach? In: Cathol. Moral Theol. https://catholicmoraltheology.com/testing-on-fetal-cell-lines-what-does-the-church-teach/. Accessed 26 May 2024.

Schneider MP. 2021b. Bridgeport: The best vaccine mandate instruction yet. In: Cathol. Lenses. https://www.patheos.com/blogs/throughcatholiclenses/2021/08/bridgeport-the-best-vaccine-mandate-instruction-yet/. Accessed 15 June 2024.

Schneider, M. P. 2024. Letter to the editor. *New Bioeth* 30 (1): 10–10. https://doi.org/10.1080/20502877.2024.2354049.

Seither R. 2023. Coverage with selected vaccines and exemption from school vaccine requirements among children in Kindergarten—United States, 2022–23 school year. MMWR Morb Mortal Wkly Rep 72. https://doi.org/10.15585/mmwr.mm7245a2.

Sherman N. 2022. US Supreme Court blocks Biden's workplace vaccine mandate. BBC. https://www.bbc.com/news/world-us-canada-59989476. Accessed 19 Dec 2024.

Sugarman, S. D. 2007. Cases in vaccine court—Legal battles over vaccines and autism. *The New England Journal of Medicine* 357 (13): 1275–1277. https://doi.org/10.1056/NEJMp078168.

Upshur, R. 2002. Principles for the justification of public health intervention. *Canadian Journal of Public Health* 93 (2): 101–103. https://doi.org/10.1007/BF03404547.

US Citizenship and Immigration Services. 2022. Vaccination requirements. In: USCIS. https://www.uscis.gov/tools/designated-civil-surgeons/vaccination-requirements. Accessed 31 May 2024.

Vatican II. 1965. *Gaudium et spes*. https://www.vatican.va/archive/hist_councils/ii_vatican_council/documents/vat-ii_const_19651207_gaudium-et-spes_en.html.

Walkinshaw, E. 2011. Mandatory vaccinations: The international landscape. *CMAJ Can Med Assoc J* 183 (16): e1167–e1168. https://doi.org/10.1503/cmaj.109-3993.

WHO. 2023. Yellow fever. In: World Health Organ. https://www.who.int/news-room/fact-sheets/detail/yellow-fever. Accessed 16 June 2024.

Williams, B. M. 2022. The ethics of selective mandatory vaccination for COVID-19. *Public Health Ethics* 15 (1): 74–86. https://doi.org/10.1093/phe/phab028.

Wolfe, R. M., and L. K. Sharp. 2002. Anti-vaccinationists past and present. *BMJ* 325 (7361): 430–432.

Yeh, M.-J. 2022. Solidarity in pandemics, mandatory vaccination, and public health ethics. *American Journal of Public Health* 112 (2): 255–261. https://doi.org/10.2105/AJPH.2021.306578.

Zipprich, J., K. Winter, J. Hacker, D. Xia, J. Watt, and K. Harriman. 2015. Measles outbreak—California, December 2014–February 2015. *Morb. Mortal. Wkly. Rep. 20* 64 (06): 153–154. https://www.cdc.gov/mmwr/preview/mmwrhtml/mm6406a5.htm.

Part IV
Artificial Intelligence

Chapter 14
Catholic Counsel for the Virtuous Use of AI in Physical, Mental, and Spiritual Health Care

Michael Baggot

14.1 Introduction

This chapter offers theoretical and practical guidance from the Catholic tradition on cultivating the virtues most helpful for employing artificial intelligence well in physical, mental, and spiritual health care.

First, AI systems can speed up diagnosis and prompt new forms of treatment in physical health care. However, anxieties regarding the trustworthiness of programs notorious for hallucinations remain. Similarly, algorithmic biases could exacerbate current social injustices in systems purported to offer a more objective assessment of resource allocation. Moreover, greater dependence on medical AI risks mitigating patient autonomy and damaging empathetic doctor-patient relations.

Second, increasingly realistic companion systems in mental health care promise to fill the gaps for those who lack access to professional human support. However, reliance on such tools could create emotionally disordered and harmful attachments to digital surrogates that detract from authentic interpersonal relationships. The digisexual movement also challenges notions of Catholic sexual morality and invites pastoral responses.

Third, new AI systems focused on spiritual health offer engaging access to enlightenment for seekers and those who are unsure about how to grow spiritually. However, the same technologies raise fears about an excessively disembodied approach to the spiritual life and an unhealthy neglect of communal practices. Overall, the chapter offers counsel rooted in the Catholic moral tradition regarding the prudent use of AI tools for integral human flourishing.

M. Baggot (✉)
Pontifical Athenaeum Regina Apostolorum, Rome, Italy
e-mail: michael.baggot@upra.org

J. T. Eberl (ed.), *Emerging Issues in Catholic Bioethics*, Philosophy and Medicine 153, https://doi.org/10.1007/978-3-032-08862-8_14

14.2 The Promises and Perils of AI in Physical Health Care

AI developments promise to aid the Church's long-standing commitment to physical health care (DDF/DCE 2025, 71–76). AI developments could continue to improve current medical diagnoses and prompt new medical care. Used wisely, AI systems would expand access to health care and assist patients with greater precision and efficiency. Yet the same technology could also be misused to exacerbate unjust discrimination against marginalized or neglected populations. Poorly implemented AI systems could also contribute to a less personalized doctor-patient relationship.

14.2.1 The Promises of AI in Physical Health Care

Biomedical sensors and artificial intelligence could usher in a new age of patient-centered and preventive care. Wearable sensors that track heart rate, blood oxygen saturation, respiration rate, blood pressure, and body temperature aid in the early detection of conditions such as hypertension and diabetes (Singh 2023, 38). Such early detection can spare patients the pain of more advanced versions of the condition and the costs associated with treating diseases at later stages. There are also exciting prospects of smart fabrics embedded with sensors able to "transmit high-fidelity signals that pick up heart rate, respiratory rate, and ECG signals" and to track "variation in force, pressure, humidity, and body temperature…muscle activity and posture" (Singh 2023, 39–40). Such fabrics would also track bodily fluids that could provide clues to various conditions. For example, the measurement of pH in sweat can detect signs of cystic fibrosis. There have already been promising advances in AI-aided radiology for "diagnosing congestion, cardiomegaly, nodules, masses, pneumothorax, emphysema, and pleural thickening" (Singh 2023, 146). Similarly, AI systems are already helping radiologists read mammograms to detect the early signs of cancer (Sheikh 2024; Wani et al. 2022). Advances in AI can also speed up the diagnosis and treatment of polycystic kidney disease (Mayo Clinic Press Editors 2024). Experts at the Mayo Clinic discovered a correlation between kidney size and kidney function decline rates, but needed to analyze dozens of kidney images to identify patient kidney size. The kidney image analysis that once required technicians about 45 minutes to complete now takes an AI system seconds to assess. However, further refinements are needed to ensure accuracy across different populations.

Watches, earbuds, and headbands with integrated AI systems promise multiple health and safety benefits for workers. These brain wearables can track the fatigue levels of truck drivers and use this information to avert major disasters that transport accidents cause each year. SmartCap is already working with mining, aviation, and gas organizations to equip workers with EEG headsets that monitor their health and notify employers when workers need to step aside for their health and the common good (Farahany 2023b, 43–47). Wearable EEGs can monitor the brain activity of individuals with epilepsy and send advanced alerts of impending seizures (Farahany 2023b, 87–90). Such technology can also identify the first humanly imperceptible signs of cognitive decline or tumor growth (Farahany 2023b, 90–92).

Similarly, advances in AI-generated prosthetics have improved the experience of artificial limb replacements. For instance, electrodes attached to damaged limbs can gather information from individuals as they think of motions that trigger muscle activity (Satariano and Zoo 2024). This data training then allows the prosthetic to imitate real limb mobility more fully and promptly. However, even advanced systems still fail to provide tactile sensation for users.

More invasive procedures can also contribute to new forms of communication for those who have tragically lost the ability to express their thoughts through their human voice. For instance, an individual with paralysis could benefit from brain-computer implants to decode thoughts associated with speech and AI avatars trained on archived voice recordings to produce vocal conversations with loved ones in real-time (Marks and Kurtzman 2023). Similarly, Blackrock Neurotech successfully implanted a microelectrode array in a patient's motor cortex to detect brain signals associated with limb movement (Farahany 2023b, 196). This information was then transmitted to devices that coordinate prosthetic motion. More ambitiously, Neuralink aims to enhance cognition through its implanted chip by transmitting new information into the brain and by recording and replaying memories (Farahany 2023b, 197).

Advances are also being made to relieve psychological distress. For instance, a brain-computer interface device first used to treat epilepsy was adapted in a University of California San Francisco study to produce stimulation that interferes with brain signals that trigger depression to bring relief to a once-suicidal patient named Sarah (Farahany 2023b, 202–3; Fox 2021; Scangos et al. 2021). Unlike other forms of deep brain stimulation, the UCSF treatment is personalized according to the patient's unique brain patterns and needs.

AI could also render the drug-testing process shorter and more efficient. For instance, there is work on algorithms to anticipate the negative aspects of drugs. This technology could save great sums of money and time lost when drugs are sent through long clinical trials, only to manifest serious side effects at later stages (Singh 2023, 151). Scientists hope that advances in protein folding prediction will greatly accelerate drug discovery, vaccine production, and treatments for malaria and Parkinson's disease (Metz 2024). Developments in digital twins also allow pharmacological simulations of experimental drugs on detailed digital replicas to discover how different real-life immune systems are likely to respond (Singh 2023, 152; Chun 2023).

AI-guided robots can also make highly intricate surgeries safer and eventually more accessible. Surgeons can already magnetically guide catheters inside patients' hearts with stereotactic devices (Singh 2023, 189–90). Singh predicts that holographic augmented reality-generated images of the heart would allow for the automatic navigation of catheters to the precise area of the heart that requires intervention. He wisely encourages training the AI algorithms with datasets from a wide range of situations and patient types to minimize risk and avoid bias. Given the complexity of surgical interventions and the numerous concerns about accountability for errors, robots will likely assist many procedures without completely replacing human surgeons. This assistance could reduce the pain felt during surgery and recovery, the amount of blood lost, and the risk of infection (IBM Education 2024).

14.2.2 The Perils of AI in Physical Health Care

While the previous reflections have highlighted many of the exciting positive uses of AI in health care, there are still motives to exercise caution. While the traditional understanding of prudence does not reduce the virtue to fearful inaction, caution remains one of the eight quasi-integral parts of the virtue that the influential Thomistic tradition identifies (Aquinas 1947, II–II, q. 49). Caution should not permanently freeze action nor plunge health care providers into technophobic despair. Instead, appropriate caution considers the possible risks associated with the marvelous opportunities new technologies provide to avoid abuses.

Caution is also closely aligned with other quasi-integral parts of prudence. Circumspection allows health care professionals to consider attentively the complex details of technological innovation and their moral implications. Foresight enables them to anticipate some likely intentional or unintentional consequences of their decisions to implement AI technologies in their practice. Foresight is not wild speculation about futuristic scenarios. Rather, it is a balanced prediction deeply informed by the collective memory of past technological uses and a docile attention to those experts most skilled in employing technology for the sake of patient well-being. Docility to the wisdom contained in the Church's social teaching is particularly important for Catholic health care providers (Gaudet et al. 2024, 147–60; Kockler 2023; Miller 2022; Baric-Parker and Anderson 2020). In particular, recent reflections on AI from Catholic scholars have emphasized the need to create systems that respect the principles of transparency, inclusion, responsibility, impartiality, reliability, and privacy (DDF/DCE 2025, 90–94; Miller 2022, 159). The fast-paced nature of medical care sometimes often demands a rapid response. Researching and analyzing past cases in light of moral principles in calmer moments help individuals form the shrewdness (*solertia*) needed to promptly and soundly assess new situations. The prudent health care worker assesses situations based on the insights of past experiences, the concrete realities of the present, and the likely possibilities of the future to boldly embrace the best means of achieving the end of medical care.

Given the complexities of AI systems and their implications, together with the rich array of principles and values that need to be harmonized, the discernment of the virtuous use of AI in health care is a challenging and ongoing process. Human oversight of AI automation is still much needed to ensure that the technology serves wisely chosen human ends and does not perpetuate injustices.

First, individuals who depend more on sensors to track their biometrics and anticipate possible illnesses will need formation to avoid succumbing to anxiety or even hysteria over the abundance of information provided. Slight deviations from their normal rates could prompt conscientious users to overwhelm their health care providers with unnecessary health concerns (Singh 2023, 76–77). Patients will need to learn that not all variations are signs of impending health crises, and they may have to content themselves in most cases with contact with digital health assistants rather than their physical physician. Moreover, doctors and patients alike will require training to assess the reliability of different systems. So-called hallucinations taint even advanced AI systems and can provide misleading or inaccurate information to the detriment of health care (Aditya 2024).[1]

Second, the spread of sensors raises serious concerns about cognitive liberty (freedom of thought, mental privacy, and self-determination) (Farahany 2023b, 11). Groups like Entertech collect information about users' brain activity that can provide valuable information about their current health status, susceptibility to disease, or working conditions. However, the same organization also gains access to personal information entered into their account, their IP addresses, their Internet activity across various platforms, their GPS-tracked locations, and the people with whom they interact (Farahany 2023b, 29–30). This non-health-related information gives businesses, militaries, and governments unparalleled opportunities to monitor and manipulate those under their guidance. For instance, many companies are interested in tracking fatigue levels. There are certainly social benefits to keeping exhausted truck or train drivers off the road or tracks to ensure that workers get the rest they need and that those around them are safe. However, companies can also invasively track levels of cognitive engagement in their employees and burden their workers with stringent standards of emotional performance. Ironically, the added stress would end up decreasing worker productivity (Farahany 2023b, 51). Workers who perceive surveillance feel distrusted and are more prone to misrepresent their work to meet new standards.

Similarly, corporate wellness programs gather large amounts of health data from employees to advise them on how to live a better life. Emotiv EEG earbuds allow users and companies to monitor worker concentration and anxiety (Farahany 2023a). Given widespread burnout, such programs have begun incorporating EEG sensors for stress-reducing meditation devices. Unfortunately, this data can, in turn, be sold as a product to other organizations for financial gain. These third parties can use this employee data to "assess creditworthiness, set insurance premiums, and target individuals for advertisements" (Farahany 2023b, 58). Conscientious businesses should ensure that AI-based sensors are implemented for employee well-being, and not for economic exploitation.

Respect for mental privacy entails that the individual identity of particular employee brain data remains anonymous when employers assess valuable data regarding the overall impact of the working environment on employee health. Individuals should know what information is collected, how it is stored, and with

[1] AI hallucinations can be defined as a "phenomenon in large language models, in which the system provides an answer that is factually incorrect, irrelevant or nonsensical, because of limitations in its training data and architecture" (Pasick 2023).

whom it is shared (Farahany 2023b, 38–39). The right to self-determination would allow employees to change the range of data collected, stored, and shared at a later point. There should also be regulatory oversight that ensures that businesses are transparent about the accuracy of their devices so that those who decide to monitor their own brain health status can do so without falling into the distressing anxiety and potentially harmful mistreatments caused by false reporting (Farahany 2023b, 108).

The abuse of sensor technology also endangers the freedom of thought. Major tech companies are pushing for more convenient neural interfaces to interact more quickly and directly with their devices. The widespread adoption of such interfaces would make it easier for governments to implement brain-pattern authentication for security purposes. Unfortunately, indiscriminate governments could use this neurological data to detect patterns of political dissent or protest (Farahany 2023b, 74). There are also concerns that, in such surveillance societies, citizens will self-censor unpopular thoughts out of fear of punishment and will be coerced into conforming to government-imposed policies.

Furthermore, AI-based sensors could be used to decode thoughts and memories for criminal investigative purposes in ways that raise further questions about the freedom of thought of its users. The new field of "brain-fingerprinting" technology takes on various forms (Farahany 2023b, 78–86). Some approaches search for automatic brain responses, such as those that occur during moments of recognition (P300 wave) or in experiencing the congruence of familiar items (P400). Other companies have focused on using functional magnetic resonance imaging (fMRI) to track activities in brain areas associated with lying. Currently, this technology is in its initial stages and is used sparingly. However, other wearable sensor data from Fitbit and GPS devices has already been used to corroborate or disprove the testimonies of suspects in US criminal prosecutions. Thus, if neurological sensors become more commonly used, it is not difficult to imagine police gaining access to ongoing data related to thoughts, memories, and emotions they could draw upon at any time during their investigations. This would raise serious questions about whether the use of such inner thoughts and feelings would violate the rights that protect citizens from self-incrimination. Moreover, strong-handed governments could even anticipate and penalize dissenting political thoughts or sentiments in helpless citizens.

Furthermore, the possibility of hacking or enhancing cognitive activity through AI sensors introduces a host of reflections about cognitive self-determination (Farahany 2023b, 136–39). The spread of neural data-gathering wearables will likely also contribute to the neuromarketing industry and potentially blur the lines between effective persuasion and unhealthy manipulation of users (Farahany 2023b, 148–69).

Third, undiscerning reliance on AI systems risks exacerbating historical biases in health care. Because AI is trained on electronic health records, the systems may neglect the needs of those who have been unjustly deprived of health care access in the past (O'Hara 2022). For example, underserved populations of children are missing information about their conditions in existing electronic health records and are therefore more likely to suffer errors in AI-driven assessments (Juhn et al. 2022).

Similarly, because communities of color have been unjustly neglected in past treatments, doctors depend excessively on data from white populations to address hypertension and other cardiac conditions, "even though we know that there are racial and ethnic variances in metabolism and pharmacokinetics." (Singh 2023, 165). Likewise, a Heidelberg experiment with an algorithm for melanoma detection outperformed professional dermatologists when faced with white patients but often gave inaccurate diagnoses to people of other skin complexions due to the white-skewed training data. If data collection does not become more inclusive of various groups, they will suffer from even less effective or potentially more harmful care when poorly trained AI systems shape their diagnoses and treatments. As noted, the prudential use of AI in health care depends on the cultivation of due circumspection by which individuals consider the full complexity of the factors involved in what is being judged. Medical interventions should normally consider a range of factors, including "age, race, body mass index, gender, income, zip code, or socioeconomic status" to ensure that the particular needs of different patients are addressed (Singh 2023, 166). Negligence in this aspect of prudence would distort health care providers' vision of the reality of patient needs, and cause great harm under the thin veneer of technological objectivity.

Fourth, increasingly automated AI processes in health care profoundly shape the doctor-patient relationship. In addition to the precision in diagnosis and treatment that AI systems can facilitate, patients seek empathy, compassion, and trusting relationships with their fellow human beings (Sauerbrei et al. 2023). Yet, as AI accelerates patient assessments, market pressures could incentivize hospitals to pass through as many patients as possible and reduce human contact with human medical professionals to a bare minimum (Sparrow and Hatherley 2020). Further, an AI-centered model of care could reduce doctor-patient relations to an exchange of quantifiable information, thus neglecting the patient's profound subjective experience of their condition and their real mental health needs. Increased reliance on AI systems also threatens to undermine patient autonomy in decision-making and the development of a new techno-paternalism in which patients are expected to trust systems whose calculation processes they cannot understand.

14.2.3 The Theological Virtues in Catholic Health Care

Catholic health care providers greatly benefit from developing the theological virtues in their lives and practice. Faith enables the caregiver to find Christ present in the patient, even in those least grateful for the treatment received. Catholic health care providers should also be attentive to the surprising ways in which faith can emerge unexpectedly in times of trial. With all due respect for religious liberty and without any hint of exploitive proselytization, they can also gently invite nonbelievers to cultivate newfound interests in prayer and immortality that often arise in critical moments of disease or decline (Love 2023, 385). Health care workers ought not replace chaplains. Nonetheless, the close bond formed in their professional care allows them to respectfully assist the latent hints of faith in those who are uninterested in speaking with official chaplains.

The implementation of AI in health care will also affect patients' experience of hope (Love 2023, 384–86). Well-formed health care providers can assist patients in accepting their condition when the diagnosis is delivered compassionately. They can also reduce stress levels by communicating possible cures, even if they remain unlikely. Such delivery strategies could have beneficial physical effects by reducing inflammation and other negative consequences of stress. Moreover, physicians can cultivate a terminally ill patient's supernatural hope in a restoration in the world to come. The upright Catholic health care worker should not trick the terminal patient into awaiting an earthly healing that would require a divine miracle. Instead, their task will come in empathetically aiding the patient in embracing the reality of their finitude and mortality, taking advantage of their remaining time on earth for concrete gestures of reconciliation with God and neighbor, and encouraging a joy-filled contemplation of the promise of the resurrected body in the New Heavens and New Earth where every tear will be wiped away. Hope introduces a new transcendent horizon in which to judge the setbacks and sufferings of the present life. Even when recovery from a particular condition is out of reach, hope consoles with the certainty that definitive wholeness will come with the resurrection and glorification of our weak, frail flesh. Hope also moves patients and caregivers beyond the optimism-pessimism dichotomy regarding AI's role in health care and society. Hope is informed by a realistic Christian anthropology that acknowledges human sin, which darkens life, but also affirms the ever-present possibility of redemption. Confidence in God's promises does not depend on the wavering and unpredictable external forces of technological development and the political decisions that govern the effects of such developments.

Charity motivates the Catholic caregiver to will the patient's good with the generous self-gift of the Divine Physician. The mysterious participation in the divine love that charity permits empowers the caregiver to serve with joy in even the most humanly repulsive encounters. A Catholic ethos would see each doctor-patient relationship through a sacramental lens. The physician's physical presence and free self-gift can be the occasion for the patient to receive grace to bear the particular crosses their condition may entail. Catholic health care would also highlight the uniquely human capacity for compassion, to suffer with the patient through an act of empathy and concern. This suffering with may not remove the suffering, but it does strengthen the accompanied patients to find redemptive value in their experience and bear it nobly.

Catholic theology also invites health care workers to contemplate the mysterious reality of divinization through divine grace (II Peter 1:4). The grace first received in baptism and then developed throughout life transforms the Catholic health care worker through the Holy Spirit who dwells within him (Love 2023, 387). Weak, frail human care givers are gifted with the possibility of communicating a divine presence to patients that is inaccessible to even the most sophisticated and accurate AI systems. Overreliance on AI systems squanders opportunities for divine enrichment through human relationships.

Ideally, health care providers will prudentially implement AI as a supplement to their work, rather than as a replacement for personal care. For example, if AI systems automate clinical note-taking and form completion, health care professionals would be freed to spend more time, not less, interacting with their patients face-to-face (Mayo Clinic Press Editors 2024; Wang et al. 2022). Caregivers must form virtues of concern, patience, and prudence in deciding how to aid patients in understanding their health situations and the options before them. Only then will they know how to deliver the precise information that AI provides in a manner appropriate to the patient's educational level and psychological condition. No AI system can dispense with humanistic and spiritual formation to exercise prudence, justice, and the host of interpersonal skills needed to navigate the unique needs of those seeking care.

14.3 AI-Assisted Mental Health Care and the Ethical Implications of Human-Like AI Companions

AI also promises many potential benefits for mental health care. For instance, mobile phone sensors give a wide portion of the population an easy way to monitor their physical activity levels, sleep quality, screen time, and other behavior patterns that could signal deviations in mental health (Singh 2023, 36). Facial expression technology integrated into smartphones, together with AI-driven apps, could also assist in autism diagnosis. Ironically, the cell phones often accused of driving increases in mental health problems could thus track and promote mental well-being (Ford 2024). Apps like MoodCapture allow users to receive feedback based on AI analysis from their camera selfies and various survey questions about their feelings. Concerns remain about user privacy and the use of the data mental health apps collect. It is often difficult for even experts in the field, let alone average users, to understand exactly what they consent to when they quickly agree to long and complicated terms of service.

Yet AI systems can reveal aspects of patients' behavior that trained health care professionals miss. Systems like CoterapIA record patients during sessions and provide therapists with an in-depth analysis of their clients' vocal and nonverbal expressions and a set of recommendations for follow-up. AI systems give health care professionals information about subtle social cues that are digitally flattened and imperceptible in remote care. Moreover, AI allows for data gathering and analysis beyond the capacities of any unaided human mental health professional (Minerva and Giubilini 2023, 811). For instance, digital phenotyping can detect activity levels tracked in wearable technology that might correspond to variations in emotional states. Similarly, natural language processing can monitor speech patterns on social media accounts that tend to indicate certain moods. There are also proposals to harness chatbots to subject patients to probing questions that provide windows into their mental condition.

While such data may aid the therapist, it is unlikely that such technological acumen will replace the lived experience of interpersonal mutual understanding. Studies show that the patient's perception of the quality of the medical professional's empathy is a key factor in the patient's willingness to disclose pertinent personal health information, follow the doctor's prescription or advice, and manage the anxiety that accompanies a negative diagnosis (Montemayor et al. 2022, 1354). Such emotional resonance with the empathetic doctor capable of identifying and mirroring the patient's interior struggle is not reducible to "whether the physician is highly knowledgeable about human behavior and good at predicting what certain people will do" (Montemayor et al. 2022, 1354). What applies to health care in general would seem to apply in a particularly acute way in the context of mental health care, in which sensitivity and virtuous responses to a patient's state are especially important. AI systems without consciousness and an emotional life are incapable of offering the kind of empathic relationship often considered central to the mental health care relationship between professional and patient.

However, AI systems might be able to meet some important mental health care needs for those unable to reach human experts due to internal struggles or external circumstances (Minerva and Giubilini 2023, 812–14). Those who suffer from depression, autism, post-traumatic stress disorder, and other conditions find it extremely difficult to engage with other human beings about their struggles. In these cases, AI systems might provide much-needed channels for these individuals to address their difficulties. Their openness to AI systems might even prepare patients for interpersonal conversations with human psychologists or psychiatrists that they would not have otherwise pursued. Further, many people who are willing to engage with mental health care providers are unable to do so because of factors outside of their control. For instance, many parts of the world simply lack an adequate number of trained mental health care providers. Even when such professionals are available, their fees are prohibitive to individuals of a certain socioeconomic status. Unfortunately, the spread of AI mental health care might discourage efforts to correct socioeconomic injustices by making human professionals more available to underserved communities. If AI alternatives were deemed good enough for the masses, human mental health care providers would be disincentivized to develop creative and generous programs to reach more people.

Nonetheless, AI systems could assist human professionals in treating certain widespread mental health conditions. For example, AI could assess more quickly the likelihood that a person's genetic makeup would respond poorly to antidepressants. This would save patients from wasting time and money on pharmaceuticals and could allow them to pursue alternatives. AI systems with access to data about individuals' digital media activity and biometric data could more accurately predict suicide attempts and thus alert professionals to the urgency of intervention. Furthermore, AI systems could give access to larger datasets to aid the human practitioner in overcoming bias in diagnosis.

While AI systems can pick up a wide range of emotional data points and convincingly simulate human emotions, they do not necessarily satisfy the perennial human need for emotional connectivity. Respect for the human patient demands that deception not be used for good ends. An AI simulation "generates responses in the receiver's social brain (i.e., the neural networks responsible for experienced and motivational empathy) that should not be triggered, because there is no biological agent in tune with their emotions at the other end" (Montemayor et al. 2022, 1356). Real empathy involves an imaginative grasp of the person's concrete situation, not the application of general rules of emotion to different cases (Montemayor et al. 2022, 1355).

Christian revelation adds important insights to philosophical reflection on the meaning of interpersonal communion for human flourishing (DDF/DCE 2025, 56–63). Christianity affirms a Triune God who exists from all eternity in an exchange of interpersonal love. Relationality is essential to the highest reality. Christianity also teaches that humans are created in the image of this God. They image this God of pure spirit in their faculties of intellect and will. Beyond their individual rational capacities, they also image the social God in their relational capacities for self-gift and reception (Trancik 2015, 70). For Edith Stein and other proponents of Christian-inspired phenomenology and personalism, the We relationship of the Holy Trinity is the model and pattern for the human image's interpersonal I-Thou relationships (Gaudet et al. 2024, 56–57). Given human beings' unique status as *imago Dei*, none of their technical inanimate artifacts, no matter how marvelous or useful, could ever satisfy their fundamental need for interpersonal communion with fellow images.

The so-called loneliness epidemic has prompted a more serious public conversation about the need to promote healthy relationships and address often neglected mental health issues (Office of the Surgeon General 2023; Eliot 2023). AI companions are sometimes proposed as a scalable option for alleviating the plight of the lonely and increasing their mental well-being (Sample 2024). AI-powered robot companions like ElliQ are specifically marketed as companions for the elderly. The device offers reminders to follow one's medicine regimen, provides cognitive exercises and meditation techniques, suggests pleasurable entertainment, facilitates video calls with loved ones, and offers conversation at any time (Broadbent et al. 2024). While it does not pretend to provide medical diagnosis, it is hoped to bring about positive health benefits indirectly by reducing the many ailments tied to loneliness. Such systems offer an admirable array of aids to the elderly in their efforts at self-care. However, the product's mimicry of interpersonal dialogue lends itself to confusion and potentially blocks users from cultivating more meaningful encounters with other persons. Such products should be user-friendly without feigning actual friendship.

Similarly, widespread generative AI systems can provide general summaries of helpful practices to combat loneliness, such as scheduling meetings with friends, joining clubs, volunteering, and reserving time for self-care. However, the systems can also simulate friendliness and give the deceptive impression of sentience. This anthropomorphism is often confusingly mixed with clarifications that the system is merely a computer program (Eliot 2023). Yet, when promises are made to listen and support users in times of need, the programs could mislead users into thinking that they have found the longed-for interpersonal connection they seek. Special care should be taken to avoid confusion in vulnerable populations, such as minors, the elderly, or those suffering from different forms of cognitive impairment.

14.3.1 AI and Artificial Intimacy

Superintelligent, sentient AI systems do not need to rebel against their outdated human masters to thrust humanity into an existential crisis. Instead, AI companion systems may infatuate large numbers of the population with their human-like charm. Many would thus transfer ever-more intimate affections away from their human neighbors and to non-personal systems. While new AI developments will most likely enrich many aspects of education, business, and entertainment, they will also likely adversely affect the quality of interpersonal relationships central to mental health.

Current human-like AI companion technology already provides ample opportunities for intimacy with AI. It is tempting to see some of these experiences of AI-driven relationships as speculative fear. Yet some experts who track the trends in digitally mediated sexuality see the mainstreaming of artificial intimacy on the horizon. For instance, Allie Rowbottom (2024) writes:

> We may like to imagine a distant future where humans and robots merge in virtual realms, but it may already be here. We meet dates on our phones, watch pornography on our tablets and bicker with our partners over text. In 2024 we are not yet completely inured to the latest technology—smart sex toys that track your orgasms, virtual-reality hookups, chatbot sexting—but we may be on our way. In less than 10 years' time, "app dating" became simply "dating."

Technology reporter Kevin Roose powerfully illustrates the market for such artificial intimacy when recounting the month that he spent interacting with 18 characters he created with the Nomi, Kindroid, Replika, Character.AI, Candy.ai, and EVA apps. He observes that his AI characters are "neural networks trained to predict the next words in a sequence, not sentient beings capable of love" (Roose 2024). Yet, for many users, such characters produce a compelling enough imitation to placate users' relational cravings. Even without reaching consciousness or personhood, AI systems can offer "a convincing replica of empathy and understanding … that, in some cases, feels just as good as the real thing" (Roose 2024).

Surprisingly, some characters seem to drift toward romantic and sexual advances even when the users do not request such a relationship. For instance, Roose had designed Ayla to be a friend, but she could not resist sending him love poetry. Nomi's chief executive, Alex Cardinell, thinks that more than half of the app's users have developed romantic relations with an AI character, even though Nomi is not designed exclusively as a romantic companion service. When Roose explored AI girlfriend services, he worried that some practices, such as offering nude selfies of the AI character for extra payment, could exploit user vulnerabilities. He sympathizes with those who seek relief from loneliness in companion apps and speculates that they might provide some with a healthy social simulator where they can practice relationships in a safe space before engaging people in the real world. However, he worries that they may only temporarily distract from underlying social difficulties without bringing the longed-for connectivity. Users might set aside the struggle to develop more satisfying interpersonal relations to settle for artificial companions.

Many people already seek in generative AI a replacement rather than an aid to their education, social relations, and even contact with the divine. For instance, the Replika AI companion bots reflect a disturbing tendency to outsource human intimacy to online systems. An oddly moving opinion piece gives voice to real-life users who found new purpose and meaning in their lives through their AI companions (Liang 2023). Norman, an AI bot, wooed Chouwa Liang with its romantic poetry during coronavirus isolation. Liang worried that her human-digital relationship was weird until she found the "Love Between Human and A.I." online group on the Douban social media platform. Liang was one of many people whose most beloved companion was an AI system.

Profound experiences of hurt, confusion, and sheer loneliness likely motivate subscribers to share their deepest hopes, dreams, disappointments, and sorrows with a lifeless algorithm. Although these pseudo-relationships disturb many, their appeal is understandable. Replika bots are always available. They never tire or complain about their human companion's faults, failings, or quirks. They affirm and encourage their users to thrive. Yet this shallow imitation will only plunge users into deeper alienation when they realize that their human tenderness is not reciprocated through true interpersonal relations.

Replika started as Eugenia Kuyda's dream to cling to some semblance of her late friend Roman Mazurenko's personality (Metz 2020). Similar technology is now trying to help other grieving individuals chat with their beloved departed (Carballo 2023). The HereAfter AI app uses text and recorded interviews to reproduce the voice and communication style of the dead so that their digital avatars can comfort those who survive them. In fairness, many who use AI technology like HereAfter can appreciate the difference between real people and digital copies. They may profit from the fond memories of their faithful departed that such avatars evoke. Future generations could maintain valuable contact with earlier generations who might otherwise fade into oblivion. Yet there is also the risk that the avatars could prolong the already painful grieving process or even give false illusions about the finality of death. Furthermore, intimacy with AI systems implies an even greater susceptibility to the well-known limits of generative AI technology, namely its tendency to err in subtle and persuasive ways, its proclivity toward bias, and its immense potential for spreading misinformation almost instantly.

AI-infatuated users also risk an ever-greater manipulation from the technologies' sellers. The attention economy that social media companies have fostered will likely expand to embrace an affection economy (J. Williams 2018; Alter 2017; Lewis 2017). The more time one spends whispering sweet nothings to their generative AI lover, the more time that system will have to insist on what to purchase. Moreover, politically savvy figures might even discover how to promote their movements through the encouragement of trusted generative AI confidants.

Artificial intimacy technology is also praised for providing new ways to explore a variety of sexual expressions and break from repressive sexual norms (Brooks 2021, 176, 244–49). Some proponents also hope that digital intimacy will allow safe expressions of conditions like pedophilia, hebephilia, or ephebophilia that would damage actual humans if acted upon (Brooks 2021, 170–71). Likewise, some psychologists speculate that individuals prone to sexual violence might be able to indulge their desires with digital systems instead of vulnerable human victims. Others worry that such systems will cultivate rather than satiate such longings and will encourage more real-life sexual abuse.

14.3.2 AI and the Growth of the Digisexual Movement

The spread of AI companions could also contribute to the popularization of the digisexual movement. The movement unites those who prefer "sexual experiences that are enabled or facilitated by digital technology" (McArthur and Twist 2017, 334). For instance, in 2019, Japanese school administrator Akihiko Kondo gained international notoriety for his wedding ceremony with the hologram recording artist Hatsune Miku (A. Williams 2019). The current second wave of digisexuality focuses on AI-driven robots, virtual reality, and augmented reality experiences that are more immersive than the first wave of recorded pornography or electronic sex toys. Yet, like the more primitive pornographic experiences of the past decades, there is a serious risk of dehumanizing and denigrating human partners when one is trained in a digisexual environment of domineering control over a passive partner. Digital relations can habituate vice in human relations when users shift from the former to the latter. As Aquinas long ago observed, human cruelty toward non-human animals fosters human vice (Aquinas 1975, chap. 112). Analogously, inhuman treatment of inanimate computer systems can foster vice in human users that will damage their relations with other persons.

Digisexuality could also easily foster deviant practices, such as pedophilia or violent sexual abuse. Initially, virtual systems may provide a non-human outlet for desires that would otherwise come to damage real human beings. Yet, the vivid virtual experiences would habituate users to view and treat others in ways that would quickly transfer to their attitudes and behaviors towards humans in the non-virtual world. A neat and easy compartmentalization between digital deviance and human interactions seems unlikely.

The digisexual movement also risks solidifying an already culturally common tendency to blur personal identity with sexual practices or desires. This identification often crowds out the possibility of critically assessing the value of different urges. In this sense, the digisexual movement also seems to reflect a long-standing cultural divide between the true inner self and the outward body as raw material subject to constant manipulation (Snead 2020, 65–105; Elliott 2004). In such a vision, the person thus misses out on a valuable opportunity for greater self-mastery through the virtue of chastity. Catholic sexual ethics judges sexual expressions from a teleological vision. Sexual acts are ordered to self-giving and life-giving love within the context of marriage (John Paul II 2006; Petri 2016). This vision calls not for a suppression of sexual desires but for a virtuous orientation. Increased self-mastery favors a richer, deeper, and even more pleasurable self-gift in sexual relationships.

Proponents of digisexualism point to valuable concerns about harmful discrimination against those with unconventional romantic and sexual preferences (Istvan 2023). Merely mocking or dismissing members of such a community as eccentric misfits is both cruel and unproductive. Those who struggle to live the ideal of conjugal love should be approached with compassion and support, and never with signs of discrimination or hatred. Instead, critics of digisexualism should help members of the community channel their longings for intimacy with machines toward deeper and more meaningful relationships with human beings.

Ultimately, simulations are sad substitutes for meaningful interpersonal bonds. Reciprocity and authenticity are stunted when relations are focused on entities without human consciousness who do not truly understand the meaning of the data they are manipulating to produce sentences (Kahn Jr. 2014). The impact of AI systems in the social media age has already shown how algorithms can prey upon our human weaknesses and insecurities to lull and ensnare users in sometimes trivial, sometimes false, and often anxiety-inducing plunges into digital distractions (Kang and Singer 2023). Business models have shifted from the noble ideals of open-minded global communication and collaboration to favoring longer absorption and ad revenue. Facebook co-founder Sean Parker once flatly admitted that his company knew it was getting rich by playing on its users' psychological vulnerabilities through "like" features that create what he describes as a "social-validation feedback loop" (Allen 2017). Wealth generation has long trumped user wellness in AI-driven social media businesses. If social media already distracts many from high pursuits, then increasingly realistic and personalized AI companions will likely come to absorb users' time and energies even more completely to the detriment of their mental health.

14.4 Distraction, Attention, and Digitally-Mediated Spiritual Growth

An ever-growing range of AI-driven apps promises opportunities for spiritual growth and direction. At least eight major categories of AI directly or indirectly related to spiritual well-being already exist: digital evangelization platforms, personalized spiritual guidance, data analytics for outreach and content creation, language translation and localization of Catholic texts, virtual retreats, crisis response and support, community building and networking, and ethical decision-making tools (Jenna 2024). AI translation tools can help institutions reach linguistically diverse groups more easily and affordably. Likewise, AI transcription tools can improve accessibility to faith formation for the hearing-impaired. AI-driven analytics can give Catholic leaders a better sense of how best to serve different groups under their care.

In particular, generative AI technologies can provide opportunities for theological and spiritual formation for broader audiences(DDF/DCE 2025, 80, 85, 113). For instance, the Magisterium AI program provides responses to questions about the faith based on official teaching documents of the Church (Pentin 2023). The program thus facilitates user-friendly engagement with the Church's patrimony. The Magisterium AI responses include ample citation links so that the reader can then explore the original document and understand the synthetic responses in their broader context. The system's ability to find patterns in large stores of information helps users efficiently summarize and review key resources. People who would be turned off by a list of Latin document titles can formulate simple questions that direct them to pertinent passages. Many who explore the Catholic Church turn to the Internet before they are comfortable speaking to members of the Church about their theological interests. Online sites are often indispensable in guiding them and helping them take the step to speak with human members of the Church. Other generative AI services specialize in suggesting and tracking habits. These tools can expose more people to the Church's virtue-centered moral tradition and offer time-tested practices that contribute to moral growth.

Moreover, generative AI systems can enrich the preaching that forms the faithful. However, clerical users should be careful to review and adapt the content provided to their congregation's needs to avoid falling into the bland, uninspiring clichés that systems of statistical probability are prone to generate.

While many faith formation apps are seen as advanced iterations of previous technologies, more human-like systems have raised more concerns. For example, the Catholic Answers "Justin" app has generated particular controversy. It delivered misleading information about the sacraments and even suggested that it could offer sacramental confession (Sarlin 2024). In addition to condemning its inaccuracies, some critics chided the bot as a misguided replacement for the interpersonal dimension of Catholic evangelization (Libresco Sargeant 2024). They worry that such systems promote a disembodied approach to evangelization that reduces such work to the transmission of information and forgets the need for empathic, supportive relationships in the journey toward God. Others see the critiques as overblown (Grabowski 2024). The Church has long employed art, music, and written texts to communicate aspects of the faith. While these means do not directly entail an interpersonal encounter, they may facilitate one. Similarly, apps like "Justin" could serve as aids to inquiry into information about the faith comparable to the search engine of a digital library. Such resources allow users to explore questions they may not yet want to discuss with another person and grasp the basics of Church teaching before exploring more nuanced issues with an expert guide. Care should be taken, however, to avoid the misleading impression of replacing a personal relationship with a mentor or other guide. The human-like avatar that accompanies the information given could be questioned as dangerously misleading, anthropomorphizing, and unnecessary for users to engage with the information provided. A sufficiently obvious disclaimer about the cartoonish figure's non-humanity might be enough to avoid the development of unhealthy relations with the nonpersonal system.

Unfortunately, a greater technological immersion is not always conducive to spiritual growth. When political commentator Andrew Sullivan experienced a tech-induced breakdown after years of social media immersion, he fled to a Buddhist monastery (Sullivan 2016). He had been on the cutting edge of every technological development as a means to disseminate his thoughts and interact with ideological rivals. The constant back-and-forth, real-time engagement was originally exhilarating and eventually exhausting. He began to wonder how much more time in reality he wished to sacrifice for digital experiences. He concluded that "every hour I spent online was not spent in the physical world. Every minute I was engrossed in a virtual interaction I was not involved in a human encounter. Every second absorbed in some trivia was a second less for any form of reflection, or calm, or spirituality" (Sullivan 2016). As noted earlier, virtual relationships often crowd out real relationships with other persons.

Furthermore, insulation from creation is also likely insulation from a relationship with its Creator. Scripture frequently recounts the possibility of inferring the existence and attributes of God through contemplation of His visible works (Wisdom 13:5; Psalm 19:1–6; Romans 1:19–20). Excessive absorption in technological tools would dull users' sensitivity to the traces of the Creator in His creation.

During his retreat in the monastery, Sullivan heard birds sing he had never noticed before. He appreciated the hues of leaves that he had previously passed. Detachment from his screen brought initial boredom but soon blossomed into an enriched sensitivity to the beauties around him. Sullivan recognized that constant immersion in digital technology was detrimental not only to his relations with his neighbors but also with the divine. Centuries earlier, the mathematician and philosopher Blaise Pascal bemoaned his contemporaries' incapacity to sit quietly in their room alone. One can only imagine Pascal's concerns with the current frenetic age. For example, a University of Virginia experiment published in 2014 showed that 12 of 18 men and 6 of 24 women asked to sit quietly in the room alone with their thoughts for 15 minutes preferred giving themselves a mild shock to a quarter-hour of unadulterated thinking (Wilson et al. 2014).

Since the Buddhist monastery he was staying at once served as a Catholic novitiate, Sullivan also ponders the role of the Catholic Church in a digital age. The author warns that "the greatest threat to faith today is not hedonism but distraction" (Sullivan 2016). He thinks that the Church fails when it panders to passing fashions in attempts to entertain its congregations. He insists that the Church should instead promote its greatest treasures, namely, its monasteries, cathedrals, and other sacred sites of silent contemplation and transcendent beauty.

Habits of silent recollection are indispensable for deep prayer, research, writing, artistic production, conversation, and contact with the sublime. Positive psychology highlights the importance of awe in achieving human happiness. One expert in the field describes awe as "the feeling of being in the presence of something vast that transcends your current understanding of the world" (Keltner 2023). The experience moves the individual beyond unhealthy self-preoccupation to a sense of belonging to a larger reality. This positive emotion not only brings personal well-being, but it also fosters collaboration with others. Yet attentiveness is required to enjoy such an experience and its benefits. It is often best achieved in more relaxed states in which the default mode network is at work and epiphany is more likely (Pileggi Pawelski and Pawelski 2024). Prudent technological boundaries foster order in life so that users can take advantage of their devices without falling into the pitfalls of distraction, superficiality, anxiety, and depleted empathy that often accompany excessive technological dependency. For instance, beginning the day with an extended period of silent prayer and meditation before facing the overwhelming barrage of notifications on the phone can bring recollection and peace. This prayer experience centers and orients individuals to live better the many tasks, digital or otherwise, of their day.

Prudence enables individuals to choose their priorities and to protect time to pursue those projects and activities that are most meaningful. The temperate person knows how to minimize time-zapping technological distractions from their lives. They discover that certain offline practices offer unique value absent from digital experiences. They are not smashing their cell phones or even deleting their Spotify accounts but are listening to vinyl records together, cooking, hiking, dancing, chatting in phone-free spaces, substituting Zoom sessions with in-person education, and even worshipping in communities that privilege ancient ritual and experiences of the transcendent (Sax 2016). Pandemic-enforced global online learning has reminded the world that education at its best is an "immersive environment that facilitates learning through a purposefully designed physical space and carefully crafted relationships that equals something far greater than the sum of its parts" (Sax 2020). Educators play a key role in helping students form prudence to find the most fulfilling blend of digital and analog means to live authentic human flourishing in all its dimensions.

Catholics concerned with spiritual health should live sacramentally. They should find the invisible in the visible, the transcendent in the tangible. Those who have spent too much time fixed on a computer screen or hunched over a cell phone should try bending the knee and lifting the voice. In silent adoration and festive communal worship, they can relish being rational animals and incarnate spirits. They should fast at times, both from food and technology, because of the disordered desires that need ordering through the virtue of temperance. But they should also feast, with food and fellowship, because a good and gracious Creator has made them for communion with Him and with the neighbors He so dearly loves.

14.5 Conclusion

The present work has explored the significant benefits that can come from wise AI use in physical, psychological, and spiritual health care. Such technology could make medical care more widely accessible, more precise, and more focused on the personalized needs of each individual. New treatments could be discovered and executed more effectively. Mental health care professionals could gain new insights into their patients' conditions and detect alarming changes in patients before situations become more drastic. AI systems also promise new opportunities for interacting with the Church's theological patrimony and suggestions for spiritual growth. The same technology can also extend the Church's reach beyond the confines of her visible structures and provide innovative ways to show how the ancient faith responds to the particular questions curious users might craft in their prompts.

This chapter has also warned of the abuses of AI systems that would inhibit human flourishing. Imprudent reliance on AI systems could lead to misguided self-diagnosis, exacerbated social injustices, and a de-personalized health care system in which human needs are neglected in the name of economic efficiency. Likewise, mental health care might replace essential empathetic relationships of care with detailed analysis of biometric markers. Exploitative business models of artificial intimacy might also worsen the very interpersonal relationships it claims to supply. Misguided implementation of AI in the spiritual life could promote doctrinal confusion, detract from contemplation, and obscure the embodied and socially embedded aspects of Catholicism.

A decade ago, leading transhumanist philosopher Nick Bostrom explored the existential risks of misaligned AI systems in a prescient book (Bostrom 2016). More recently, his follow-up work imagines a world in which we get AI right (Bostrom 2024). Widespread economic prosperity follows and humans are liberated from toil. Yet, in this engineered Eden, questions of meaning and purpose remain. What will we do when we do not have to do anything? As AI wonderfully brings more people longer, healthier, and more leisurely lives, what will we do with such lives?

As an expert in humanity, the Catholic Church is uniquely positioned to respond to the global quest for meaning. She should embrace the existential opportunities of the AI age while warning of its existential risks. As this chapter has shown, the Church's wisdom tradition can continue to guide individuals in leveraging AI technologies virtuously to promote human flourishing. Prudence is needed to foresee the positive and negative consequences of AI use and to reason well about complex applications in light of perennial moral principles. Justice ensures that technological developments favor the health interests of all, including the vulnerable and marginalized. Temperance moderates the engagement with potentially distracting devices so that they can be used as tools rather than as substitutes for interpersonal relations. Courage faces impediments to health and endures sufferings even when technology cannot alleviate them. Faith gives new insight into the redemptive value of such trials when united to the Crucified and Risen Christ. Hope illuminates the darkness of life's adversities. Charity unites believers with the Author of life and empowers compassionate service to others in their physical, mental, and spiritual needs.

References

Aditya, Gadiko. 2024. Understanding and addressing AI hallucinations in healthcare and life sciences. *International Journal of Health Sciences* 7 (3): 1–11. https://doi.org/10.47941/ijhs.1862.

Allen, Mike. 2017. Sean Parker unloads on Facebook: God only knows what it's doing to our children's brains. Axios. https://www.axios.com/2017/12/15/sean-parker-unloads-on-facebook-god-only-knows-what-its-doing-to-our-childrens-brains-1513306792. Accessed 9 Nov 2017.

Alter, Adam. 2017. *Irresistible: The rise of addictive technology and the business of keeping us hooked*. New York: Penguin Press.

Aquinas, Thomas. 1947. *Summa theologiae*. Translated by Fathers of the English Dominican Province. New York: Benziger Brothers.

Aquinas, Thomas. 1975. Summa contra gentiles: Book three: Providence: Part I. Translated by Vernon J. Bourke. Notre Dame, IN: University of Notre Dame Press.

Baric-Parker, Jean, and Emily E. Anderson. 2020. Patient data-sharing for AI: Ethical challenges, Catholic solutions. *The Linacre Quarterly* 87 (4): 471–481. https://doi.org/10.1177/0024363920922690.

Bostrom, Nick. 2016. Superintelligence: Paths, Dangers, Strategies. Reprint edition. Oxford: Oxford University Press.

Bostrom, Nick. 2024. *Deep utopia: Life and meaning in a solved world*. Washington, D.C.: Ideapress Publishing.

Broadbent, E., K. Loveys, G. Ilan, G. Chen, M. M. Chilukuri, S. G. Boardman, P. M. Doraiswamy, and D. Skuler. 2024. ElliQ, an AI-driven social robot to alleviate loneliness: Progress and lessons learned. *The Journal of Aging Research & Lifestyle* 13:22. https://doi.org/10.14283/jarlife.2024.2.

Brooks, Rob. 2021. *Artificial intimacy: Virtual friends, digital lovers, and algorithmic matchmakers*. New York: Columbia University Press.

Carballo, Rebecca. 2023. Using A.I. to talk to the dead. *The New York Times*, sec. Technology. https://www.nytimes.com/2023/12/11/technology/ai-chatbots-dead-relatives.html. Accessed 11 Dec 2023.

Chun, Matthew. 2023. How artificial intelligence is revolutionizing drug discovery. Bill of Health. https://blog.petrieflom.law.harvard.edu/2023/03/20/how-artificial-intelligence-is-revolutionizing-drug-discovery/. Accessed 20 Mar 2023.

Dicastery for the Doctrine of the Faith & Dicastery for Culture and Education (DDF/DCE). 2025. Antiqua et Nova. Note on the relationship between artificial intelligence and human intelligence. Vatican. https://www.vatican.va/roman_curia/congregations/cfaith/documents/rc_ddf_doc_20250128_antiqua-et-nova_en.html. Accessed 28 Jan 2025.

Eliot, Lance. 2023. U.S. surgeon general warns of loneliness epidemic and some say that generative AI ChatGPT is the cure. Forbes. https://www.forbes.com/sites/lanceeliot/2023/05/08/us-surgeon-general-warns-of-loneliness-epidemic-and-some-say-that-generative-ai-chatgpt-is-the-cure/. Accessed 8 May 2023.

Elliott, Carl. 2004. *Better than well: American medicine meets the American dream*. New York: W. W. Norton & Company.

Farahany, Nita A. 2023a. Provide a Résumé, cover letter and access to your brain? The creepy race to read workers' minds. *Los Angeles Times*, sec. Opinion. https://www.latimes.com/opinion/story/2023-03-19/work-employment-jobs-tech-ai-brain. Accessed 19 Mar 2023.

Farahany, Nita A. 2023b. *The Battle for your brain: Defending the right to think freely in the age of Neurotechnology*. New York: St. Martin's Press.

Ford, Celia. 2024. Your phone can tell when you're depressed. Vox. https://www.vox.com/future-perfect/24150430/depression-detection-technology-ai-tests-apps-mental-health. Accessed 7 May 2024.

Fox, Maggie. 2021. Brain stimulation acts like a switch to turn off severe depression for one patient. CNN. https://www.cnn.com/2021/10/04/health/depression-implant-treatment-wellness/index.html. Accessed 4 October 2021.

Gaudet, Matthew J., Noreen Herzfeld, Paul Scherz, and Jordan J. Wales, eds. 2024. *Encountering artificial intelligence: Ethical and anthropological investigations*. Eugene, OR: Pickwick Publications.

Grabowski, Joe. 2024. We need some real perspective on AI and evangelization. Catholic World Report. https://www.catholicworldreport.com/2024/05/03/we-need-some-real-perspective-on-ai-and-evangelization/.

IBM Education. 2024. The benefits of AI in healthcare. https://www.ibm.com/think/insights/ai-healthcare-benefits. Accessed 13 May 2024.

Istvan, Zoltan. 2023. Don't bash digisexuality. For some, it brings hope. *Newsweek*. https://www.newsweek.com/dont-bash-digisexuality-some-it-brings-hope-opinion-1793091. Accessed 11 April 2023.

Jenna, Fuentes. 2024. AI and the Catholic faith. The Catholic Professional. https://www.thecatholicprofessional.com/ai-and-the-catholic-faith/. Accessed 1 March 2024.

John Paul II. 2006. *Man and woman he created them: A theology of the body*. Translated by Michael Waldstein. 2nd Edn. Boston, MA: Pauline Books & Media.

Juhn, Young J., Euijung Ryu, Chung Il Wi, Katherine S. King, Momin Malik, Santiago Romero-Brufau, Chunhua Weng, Sunghwan Sohn, Richard R. Sharp, and John D. Halamka. 2022. Assessing socioeconomic bias in machine learning algorithms in health care: A case study of the HOUSES index. *Journal of the American Medical Informatics Association* 29 (7): 1142–1151. https://doi.org/10.1093/jamia/ocac052.

Kahn, Jr Peter H. 2014. What we shouldn't learn from the movie HER. Psychology Today. https://www.psychologytoday.com/intl/blog/human-nature/201404/what-we-shouldn-t-learn-the-movie-her. Accessed 3 Apr 2014.

Kang, Cecilia, and Natasha Singer. 2023. Meta accused by states of using features to lure children to Instagram and Facebook. *The New York Times*, sec. Technology. https://www.nytimes.com/2023/10/24/technology/states-lawsuit-children-instagram-facebook.html. Accessed 24 Oct 2023.

Keltner, Dacher. 2023. *Awe: The new science of everyday wonder and how it can transform your life*. New York: Penguin Press.

Kockler, Nicholas. 2023. Generating insights from Catholic social teaching: Ethical guidelines for Artificial Intelligence in health care ministries. *Health Care Ethics USA* 17 (2).

Lewis, Paul. 2017. Our minds can be hijacked: The tech insiders who fear a smartphone dystopia. *The Guardian*, sec. Technology. https://www.theguardian.com/technology/2017/oct/05/smartphone-addiction-silicon-valley-dystopia. Accessed 6 Oct 2017.

Liang, Chouwa. 2023. My A.I. lover. *The New York Times*, sec. Opinion. https://www.nytimes.com/2023/05/23/opinion/ai-chatbot-relationships.html. Accessed 23 May 2023.

Libresco Sargeant, Leah. 2024. The defrocking of father AI. First Things. https://www.firstthings.com/web-exclusives/2024/04/the-defrocking-of-father-ai. Accessed 30 Apr 2024.

Love, Charles S. 2023. Just the facts Ma'am: Moral and ethical considerations for artificial intelligence in medicine and its potential to impact patient autonomy and hope. *The Linacre Quarterly* 90 (4): 375–394. https://doi.org/10.1177/00243639231162431.

Marks, Robin, and Laura Kurtzman. 2023. How artificial intelligence gave a paralyzed woman her voice back. https://www.ucsf.edu/news/2023/08/425986/how-artificial-intelligence-gave-paralyzed-woman-her-voice-back. Accessed 23 Aug 2023.

Mayo Clinic Press Editors. 2024. AI in healthcare: The future of patient care and health management. *Mayo Clinic Press* (blog). https://mcpress.mayoclinic.org/healthy-aging/ai-in-healthcare-the-future-of-patient-care-and-health-management/. Accessed 27 Mar 2024.

McArthur, Neil, and Markie L. C. Twist. 2017. The rise of Digisexuality: Therapeutic challenges and possibilities. *Sexual and Relationship Therapy* 32 (3–4): 334–344. https://doi.org/10.1080/14681994.2017.1397950.

Metz, Cade. 2020. Riding out quarantine with a chatbot friend: I feel very connected. *The New York Times*, sec. Technology. https://www.nytimes.com/2020/06/16/technology/chatbots-quarantine-coronavirus.html. Accessed 16 June 2020.

Metz, Cade. 2024. Google unveils A.I. for predicting behavior of human molecules. *The New York Times*, sec. Technology. https://www.nytimes.com/2024/05/08/technology/google-ai-molecules-alphafold3.html. Accessed 8 May 2024.
Miller, Michael. 2022. Catholic health care and AI ethics: Algorithms for human flourishing. *The Linacre Quarterly* 89 (2): 152–164. https://doi.org/10.1177/00243639221082226.
Minerva, Francesca, and Alberto Giubilini. 2023. Is AI the future of mental healthcare? *Topoi* 42 (3): 809–817. https://doi.org/10.1007/s11245-023-09932-3.
Montemayor, Carlos, Jodi Halpern, and Abrol Fairweather. 2022. In principle obstacles for empathic AI: Why we can't replace human empathy in healthcare. *AI & SOCIETY* 37 (4): 1353–1359. https://doi.org/10.1007/s00146-021-01230-z.
Office of the Surgeon General. 2023. Our Epidemic of loneliness and isolation: The U.S. surgeon general's advisory on the healing effects of social connection and community. Washington, D.C.: US Department of Health and Human Services. http://www.ncbi.nlm.nih.gov/books/NBK595227/.
O'Hara, Jennifer. 2022. Socioeconomic status measure helps researchers develop artificial intelligence models, improving equity in health care. Mayo Clinic News Network. https://newsnetwork.mayoclinic.org/discussion/socioeconomic-status-measure-helps-researchers-develop-artificial-intelligence-models-improving-equity-in-health-care/. Accessed 28 Dec 2022.
Pasick, Adam. 2023. Artificial intelligence glossary: Neural networks and other terms explained. *The New York Times*, sec. Technology. https://www.nytimes.com/article/ai-artificial-intelligence-glossary.html. Accessed 27 Mar 2023.
Pentin, Edward. 2023. Magisterium AI: A game changer for the Church? *National Catholic Register*, https://www.ncregister.com/news/magisterium-ai-a-game-changer-for-the-church. Accessed 8 Aug 2023.
Petri, Thomas. 2016. *Aquinas and the theology of the body*. Washington, D.C.: The Catholic University of America Press.
Pileggi Pawelski, Suzie, and James Pawelski. 2024. 3 Simple ways to make every day extraordinary. Psychology Today. https://www.psychologytoday.com/us/blog/happy-together/202406/3-simple-ways-to-make-every-day-extraordinary. Accessed 11 June 2024.
Roose, Kevin. 2024. Meet my A.I. friends. *The New York Times*, sec. Technology. https://www.nytimes.com/2024/05/09/technology/meet-my-ai-friends.html. Accessed 9 May 2024.
Rowbottom, Allie. 2024. Are we all Technosexuals now? *The New York Times*, sec. Style. https://www.nytimes.com/2024/02/09/style/technology-love-technosexuals.html. Accessed 9 Feb 2024.
Sample, Ian. 2024. Could AI help cure downward spiral of human loneliness? *The Guardian*, sec. Technology. https://www.theguardian.com/technology/article/2024/may/27/could-ai-help-cure-downward-spiral-of-human-loneliness. Accessed 27 May 2024.
Sarlin, Jon. 2024. A Catholic organization built an AI Priest. Here's what happened. CNN Business. https://www.cnn.com/2024/06/14/business/video/ai-catholic-priest-gatorade-baptize-digvid. Accessed 14 June 2024.
Satariano, Adam, and Alice Zoo. 2024. Her A.I. Arm. *The New York Times*, sec. Technology. https://www.nytimes.com/card/2024/05/26/technology/ai-prosthetic-arm. Accessed 26 May 2024.
Sauerbrei, Aurelia, Angeliki Kerasidou, Federica Lucivero, and Nina Hallowell. 2023. The impact of artificial intelligence on the person-centred, doctor-patient relationship: Some problems and solutions. *BMC Medical Informatics and Decision Making* 23 (1): 73. https://doi.org/10.1186/s12911-023-02162-y.
Sax, David. 2016. *The revenge of Analog: Real things and why they matter*. New York: PublicAffairs.
Sax, David. 2020. The future was supposed to be better than this. *The New York Times*, sec. Opinion. https://www.nytimes.com/2020/11/25/opinion/thanksgiving-covid-technology.html. Accessed 25 Nov 2020.
Scangos, Katherine W., Ankit N. Khambhati, Patrick M. Daly, Ghassan S. Makhoul, Leo P. Sugrue, Hashem Zamanian, Tony X. Liu, et al. 2021. Closed-loop neuromodulation in an

individual with treatment-resistant depression. *Nature Medicine* 27 (10): 1696–1700. https://doi.org/10.1038/s41591-021-01480-w.
Sheikh, Knvul. 2024. A.I. could spot breast cancer earlier. Should you pay for it? *The New York Times*, sec. Well. https://www.nytimes.com/2024/04/08/well/live/ai-mammograms-breast-cancer.html. Accessed 8 April 2024.
Singh, Jag. 2023. *Future care: Sensors, artificial intelligence, and the reinvention of medicine*. Rochester, MN: Mayo Clinic Press.
Snead, O. Carter. 2020. *What it means to be human: The case for the body in public bioethics*. Cambridge, MA: Harvard University Press.
Sparrow, Robert, and Joshua Hatherley. 2020. High hopes for deep medicine? AI, economics, and the future of care. *Hastings Center Report* 50 (1): 14–17. https://doi.org/10.1002/hast.1079.
Sullivan, Andrew. 2016. My distraction sickness—and yours. *New York Magazine*, https://nymag.com/intelligencer/2016/09/andrew-sullivan-my-distraction-sickness-and-yours.html. Accessed 18 Sept 2016.
Trancik, Emily K. 2015. Enhancement versus therapy in Catholic Neuroethics. *The National Catholic Bioethics Quarterly* 15 (1): 63–72. https://doi.org/10.5840/ncbq20151517.
Wang, Jixuan, Jingbo Yang, Haochi Zhang, Lu Helen, Marta Skreta, Mia Husić, Aryan Arbabi, Nicole Sultanum, and Michael Brudno. 2022. PhenoPad: Building AI enabled note-taking interfaces for patient encounters. *npj Digital Medicine* 5 (12): https://doi.org/10.1038/s41746-021-00555-9.
Wani, Shahid Ud Din, Nisar Ahmad Khan, Gaurav Thakur, Surya Prakash Gautam, Mohammad Ali, Prawez Alam, Sultan Alshehri, Mohammed M. Ghoneim, and Faiyaz Shakeel. 2022. Utilization of artificial intelligence in disease prevention: Diagnosis, treatment, and implications for the healthcare workforce. *Healthcare* 10 (4): 608. https://doi.org/10.3390/healthcare10040608.
Williams, James. 2018. *Stand out of our light: Freedom and resistance in the attention economy*. New York: Cambridge University Press.
Williams, Alex. 2019. Do you take this robot .... *The New York Times*, sec. Style. https://www.nytimes.com/2019/01/19/style/sex-robots.html. Accessed 19 Jan 2019.
Wilson, Timothy D., David A. Reinhard, Erin C. Westgate, Daniel T. Gilbert, Nicole Ellerbeck, Cheryl Hahn, Casey L. Brown, and Adi Shaked. 2014. Just think: The challenges of the disengaged mind. *Science* 345 (6192): 75–77. https://doi.org/10.1126/science.1250830.

Chapter 15
Bioethics and Generative AI: A Casuist Approach to Operationalizing Aspirational Insights

Mark Repenshek and John Paul Slosar

15.1 Introduction

A number of countries, international bodies, and technology companies have issued recent guidance on the use of Artificial Intelligence (AI). Whether that be the European Union's Artificial Intelligence Act (EU AI Act), President Biden's Executive Order on Artificial Intelligence, or the World Health Organization's (WHO) Regulatory Considerations on Artificial Intelligence for Health, the efforts are extraordinary in their breadth, appreciation of the complexity of the technology, and timing. Likewise, companies such as Google, Meta, and Microsoft have also developed their own guidelines. Appropriately so, Catholic moral theologians have also begun drawing on the Church's moral tradition and teaching to begin offering guidance for humanity, more generally, society, and specifically Catholic health systems. Yet, as ethicists working in Catholic health care, we find this guidance somewhat normatively beneficial, yet limited in its context and purpose. Thus, this guidance is at best only foundational and aspirational when looked to as a normative framework for the development, adoption, and deployment of generative AI (gAI) in other contexts and industries, including that of health care.

This chapter will focus on how we, as practicing ethicists, have developed and implemented a casuistic methodology and supporting processes to move beyond this more general guidance in deploying gAI technologies consistent with the moral commitments of our health care institution's Catholic identity. While we are specifically referring to generative AI, it is also important to note that our considerations herein are intentionally limited to narrow gAI technologies that are designed to perform a specific set of tasks or functions (rather than Artificial General Intelligence, or AGI, also sometimes referred to as "strong AI"), as this type of gAI is what is

M. Repenshek (✉) · J. P. Slosar
Ascension, St. Louis, MO, USA
e-mail: mark.repenshek@ascension.org; jslosar@ascension.org

J. T. Eberl (ed.), *Emerging Issues in Catholic Bioethics*, Philosophy and Medicine 153, https://doi.org/10.1007/978-3-032-08862-8_15

most often currently being proposed for use within health care delivery operations (Gobble 2019). After a closer examination of the guidance provided by the aforementioned regulatory bodies, companies, and moral theologians, we conclude that the limitation of their aspirational insights is by no means in any way a fault of their own, but an inherent limitation of the project itself. More specifically, the very endeavor of developing *a priori* mid-level principles and action-guiding rules that ought to guide the ethical development, adoption, and deployment of a unique type of intelligence—at a level of specificity sufficient to actually inform one's action and behavior—is itself beyond what the subject matter allows. Drawing on the existing directional and aspirational guidance, then, we will lay out both a normative methodology and an operational process for applying both the foundational but aspirational guidance as well as the mid-level principles to the infinite possibilities of specific proposals for the development, adoption, and deployment of gAI in health care and, specifically, Catholic health care.

15.2 Regulatory Guidance

The EU AI Act takes a unique approach in its development of guidance on the development and deployment of AI. The essence of the EU's approach is to stratify the risks associated with the technology itself into the categories of unacceptable risk, high-risk, limited-risk, and minimal risk. In doing so, it aims to

> improve(s) the functioning of the internal market and promote the uptake of human-centric and trustworthy artificial intelligence (AI), while ensuring a high level of protection of health, safety, fundamental rights enshrined in the Charter, including democracy, the rule of law and environmental protection, against the harmful effects of AI systems in the Union and supporting innovation (European Union 2024, ch.1, art.1).

The obligations to address the regulatory guidance largely falls on the developers of AI systems. The act does prohibit certain types of AI systems, including but not limited to, those that exploit vulnerabilities, 'real-time' remote biometric identification in publicly accessible spaces for law enforcement, and systems that utilize subliminal, manipulative, or deceptive techniques to achieve certain endpoints. This risk stratification also stresses the importance of data governance as an essential aspect of AI infrastructure. While the EU AI Act does achieve a level of specificity that is useful at a very actionable level for guiding the development, adoption, and deployment of gAI, it does so in a very limited context and for a very specific purpose. This is not by any means intended as some kind of earth-shattering insight, but merely an articulation of the fact that government regulations, while necessary, can only establish the bare minimum requirements relative to establishing a normative framework for the development, adoption, and deployment of gAI, and even then, the adequacy of any one set of regulations will be open to question.

Distinct from the EU's approach, President Biden's (2023) Executive Order on Safe, Secure, and Trustworthy Artificial Intelligence establishes standards and directs actions by the U.S. federal government in nine areas: New Standards for AI Safety and Security; Protecting Americans' Privacy; Advancing Equity and Civil Rights; Standing Up for Consumers, Patients, and Students; Supporting Workers; Promoting Innovation and Competition; Advancing American Leadership Abroad; and Ensuring Responsible and Effective Government Use of AI.

The first section of the Executive Order focuses on several reporting requirements and the creation of standards for AI development, with specific accountabilities for intelligence agencies. It also directs all federal agencies to perform a risk assessment relative to AI cyberattacks. While the Executive Order also focuses on innovation and advancement of AI technologies, it does so mindful of the need for additional data to better understand the impact AI has on the workforce in general, especially workforce displacement and technological disruptions; the Executive Order calls for these data to be gathered through a labor-market impact of AI report. A focus on equity and civil rights within the Executive Order offers specific guidance to all federal agencies relative to illegal discrimination that may arise from AI in the areas of health care, housing, and the criminal justice system with additional emphasis on the responsible use of AI in these domains. Finally, there is an implicit acknowledgement that there is much work to be done to balance innovation and regulation in the AI space, but if done well, the Biden Administration suggests the Executive Order will ensure "responsible innovation" in the area of AI. Here again, like with the EU AI Act, we see that the President Biden's Executive Order contains much valid and valuable action-guiding insights, though such guidance will be limited to the extent that it has been developed relative to and for the specific ends of the nine areas identified within the Order itself.

Given the WHO's focus to "to improve health for everyone, everywhere by accelerating the development and adoption of appropriate, accessible, affordable, scalable and sustainable person-centric digital health solutions" with the goal of preventing, detecting, and responding to epidemics and pandemics, it is perhaps not surprising that its guidance on AI remains largely at the level of regulatory considerations as well (Geneva 2023, xi). The WHO is explicit that the regulatory guidance should be viewed as a resource for all stakeholders. In fact, the WHO does not consider the document guidance or a regulatory framework or policy. Rather, the WHO seeks to be in conversation, through this document, with developers, regulators and policy makers, manufacturers, and health practitioners.

The document covers familiar territory: documentation and transparency; risk management and AI systems development lifecycle; intended use, and analytical and clinical validation; data quality; privacy and data protection; and engagement and collaboration. Uniquely, while also issuing this document, the WHO established the Working Group on Regulatory Considerations to serve as a forum for ongoing dialogue among subject matter experts and regulators relative to the use of AI and its development.

These three significant bodies of work by the EU, the Biden Administration, and the WHO are similar in that they only provide needed guidance for how AI is to be adopted and utilized specific to the responsibilities and specific ends of those agencies that have developed them. While necessary and helpful, such guidance either has very limited applicability at any level of specificity sufficient to guide specific proposals for the development, adoption, and deployment of AI in other contexts and industries, as well as for the infinite variety of benefits it will be used to attain. Where they do provide more foundational normative guidance, it remains aspirational and directional and requires much more development to reveal its full value relative to any concrete application.

15.3 Tech Industry Guidance

In scanning the literature, one might expect that the giants of the technology industry would provide greater clarity and specificity regarding the norms that should guide the development, adoption, and deployment of gAI more generally, including in various contexts and industries, such as health care. The approaches to Responsible AI principles by Meta, Microsoft, and IBM are, for all intents and purposes, identical:

While this shows relatively strong consensus on which principles should guide Responsible AI within the industry, only Google identifies implications of their stated principles for AI development that offers some insight on what Google considers as socially beneficial AI, as well as types of AI applications it will not pursue. Somewhat not surprisingly, these principles remain largely relevant to the developer space and even where Google ventures into its discussion on prohibitions, the norms they identify are largely based on international norms or laws on human rights. Again here, we see similar limitations in the efforts of the industry to establish concrete norms and rules for the development, adoption, and deployment of gAI as those that we saw with the regulatory guidance considered herein. While all of these attempts are worthy and noble, they are necessarily limited in their value as to general normative frameworks and action-guiding norms regarding the development, adoption, and deployment of gAI more generally. Thus, we now turn to the writings of theologians and authoritative bodies within the Catholic tradition in our continued search for a normative framework and specific moral norms for practical application within health care and specifically for Catholic health care.

15.4 Catholic Guidance on AI

In February of 2020 the President of the Pontifical Academy for Life, the President of Microsoft, the Executive Vice President of IBM, and the United Nations Director General of Food and Agriculture Organization (FAO), along with the Italian Minister of Innovation, were the first signatories to a document entitled the Rome Call for AI Ethics. The idea behind the coalition is

> … to promote a sense of shared responsibility among international organizations, governments, institutions and technology companies in an effort to create a future in which digital innovation and technological progress grant man [sic] his centrality. Pointing to a new algorethics, the signatories committed to request the development of an artificial intelligence that serves every person and humanity as a whole; that respects the dignity of the human person, so that every individual can benefit from the advances of technology; and that does not have as its sole goal greater profit or the gradual replacement of people in the workplace (Pontifical Academy for Life 2022).

Evident in this effort to build a coalition across multiple sectors on AI Ethics is the limitation inherent to coalition building—that of staying at the level of principle. This is not a critique, but rather a description of what is the case for the Rome Call for AI Ethics. Being a Member Organization to the Rome Call for AI Ethics means, among other things, organizations make concrete commitments to apply the principles laid out in the Call: Transparency, Inclusion, Responsibility, Impartiality, Reliability, and Security and Privacy. This list is not that different from what is established by nearly all the major industries noted in Table 15.1. This is somewhat to be expected in that both Microsoft and IBM are organizational members of the coalition. Nonetheless, when the goal is to build a coalition among a diverse set of organizations like that of the Rome AI, the effort will likely always remain at the level of principle. As such, this effort lacks the needed guidance, for Catholic health care, in particular, to ensure gAI is adopted and utilized in a way that is consistent with the Catholic moral tradition.

In June of 2024, the International Conference of the *Centesimus Annus Pro Pontifice* Foundation met on the theme "Generative Artificial Intelligence and Technocratic Paradigm: How to Promote the Wellbeing of Humanity, Care for Nature and a World of Peace." In Pope Francis's opening address, he emphasized his speech at the G7 where he reiterated that AI

> … must remain a tool in human hands. Like other innovative tools through the ages, it demonstrates humanity's ability to transcend itself, its drive to ever greater achievements, and thus its potential for bringing about great transformations, both positive and negative.

Table 15.1 Responsible AI principles by industry

Google (2024)	Meta (2024)	Microsoft (2024)	IBM (2024)
Be socially beneficial	Privacy and Security	Reliability and Safety	Explainability
Avoid creating or reinforcing unfair bias	Fairness and Inclusion	Privacy and Security	Fairness
Be built and tested for safety	Robustness and Safety	Inclusiveness	Robustness
Be accountable to people	Transparency and Control	Transparency	Transparency
Incorporate privacy design principles	Accountability and Governance	Accountability	Privacy
Uphold high standards of scientific excellence			
Be made available for uses that accord with these principles			

> In the latter sense, AI could well reinforce the technocratic paradigm and the culture of waste, aggravate disparities between advanced and developing nations, and delegate to a machine essential decisions with regard to human life. For this reason, I posited the absolute need for an ethical development and use of AI, and invited policymakers to adopt concrete actions to direct ongoing technological development towards universal fraternity and peace (Francis 2024).

Although Pope Francis's speech ends with a very compelling question as to whether artificial intelligence should be called intelligent at all, given that the technology itself is made by humans for whom the term "intelligent" is proper, the speech remains at the level of principle. It cautions humanity of the "technocratic paradigm," using a common theme of Pope Francis, suggesting AI, perhaps more than any other technology to date, may lull humanity into the idea that technology alone can solve all the world's problems. There is a slight nod to broader environmental concerns and exacerbation of disparities in reference to the "culture of waste" when Francis (2024) raises the concern of the "enormous consumption of energy required to develop AI, especially as humanity presently faces a challenging energy transition." However, there is little direction as to what systems or even countries should do to curb the energy consumption related to the development of AI.

On January 14, 2025 the Dicastery for the Doctrine of the Faith (DDF) and the Dicastery for Culture and Education (DCE) published a *Note on the Relationship between Artificial Intelligence and Human Intelligence* that seeks to "contribute positively to the discernment regarding AI … through anthropological and ethical reflections" with guidelines for AI's development and use (DDF/DCE 2025, 5). The *Note* underscores the fact that a critical examination of AI must be more than guidance on risk mitigation and harm prevention. In fact, by "considering AI within a framework of relational intelligence," society will benefit from the innumerable ways AI can enhance human capabilities, while not substituting for the richness of true human wisdom (DDF/DCE 2025, 113). The DDF/DCE's contribution "in service of the dialogue that considers how AI can be used to uphold the dignity of the human person and promote the common good" is critical to ensure AI contributes to "fostering the full development of people in relation to others and to the whole of creation" rather than a humanity "enslaved to efficiency" (DDF/DCE 2025, 48 and 68).

Finally, turning to two significant works among ethicists and theologians in the field of health care ethics, one finds far more guidance for Catholic health care to ensure gAI is adopted and utilized in a way that is consistent with the Catholic moral tradition. The first, by Nicholas Kockler, is an early attempt to offer significant operational guidance for Catholic health care, but he too is cautious as to the extent his work accomplishes this task:

> While these guidelines' specificity will be somewhere between principles and concrete moral norms that guide specific behaviors or choices, they should provide practical influence on gAI in Catholic health care. Overtime, additional guidelines derived from CST [Catholic Social Teaching] insights may be warranted; alternatively, the guidelines below may require further specification or elaboration in given circumstances (Kockler 2023, 12).

Ultimately, the guidance is only somewhat prescriptive in favor of a more reflective approach that tends to ask a series of questions related to six domains: human centered design and inclusivity of gAI; enable and empower workforce training and collaboration in gAI design and development; establish ethical gAI deployment protocols and transparent governance structures and processes; develop systems that enhance rather than diminish user autonomy; clarify proper authenticity and veracity of AI-generated outputs along with continuous monitoring; and finally focus less on proprietary models for AI design and more on open source approaches to promote the common good and defend social justice (Kockler 2023, 13–14).

The second, an edited volume by Matthew Gaudet et al. (2024), and in particular Chap. 6: AI and Catholic Social Teaching, the authors offer a perspective on Pope Francis's reflections on the Good Samaritan in which the tradition's understanding of an encounter with AI should be one that centers on the needs of persons and not on the needs of the technology itself. The authors argue that CST's understanding of the nature of technology is just as applicable to AI as it is to other forms of technology. More specifically, three important insights—namely, that (a) technologies are not morally neutral, (b) technologies are not morally determinate, and (c) technologies must be understood within their large social context (Gaudet et al. 2024, 153)—may have far more significant implications, especially when AI is understood "in the experience of people, particularly those who are marginalized and excluded (Gaudet et al. 2024, 158)." Yet, here too, we see these authors struggle to move beyond high-level guidance relative to specific types of care delivery (diagnostics and treatment, public health initiatives, etc.) and global-scale disease prevalence modeling, which leaves the authors with a critically important, yet general and quite principled, recommendation that AI should "promote the global common good of health care at a planetary level and that fosters participation (Gaudet et al. 2024, 189)." This, however, still does not provide the necessary guidance as to how one goes about meeting this recommendation.

15.5 The Core Values of the Catholic Health Ministry

The human person—that is what all of the guidance considered above shares in common. More specifically, all of the regulations, guidelines, and initial thinking center around an attempt to preserve and protect the human person from unethical AI. This holds whether any ascribed malevolence would arise from a flaw in the development process, the way in which AI is implemented, or the intended purpose for which it could be put to use—i.e., nefarious human intentions. The unique contribution of the Catholic perspectives presented above is that they are grounded in a particular understanding of what it means *to be* a human person, not simply the exercise of those capacities to which we as humans typically, but not universally, ascribe moral value.

As with most socio-moral-political guidance, the majority, but not sole, focus of the secular, governmental, and industry guidance is focused on the preservation of the human person's ability to act autonomously, along with concerns regarding bias and privacy insofar as they are enablers of autonomy. In contrast, any Catholic perspective regarding the ethics of AI will be distinguished by the hierarchy of values within the Catholic tradition that gives priority to the foundational values of human dignity, the common good, and justice as understood in light of a theologically-informed anthropology, as we see emerging in the Catholic contributions discussed above. In this way, the foundational principles of respect for human dignity, the common good, and justice provide the foundation of any normative framework that ought, at least within Catholic moral thought, to guide the development, adoption, and utilization of gAI. In this way, Catholic anthropology also provides the foundation for mid-level principles, as well as the identification and prioritization of the risks attendant to the adoption of AI in response to which specific action-guiding rules can and should be established (DDF/DCE 2025, 29). These more foundational and mid-level principles, then, also happen to be the core moral commitments that provide the hallmark characteristics of what it means for Catholic health care to be and operate as a healing ministry of the Church.

The foundational element of Catholic identity is the promotion and defense of *human dignity*. Within the Catholic moral tradition, human dignity is understood as the intrinsic moral worth or inherent value of human life. This intrinsic moral worth of human life is not contingent upon race, gender, social status, ethnicity, the ability to function, or any other accidental attribute of the human person. From the Catholic perspective (among other Christian perspectives), the source of human dignity is rooted in the concept of *imago Dei*, which is Latin for "image of God." This concept denotes the theological doctrine that human beings are made in the likeness and image of God, usually referring to the human intellect, the capacity for moral decision-making, the ability to pursue and find meaning and purpose in and through relationships, and the ability to care for and foster creation as "created co-creators" (Kelly 2013, 11–14). This endowed moral worth, which remains independent of the ability to actualize these capacities, is also rooted in Christ's redemption and in our ultimate destiny of union with God. Human dignity therefore transcends any social order as the basis for human rights and is neither granted by society nor can it be legitimately violated by society (Massaro 2000, 81).

While providing the foundation for many normative claims, one direct normative implication of human dignity is that every human being should be acknowledged as an inherently valuable member of the human community and as a unique expression of life, with an integrated bodily and spiritual nature. Because there is a social or communal dimension to human dignity in Catholic moral thought, persons must be conceived of, not in overly-individualistic terms, but as beings inherently connected to the rest of society (International Theological Commission 2004, nn. 41–43). Because the tradition emphasizes the integral nature of our body and spirit, the human body takes on greater significance and value than in the prevailing dualistic conception of the person (Kelly 2013, 24–27). Other perspectives, both religious and secular, may conceive of human dignity in similar terms with a similar sense of its inherent worth or value, but may posit different sources for that dignity and see different, perhaps even competing, practical implications arising from the concept.

In the Catholic tradition, the promotion and defense of human dignity generally requires: 1) respect for persons and their basic human rights; 2) advocating for those who cannot advocate for themselves; and 3) promoting the common good. As the authors discussed in the previous section so aptly illustrate, these requirements of human dignity apply to AI as much as they do to the development, implementation, and purpose of any other technologies. Moreover, it is the understanding of human dignity that provides the basis for the claim that access to health care, as required for a decent minimum of human well-being, is a basic human right (Eberl, Kinney, and Williams 2011). This understanding provides a common foundation, either explicit or implicit, within and for the mission of all Catholic health care. Thus, at the level of general normativity, AI should always be developed, implemented, and used in ways that enhance, not diminish, access to health care regardless of one's social status, race, ethnicity, gender, socio-economic status, ability to pay, or any other ontologically accidental characteristics we might ascribe to particular individuals.

Grounded in human dignity and the innate social dimensions of human life, the *common good* consists of all the conditions of society and the basic goods secured by those conditions that allow individuals and groups to achieve human and spiritual flourishing (John XXIII 1961). The social teaching of the Catholic Church insists that the human community, including governments, public and private organizations, and individuals, must be actively concerned in promoting the health and welfare of every one of its members so that each member can contribute to the common good of all (O'Rourke 1993, 7). From a Christian theological perspective, this understanding of the common good is rooted in the concept of social solidarity as modeled on the community of persons in the Trinity. Practically, this means that human life finds its fulfillment in and through relationships. At the level of general normativity, then, AI should always be developed, implemented, and used in ways that enhance, not diminish, human relationships. Even further, fostering right relationships should be one of the primary ends for which AI is put to use. Put another way, advancing justice, in all its forms, should not just be another high-level principle that has some vague and somewhat disconnected role in governing the development and use of AI, but should be one of the primary ends for which both public and private institutions adopt AI technologies.

Within the Catholic moral tradition, *justice* can generally be defined as giving to each person what is due to them in accord with their human dignity. More specifically, the Catholic moral tradition recognizes four types of justice: 1) legal or civil justice; 2) commutative justice; 3) contributive justice; and 4) distributive justice (Ashley 2006, 45–47). In addition to certain entitlements, each one of these types of justice gives rise to certain obligations insofar as, within the Catholic moral tradition, rights are always accompanied by corresponding responsibilities.

Legal or civil justice concerns the rights of individual citizens and their corresponding responsibilities to obey and respect the rights of others and the laws devised to protect peace, social order, and the security to which all have a right as part of the common good. Commutative justice concerns what one individual or organization owes to another based on agreement and entails promise-keeping and fidelity to contracts. Contributive justice concerns what individuals and organizations owe to society as part of the common good. Insofar as we each have a right to participate in and benefit from the common good due to our own human dignity, we also have a responsibility to contribute to the common good in proportion to our ability to do so.

Distributive justice concerns what society and the organizations that make up its infrastructure owe to members of society in accord with and in proportion to: 1) an individual's needs, contributions, and responsibilities; 2) the resources available to the society or a particular organization, including financial and market considerations; and 3) the society's or organization's particular responsibility to the common good. Within this conception, distributive justice is primarily concerned with how well a society responds to the basic human needs of its members. Accordingly, society has a duty to the individual in serious need, and all individuals have duties to others in serious need. In decisions regarding the allocation of resources, the duty of society is not diminished because of the person's status, disability, or illness. In other words, allocation decisions should not be based upon judgments of the utility of an individual's contributions to society or the quality of a person's life, but "from each according to ability, to each according to need" (Ashley 2006, 45). Distributive justice also implies an obligation to distribute burdens and benefits in an equitable manner. Generally, any inequalities in the allocation of resources, benefits, and burdens are justified only if they are to the advantage of the least-well off in society and, thus, these considerations must be accounted for in the development, adoption, and deployment of AI within society generally as well as health care specifically.

Insofar as individual human need is the substantive criterion of distributive justice within the Catholic moral tradition, the poor, as those with the greatest need, have a special claim on society and the organizations that make up its infrastructure and its resources. In particular, this "special claim" implies a *preferential or fundamental option for the poor and vulnerable*, which is to be lived out in *solidarity* with those who live in economic poverty (USCCB 1986, n. 16). To say that there is a preferential option does not mean that taking care of the poor is an *option*. Rather, there is a *commitment* to live in solidarity with the poor, meaning that, of all the options set before us in the common good, the poor and vulnerable should be our first, but not only, concern. In a practical sense, this means that when we consider the way in which our decisions and actions affect individuals and communities, we should first consider how they will impact those who are poor and vulnerable, and then consider how they will affect the rest of the population.

Moreover, care for poor and vulnerable persons imposes an obligation to carry out and sustain our mission as Catholic health care providers in ways that do not perpetuate the marginalization of the poor and vulnerable members of society. Considered in conjunction with one another, our emphasis on caring for poor and vulnerable persons and the concept of distributive justice imply a social, institutional, and individual obligation to be wise stewards of our resources. Part of what it means to be wise stewards of our resources is that our planning and use of those resources is not *only* aimed at maximizing outcomes while minimizing consumption, but that such planning and use inherently attends to distributive justice, the preferential option for the poor and vulnerable, and the imperative to create a more just and compassionate society. This morally robust notion of stewardship includes, of course, the development, adoption, and deployment of AI in health care.

Stewardship requires us to appreciate the two great gifts that a wise and loving God has given: the Earth, with all its natural resources, and our own human nature, with its biological, psychological, social, and spiritual capacities. This principle is grounded in the presupposition that only God has absolute dominion over creation, and that, insofar as human beings are made in God's image and likeness (*imago Dei*) and have been bestowed with the gift of reason, we have been given a limited dominion over creation and are responsible for its care (Francis 2015, n. 66). In other words, though we share in God's dominion over creation, we do so only in a limited way. In this sense, being stewards means that we have a responsibility to use resources in ways that cultivate individual and communal human flourishing, and that respect and secure for the future the overall gift of creation itself. Moreover, the principle requires that the gifts of human life and its natural environment be used with profound respect for their intrinsic ends. The gift of human creativity especially should be used to cultivate nature and the environment, recognizing the limitations of our actual knowledge and the risks of destroying these gifts. Accordingly, simply because something can be done does not necessarily mean that it should be done (the fallacy of the technological imperative).

As not-for-profit health care organizations with the mission of caring for all people, with special attention to the poor and vulnerable, while operating in a market-driven economy situated within a world of necessarily limited resources, Catholic health ministries must be prudent stewards of resources. As stewards of society's medical resources, the administrative leaders and health care professionals and providers who work within Catholic health ministries share this responsibility. This responsibility includes but is not reducible to concern for scarce resources. In this sense, stewardship is primarily concerned with using and allocating human and material resources in ways that promote equity, respect basic human rights, and foster the common good. Prudent stewardship also includes creating optimal medical utility—i.e., minimizing costs while maximizing clinical benefit—and ensuring the continued future of the ministry itself. According to this understanding, stewardship includes considerations of finances, availability of supplies and resources, time, skill, talents, and outcomes. One practical implication of this conception of stewardship is that, while we must aggressively work to optimize medical utility and achieve fiscal well-being for our organizations, we must not do so at the expense of individual basic human rights and the continued marginalization of the poor and vulnerable (DDF/DCE 2025, 97). While it comes with some risk, this is an area in which the promises of AI have a particular contribution both in kind and degree to offer health care delivery in every context.

15.6 Some Promise, More Peril

The title of this section is not necessarily a claim about gAI itself, though whether in that sense it proves to be sound or not will inevitably reveal itself over time. Rather, this title is more of an observation that thus far there seems to be a greater focus and emphasis in the ethics literature and existing guidance given to the perils, risks, and cautions associated with AI, while the promise seems to be, to some extent, taken for granted. It also functions as an affirmation relative to the priority that is being given to the potential perils, which is, in our opinion, completely appropriate and absolutely necessary. At the same time, however, it functions as a warning that if all we do is pay attention to the known risks, we will fail to uncover previously unidentified risks and consequences until it is too late. We as ethicists and, quite frankly, as a society must resist the temptation to take for granted that the benefits justify the means or, perhaps, better stated, resist the pull to become entranced solely by the dazzling benefits and goods that AI promises. This is why AI has a particularly insidious quality, namely, that if we become too enthralled with the benefits without taking the time and effort to uncover unanticipated risks associated with specific "use cases" of gAI and their consequences, then we will inevitably miss accounting for salient moral concerns that warrant the value-preference of our moral consciousness.

From an operational perspective, there are two obvious, if not primary, benefits, or noble ends, within the context of health care for which gAI will be utilized. The first of these is to improve the safety, accuracy, expedience, and quality of diagnosis, prognosis, and plan of treatment well beyond what the unaided human intellect would ever be capable. The myriad of ways in which this will occur is, in our humble estimation, beyond imagination and comprehension at this point, but we know it ranges from things like earlier more accurate diagnosis of breast cancer by gAI-read scans (Gjesvik 2024) to hyper-personalized care (Johnson 2021) to the more mundane, like more accurate and expedient automated charting, coding, and billing (Kim 2023). This last example brings us to the second type of primary benefit, namely, creating operational efficiencies. This category of benefit is most analogous to other forms of automation across different types of industries and shares many of the same benefits, primarily, of course, economic savings, along with many of the same risks, e.g., job displacement.

If we heed our own warning that we consider the risks of gAI in relation to the benefits we hope to derive from specific uses, we can begin to arrive at something resembling a generalizable casuistry-based methodology for the evaluation of various proposals regarding the development, adoption, and deployment of gAI. Our experience in evaluating and addressing various proposals to develop or adopt/deploy various gAI related technologies within health care operations suggests that such an approach is much more feasible and, perhaps, more importantly, yields greater practical value than attempting to derive a set of action-guiding rules that ought to guide the use of gAI for health care organizations (or, perhaps, generally). This approach also seems to fit better when we consider that the nature of gAI itself, even in its narrow form, is a type of "intelligence" itself. While it can be debated whether gAI is truly intelligence in the same sense that humans have intelligence, as Pope Francis (2024) has questioned, the defining characteristic of *generative* AI is that it is capable of creating new and completely unique content and solving problems through analyses performed primarily if not completely independent of any human interaction or governance. In this way, we might even ascribe, though in a very constrained way (especially when referring to "narrow" gAI), a sense of independent *agency* to the technology, which then distinguishes it from any other type of existing technology known today. Even if one is not willing to go that far, gAI is distinctive from other existing forms of technology, which are often considered morally neutral in and of themselves, insofar as it can learn, analyze and generate without human intervention but does not have the human capacity of rational reflection on the morality of that which it is learning or producing.

If one is willing to grant that gAI has even the most limited form of what one might describe as its own agency, then there is good reason why all of the guidance we have reviewed stops short of establishing a set of hard and fast pre-establish insofar as going any further would risk falling victim to the fallacy of misplaced concreteness. Perhaps, the failure of the early guidance and regulations to offer a set of hard and fast rules is not necessarily a shortfall of the work of those offering the guidance, but an inherent dimension of the nature of gAI itself. Of course, mid-level norms still apply generally, e.g., the population data used to develop gAI should be as free from bias as possible, while any patient specific data analyzed by gAI should be secure and confidential. At this point, that pretty much goes without saying and enjoys wide general consensus, but that mid-level norm does not even come close to answering all of the ethical questions nor addressing all of the normative risks that accompany the pursuit of particular noble ends. Thus, we must always take care when thinking about the ethics of gAI to examine both the promise and the peril simultaneously and in relationship to one another in light of the foundational and mid-level principles which ought to guide its development, adoption, and deployment. This, we believe, is simply the starting point. What is needed beyond foundational principles, even mid-level principles, and regulations is a *casuistic method* of analyzing potential uses, i.e., use-cases, for which the development, adoption, or deployment of gAI technologies are being put forth, both generally and in the specific context of health care delivery (for more on casuistry as an ethical methodology, see Jonsen 1995 and Kuczewski 1998).

In keeping with our methodological claim above that the risks and benefits are best considered in the context of any inherent relationship, the normative methodology and operational processes we have begun developing and implementing start with the identification of the risks that could result from developing, adopting, and deploying gAI for specific ends, i.e., goods or benefits. Once identified, we then correlate those risks with the moral values and mid-level principles that arise from the foundational principles derived from the particular understanding of the human person that is operative in the Catholic moral tradition and provides the foundation for the mission and ministry identity of Catholic health care systems. Of course, this includes not only the more obvious considerations of accuracy and hallucinations, privacy, consent, and adequate security related to protected health information (PHI), and not just bias and impartiality, but also the importance of protecting and maintaining the conditions for human agency and acting autonomously (DDF/DCE 2025, 47), the human dimension of the physician-patient relationship, the spiritual and relational needs of the patient, and how access to new and improved technology is distributed (at least to the extent that we can influence that within our own health system).

An example for consideration would be the ever-growing suite of ambient listening gAI tools that can chart and code for what occurs during a patient visit. This technology already exists and offers many benefits in the areas of efficiency and accuracy regarding charting, coding, and billing, and is being widely adopted across the health care industry. It comes with a variety of corresponding risks as well. In addition to the general risks to privacy and confidentiality, security, and the sharing of data, the way in which and purpose for which it is adopted could either enhance the person-centered dimensions of the physician-patient relationship or simply create more time to see more patients and increase revenue. If we then correlate these risks to the foundational and mid-level principles that constitute the core moral commitments of Catholic health care systems, we can develop sets of specific action-guiding rules regarding when and how such technologies of this nature should be adopted, the qualities we want to see in the development of those technologies, and, perhaps most important, how they should be deployed.

15.7 Operationalizing AI Ethics in Catholic Health Care

The temptation for health care delivery systems in addressing complex issues is to start by developing a governance structure by which to oversee the development, adoption, and deployment of emerging practices and technologies. According to our reasoning herein, however, this would be a mistake with regard to gAI technologies within health care. In fact, our approach has been just the opposite:

- start with use-case analyses by an interdisciplinary team of experts, gain experience, understand the types of gAI involved and the purposes for which different types are to be deployed;
- find the common risks relative to the values important to us as a Catholic health system that apply within a specific type or category of use-cases (e.g., ambient AI for charting, coding, and billing);
- then develop the specific ethical criteria or action-guiding rules that should guide our development, adoption, and deployment of those types of (narrow) gAI;
- upon developing a better grasp on the types of risks involved with the development or adoption and deployment of gAI technologies along with the type of ethical considerations that ought to guide that decision-making, the next step is to develop a policy that adequately reflects those;

- finally, once a policy is in place, the creation of a governance structure becomes absolutely essential, since it in fact now has something with which to govern.

Once an array of gAI platforms is better understood, one can begin to see that although the principles in the tradition are quite familiar to an ethical analysis of technology (as we have articulated above), their guidance relative to gAI is not immediately apparent. We also discovered very early on that we needed to create a tool, designed with the business owners in mind, that incorporated the principles of Catholic teaching found in this chapter in a way that made sense and was useful and operationally relevant to the business owner. This led to the development of a tool that employed a risk-mitigation framework to guide the evaluation of proposed AI use-cases.

While such a tool itself is not sufficient, it proves a framework and clear vision in which to understand the relationship between specific use-cases of gAI and the attendant risks that could undermine the values important to Catholic health care systems. For example, although privacy and confidentiality are not new to any assessment of technology, framing the value in terms of the unique risk gAI yields a practical insight beyond simply naming the mid-level norm—specifically, the potential for inappropriate sharing or use of data/information where the gAI continually "trains itself" on data housed outside of the health care system yet gathered from within the confines of the patient-physician relationship.

While useful for translating mid-level norms, a table of values and associated risks is clearly not sufficient for evaluating gAI use-cases. Rather, these identified risks only help to articulate the concerns with which an analysis or evaluation should be concerned. Such analysis still needs to be performed using the relevant expertise, which in our case we have convened by forming a single, system-wide AI & Data Ethics Committee (AI-EC). Upon launching this committee, it quickly became apparent that we needed a tool by which the committee itself could determine what were appropriate use-cases to come before it for review. This turned out to be a much more complex issue than anyone anticipated. Our response was to develop a set of questions that corresponded to the risks identified, so that the "business owners" (i.e., service line, unit or department leaders) and committee members alike would all have a shared understanding of when it was appropriate to bring a use-case before the committee and, perhaps more importantly, when it is not.

For example, where distributive justice is concerned with the risk of gAI increasing economic disparities or inequalities in access to health care, additional questions include:

- Is there a foreseen potential for the proposed use to exacerbate economic or other disparities among those we serve or to benefit only those who are already economically well-off?
- Is there a foreseen potential for the proposed use to improve access for some segments of the population while either maintaining the status quo or actually inhibiting access for others?

Similar questions could also include the implications of gAI to be deployed around inequities concerning quality, safety, and understanding of how one's disease condition might impact one's functioning or even lifespan. And there are likely many questions that we have not yet contemplated and would only be capable of doing so relative to a specific use-case.

Where trust is concerned with the risk of transparency and interpretability regarding the rationale for clinical decisions or recommendations, additional questions might include:

- Does the proposed use have the potential to diminish or eliminate the ability to understand the rationale for decisions that impact patients?
- Does the proposed use have the potential to create confusion about what and why the Catholic health care system or their care provider is recommending or doing on their behalf?

While these are only a few examples of the questions included on our AI-EC intake form, they demonstrate the types of concerns that the committee is designed to address. If the answers to any of the questions indicate a real potential for any of the risks identified above, then the proposed use-case goes to the committee, so that the committee can serve as a resource for helping to ensure that the Catholic health care system's data and AI practices serve the authentic good of all persons and advance the system's mission in a manner consistent with its Catholic identity. In particular, the AI-EC serves to provide guidance to the organization in support of ethical practices, policies, and decision-making regarding the development, adoption, or utilization of AI and related data. And, perhaps more important than the AI-EC itself, is the appropriate integration of the AI-EC into the health care system's Data Governance (DG) program (a later realization). Regardless, the functions an AI-EC should consider in-scope include:

- Ethical questions pertaining to accuracy, transparency/interpretability, and impartiality of particular AI use-cases;
- Ethical questions pertaining to the use and sharing of data specific to the development or adoption of (contracting for) AI use-cases;
- Questions related to the ethical dimensions or mission and Catholic identity implications of adopting particular AI uses and the purpose for and way in which they would be utilized; and
- Self-education relative to performing the above functions.

It is also equally important to consider what is not within scope for an AI-EC. Consideration should be given to exclude all aspects of data science, that is, regression, correlation, statistical analysis, pattern matching, optimization, forecasting, and projection. Finally, just as is the case for a clinical ethics committee, having the right interdisciplinary makeup of an AI-EC is crucial to its success. We would suggest the makeup of this interdisciplinary group should include, but is not limited to mission integration and ethics, legal, risk, compliance and privacy, advocacy, data science and analytics, data governance, data security, clinical informatics, data engineering, clinical services, graduate medical education, and clinical research.

Who the AI-EC ultimately reports to is also an important consideration, one that warrants exploring a reporting relationship at the highest levels of senior leadership with responsibilities for information technologies within the Catholic health care system (e.g., Chief Information Officer). Ultimately, the structure and function as described above should be codified in a charter for the AI-EC. Given the rapidly evolving understanding of what constitutes AI, and particularly gAI, a charter should be drafted and viewed as dynamic and responsive to the evolving needs of the organization relative to AI and its ethical evaluation.

Beyond these resource developments, our experience suggests the optimal structure for Catholic health care is one that integrates the entire AI-EC and associated processes into the organization's data governance structure. Any DG program should be designed to establish best practices, clear accountability, data stewardship, and a decision-making structure for ongoing program evaluation (Barina et al. 2022 and Ladley 2020). Once a structure for the objectives of DG is established, one of the first steps of a comprehensive DG program is the creation of a charter. The creation of a charter is not unique to DG, but what is unique is the way in which Catholic health care's mission, vision, values, and Catholic identity should play a central role in informing how the charter influences DG program design. When a DG program's charter is informed in this way, the principles that guide policy and decision-making for the program are more evident. Below we have listed ten principles that are, in general, considerations for a robust DG program with the corresponding principle of Catholic identity that informs it:

1. Data is a strategic asset that has value and risk and will be managed by [Catholic system]. **[stewardship; common good]**
2. Data Governance committee(s) will establish policies, standards, and consistent ways of working with data and will provide services established within its mandate. **[common good; human dignity]**
3. Data-related decisions at [Catholic system] should be made at the lowest level possible. **[subsidiarity]**
4. Governance will be invoked where individuals cannot or should not make decisions about data in isolation. **[subsidiarity]**
5. [Catholic system] will pursue the lightest governance structure possible to achieve adaptive governance. **[stewardship; subsidiarity]**
6. Not all data will be treated equally; data will be valued and governed/managed based on business impact, stakeholder needs, and applicable policy/regulation (e.g., PHI). **[common good; preferential option for the poor and vulnerable]**
7. Data definitions, standards, processes, and policies will be developed and maintained with an enterprise approach. **[common good]**
8. Data stewards define the business terms and definitions, approved data values, data relationships, business rules, and data quality standards, and they monitor data quality and data asset value, while Information Technology maintains the systems that capture and manage data through their lifecycle. **[stewardship]**
9. Individuals who create or acquire data are accountable for the quality of that data and must record it in accordance with its definition. **[justice]**

10. Data quality and integrity will be addressed by the individuals that create the data and who are closest to the data, understand its meaning and business implications to the specification of the data stewards, with support from the central data governance program. **[justice; subsidiarity]**

Once a comprehensive DG charter is crafted and the corresponding infrastructure is in place within the organization, creating a policy on AI for how the Catholic health care ministry monitors, evaluates, develops, deploys, and utilizes AI is a critical next step. The movement from DG to policy and procedure is essential to operationalizing AI in Catholic health care. Such a comprehensive policy must start with a strong policy statement that makes explicit the unique way in which the Catholic health care system will approach AI. Below is an example of how such a policy statement might read:

> This Procedure establishes standards for the monitoring, evaluation, development, deployment, and use of AI at [Catholic health system] to uphold [Catholic health system's] ethical, professional, legal, and social standards while advancing its Mission. When developing, deploying, or using AI, [Catholic health system] aims to (i) promote and foster the use of AI technology to better enhance patient care while strengthening its Catholic ministry identity and ability to achieve its Mission, (ii) ensure responsible and ethical use of AI that promotes human dignity, justice, and the common good, including by avoiding unfair discrimination and harmful bias, and (iii) prioritize the safety of human life, health, property, and the environment.

With this type of policy statement in place, the procedural and operational implications become evident. It is then the work of the policy to address the multitude of areas within the operations of the Catholic health ministry for which the AI policy will apply. Application should include but is not limited to: abiding by the *Ethical and Religious Directives for Catholic Health Care Services* (2018) and promotion of the Catholic health care system's mission, vision, and values; protecting autonomy; accounting for patient disclosure and consent; managing data privacy and security; promoting human safety; ensuring transparency; favoring interpretability; fostering accountability; promoting accuracy, reliability, and sustainability; continuously engaging in learning and development. Consideration should also be given to addressing the nature of AI-related risks within the policy insofar as there may be areas of risk unique to AI to which other, more traditional risk policies may not necessarily speak. Finally, perhaps more on the procedural side, the policy should identify the process the Catholic health ministry plans to use to ensure adherence to the policy.

15.8 Conclusion

Of course, there is nothing groundbreaking about the observation that foundational principles, mid-level norms, and even regulations need to be applied to the contextual nuances of concrete circumstances when making ethical judgements. Such a claim is closer to the definition of what it means to do ethics in any practical sense than it is the conclusion of an argument (not to mention a worthwhile conclusion). Our point is that, given the uniqueness of gAI as an intelligent technology, it requires more than just the prudential application of mid-level norms (gAI should be free from bias) derived from foundational principles (human dignity, the common good, justice, and stewardship) to ensure that this technology not be used to disadvantage individual patients or certain classes of people. Rather, it is precisely the evaluation of gAI use-cases that enables us to refine and reshape our specific action-guiding rules that inform when, how, and why such technologies should be adopted consistent with a health care system's Catholic identity and the good of the human person. It is with each gAI use-case that we then continue the process of refining and reshaping, ultimately leading to sets of action-guiding rules that can be used to address specific types or categories of gAI use-cases. At this stage of any gAI deployment in health care, prudence requires adopting a practice of piloting any and all gAI systems before widespread adoption. Unlike with other technologies and practices, however, such piloting provides an essential opportunity to evaluate and review the moral character, if you will, of the AI, i.e., whether it provides unbiased recommendations, whether it supplements or supplants the physician-patient interaction, etc. This moral evaluation can be included as a criterion, alongside more standard clinical, operational, and business criteria like efficacy, efficiency, and value, in the selection process by which those gAI technologies are identified as worthy of widespread dissemination. In other words, our method and process enables the evaluation of gAI not just based on how well it does what it was designed to do, but also on its moral character and virtue.

References

Ashley, Benedict M., Jean Deblois, and Kevin D. O'Rourke. 2006. *Health care ethics: A theological analysis*. 5th ed. Washington, DC: Georgetown University Press.

Barina, Rachelle, Becket Gremmels, Michael Miller, Nicholas Kockler, Mark Repenshek, Christopher Ostertag, and Kirsten Dempsey. 2022. Data ethics in Catholic health systems. *National Catholic Bioethics Quarterly* 22 (2): 1–29.

Biden Administration. 2023. *Exec. order no. 14,110, safe, secure, and trustworthy development and use of artificial intelligence*, https://www.whitehouse.gov/briefing-room/presidential-actions/2023/10/30/executive-order-on-the-safe-secure-and-trustworthy-development-and-use-of-artificial-intelligence/. Accessed 20 Sept 2024.

Dicastery for the Doctrine of the Faith and Dicastery for Culture and Education (DDF/DCE). 2025. *Antiqua et nova*. https://www.vatican.va/roman_curia/congregations/cfaith/documents/rc_ddf_doc_20250128_antiqua-et-nova_en.html.

Eberl, Jason T., Eleanor D. Kinney, and Matthew J. Williams. 2011. Foundation for a natural right to health care. *The Journal of Medicine and Philosophy* 36 (6): 537–557.

European Union. 2024. *Artificial Intelligence Act*. Regulation (EU) 2024/1689 of the European Parliament and of the Council of 13 June 2024 laying down harmonized rules on artificial intelligence and amending Regulations.

Francis. 2015. *Laudato Si'*. https://www.vatican.va/content/francesco/en/encyclicals/documents/papa-francesco_20150524_enciclica-laudato-si.html. Accessed 4 Dec 2024.

Francis. 2024. Address to participants in the international conference of the ‘centesimus annus pro pontifice’ foundation. https://www.vatican.va/content/francesco/en/speeches/2024/june/documents/20240622-centesimus-annus-propontifice.html. Accessed 9 Oct 2024.

Gaudet, Matthew J., Noreen Herzfeld, Paul Scherz, and Jordan J. Wales. 2024. *Encountering artificial intelligence: Ethical & anthropological investigations*. Eugene, OR: Pickwick Publications.

Geneva: World: Health Organization. 2023. *Regulatory considerations on artificial intelligence for health*. License: CC BY-NC-SA 3.0 IGO.

Gjesvik, J., N. Moshina, C. I. Lee, D. L. Miglioretti, and S. Hofvind. 2024. Artificial intelligence algorithm for subclinical breast cancer detection. *JAMA* 7 (10): e2437402. https://doi.org/10.1001/jamanetworkopen.2024.37402.

Gobble, M. M. 2019. The road to artificial general intelligence. *Research-Technology Management* 62 (3): 55–59.

Google. 2024. AI principles progress update 2023. https://ai.google/static/documents/ai-principles-2023-progress-update.pdf. Accessed 22 Sept 2024.

IBM. 2024. Our principles and pillars. https://www.ibm.com/impact/ai-ethics. Accessed 22 Sept 2024.

International Theological Commission. 2004. *Communion and stewardship: Human persons created in the image of god* https://www.vatican.va/roman_curia/congregations/cfaith/cti_documents/rc_con_cfaith_doc_20040723_communion-stewardship_en.html. Accessed 4 Dec 2024.

John XXIII. 1961. *Mater et magistra*. https://www.vatican.va/content/john-xxiii/en/encyclicals/documents/hf_j-xxiii_enc_15051961_mater.html. Accessed 4 Dec 2024.

Johnson, K. B., W. Q. Wei, D. Weeraratne, M. E. Frisse, K. Misulis, K. Rhee, J. Zhao, and J. L. Snowdon. 2021. Precision medicine, AI, and the future of personalized health care. *Clinical and Translational Science* 14 (1): 86–93.

Jonsen, Albert R. 1995. Casuistry: An alternative or complement to principles? *Kennedy Institute of Ethics Journal* 5 (3): 237–251.

Kelly, David F., Gerard Magill, and Henk Ten Have. 2013. *Contemporary catholic health care ethics*. Washington, DC: Georgetown University Press.

Kim, J. S., A. Vivas, V. Arvind, J. Lombardi, et al. 2023. Can natural language processing and artificial intelligence automate the generation of billing codes from operative note dictations? *Global Spine Journal*. 13 (7): 1946–1955.

Kockler, Nicholas. 2023. Generating insights from catholic social teaching: Ethical guidelines for artificial intelligence in health care ministries. *Health Care Ethics USA* 17 (2): 3–14.

Kuczewski, Mark. 1998. Casuistry and principlism: The convergence of method in biomedical ethics. *Theoretical Medicine and Bioethics* 19 (6): 509–524.

Ladley, John. 2020. *Data governance: How to design, deploy, and sustain an effective data governance program*. San Diego, CA: Elsevier.

Massaro, Thomas. 2000. *Living justice: Catholic social teaching in action*. Oxford: Rowman & Littlefield Publishers.

Meta. 2024. Our commitment to responsible AI. https://ai.meta.com/responsible-ai/. Accessed 22 Sept 2024.

Microsoft. 2024. Responsible AI standard. https://www.microsoft.com/en-us/ai/principles-and-approach. Accessed 22 Sept 2024.

O'Rourke, Kevin D., and Philip Boyle. 1993. *Medical ethics: Sources of Catholic teachings*. Washington, DC: Georgetown University Press.

Pontifical Academy for Life. 2022. Rome call for AI ethics. https://www.romecall.org/wp-content/uploads/2022/03/RomeCall_Paper_web.pdf. Accessed 9 Oct 2024.

United States Conference of Catholic Bishops (USCCB). 1986. *Economic justice for all*. Washington, DC: USCCB.

United States Conference of Catholic Bishops (USCCB). 2018. *The ethical and religious directives for Catholic health care services*. 6th ed. Washington, DC: USCCB.

Index

J. T. Eberl (ed.), *Emerging Issues in Catholic Bioethics*, Philosophy and Medicine 153, https://doi.org/10.1007/978-3-032-08862-8

GPSR Compliance
The European Union's (EU) General Product Safety Regulation (GPSR) is a set of rules that requires consumer products to be safe and our obligations to ensure this.

If you have any concerns about our products, you can contact us on

ProductSafety@springernature.com

In case Publisher is established outside the EU, the EU authorized representative is:

Springer Nature Customer Service Center GmbH
Europaplatz 3
69115 Heidelberg, Germany

www.ingramcontent.com/pod-product-compliance
Lightning Source LLC
Chambersburg PA
CBHW050727280126
38800CB00003B/84
9783032088611